Pascal's Wager

In his famous Wager, Blaise Pascal (1623–62) offers the reader an argument that it is rational to strive to believe in God. Philosophical debates about this classic argument have continued until our own times. This volume provides a comprehensive examination of Pascal's Wager, including its theological framework, its place in the history of philosophy, and its importance to contemporary decision theory. The volume starts with a valuable primer on infinity and decision theory for students and non-specialists. A sequence of chapters then examines topics including the Wager's underlying theology, its influence on later philosophical figures, and contemporary analyses of the Wager including Alan Hájek's challenge to its validity, the many-gods objection, and the ethics of belief. The final five chapters explore various ways in which the Wager has inspired contemporary decision theory, including questions related to infinite utility, imprecise probabilities, and infinitesimals.

Paul Bartha is Professor and Head of the Department of Philosophy, University of British Columbia. He is the author of *By Parallel Reasoning: The Construction and Evaluation of Analogical Arguments* (2010).

Lawrence Pasternack is Professor of Philosophy at Oklahoma State University. His publications include *Kant's Religion within the Boundaries of Mere Reason: An Interpretation and Defense* (2014).

Classic Philosophical Arguments

Over the centuries, a number of individual arguments have formed a crucial part of philosophical enquiry. The volumes in this series examine these arguments, looking at the ramifications and applications which they have come to have, the challenges which they have encountered, and the ways in which they have stood the test of time.

Titles in the series

The Prisoner's Dilemma
Edited by Martin Peterson

The Original Position
Edited by Timothy Hinton

The Brain in a Vat
Edited by Sanford C. Goldberg

Pascal's Wager
Edited by Paul Bartha and Lawrence Pasternack

Pascal's Wager

Edited by

Paul Bartha
University of British Columbia, Vancouver

Lawrence Pasternack
Oklahoma State University

CAMBRIDGE
UNIVERSITY PRESS

University Printing House, Cambridge CB2 8BS, United Kingdom

One Liberty Plaza, 20th Floor, New York, NY 10006, USA

477 Williamstown Road, Port Melbourne, VIC 3207, Australia

314–321, 3rd Floor, Plot 3, Splendor Forum, Jasola District Centre, New Delhi – 110025, India

79 Anson Road, #06–04/06, Singapore 079906

Cambridge University Press is part of the University of Cambridge.

It furthers the University's mission by disseminating knowledge in the pursuit of education, learning, and research at the highest international levels of excellence.

www.cambridge.org
Information on this title: www.cambridge.org/9781107181434
DOI: 10.1017/9781316850398

© Cambridge University Press 2018

This publication is in copyright. Subject to statutory exception and to the provisions of relevant collective licensing agreements, no reproduction of any part may take place without the written permission of Cambridge University Press.

First published 2018

Printed and bound in Great Britain by Clays Ltd, Elcograf S.p.A.

A catalogue record for this publication is available from the British Library.

Library of Congress Cataloging-in-Publication Data
Names: Bartha, Paul F. A., 1964– and Pasternack, Lawrence R., 1967– editors.
Title: Pascal's wager / edited by Paul Bartha, University of British Columbia, Vancouver, and Lawrence Pasternack, Oklahoma State University.
Description: New York : Cambridge University Press, 2018. | Series: Classic philosophical arguments | Includes bibliographical references and index.
Identifiers: LCCN 2018026134 | ISBN 9781107181434
Subjects: LCSH: Faith and reason – Christianity. | Apologetics. | Pascal, Blaise, 1623–1662. | Philosophical theology.
Classification: LCC BT50 .P38 2018 | DDC 212/.1–dc23
LC record available at https://lccn.loc.gov/2018026134

ISBN 978-1-107-18143-4 Hardback
ISBN 978-1-316-63265-9 Paperback

Cambridge University Press has no responsibility for the persistence or accuracy of URLs for external or third-party internet websites referred to in this publication and does not guarantee that any content on such websites is, or will remain, accurate or appropriate.

Contents

List of Figures and Tables	*page* vii
List of Contributors	ix
Acknowledgments	x
Introduction Paul Bartha and Lawrence Pasternack	1

Part I Historical Context and Influence — 25

1. Pascal's Wager and the Origins of Decision Theory: Decision-Making by Real Decision-Makers — 27
   James Franklin

2. The Wager and Pascal's Theology — 45
   William Wood

3. Pascal's Wager and the Ethics for Inquiry about God — 64
   Paul Moser

4. Pascal and His Wager in the Eighteenth and Nineteenth Centuries — 84
   Adam Buben

5. The Wager and William James — 101
   Jeffrey Jordan

Part II Assessment — 121

6. The (In)validity of Pascal's Wager — 123
   Alan Hájek

7. The Many-Gods Objection to Pascal's Wager: A Defeat, then a Resurrection — 148
   Craig Duncan

8	The Wager as Decision under Ignorance: Decision-Theoretic Responses to the Many-Gods Objection Lawrence Pasternack	168
9	Rationality and the Wager Paul Saka	187
10	The Role of Pascal's Wager in Authentic Religious Commitment Joshua Golding	209
Part III	**Extensions**	223
11	The Arbitrary Prudentialism of Pascal's Wager and How to Overcome It by Using Game Theory Elliott Sober	225
12	Pascal's Wager and the Dynamics of Rational Deliberation Paul Bartha	236
13	Infinity in Pascal's Wager Graham Oppy	260
14	Pascal's Wager and Imprecise Probability Susanna Rinard	278
15	Do Infinitesimal Probabilities Neutralize the Infinite Utility in Pascal's Wager? Sylvia Wenmackers	293
	Bibliography	315
	Index	329

Figures and Tables

Figures
11.1	Winding Road	*page* 231
11.2	Hill Climbing	232
12.1	Infinite Relative Utility	243
12.2	Equal Relative Utility	243

Tables
I.1	Dominance Reasoning	3
I.2	EMV Reasoning	4
I.3	EU Reasoning	5
I.4	Pascal's Wager	7
I.5	Pascal's Wager	10
I.6	Pascal's Wager with Probabilities	14
1.1	Pascal's Wager	37
1.2	Pascal's Wager with Huge Payoff	40
1.3	Contemporary Wager	42
6.1	(Super)Dominance Wager	124
6.2	Wager with Probability ½	127
6.3	Damnation Has Negative Infinite Utility	133
6.4	Infinite Surreal Utility	137
6.5	Vector-Valued Utilities	139
6.6	Long-Run Average Utilities	140
6.7	Finite Utility	141
6.8	Negative Infinite and Positive Finite Utilities	142
6.9	Relative Utilities	144
6.10	Superduperdominance	146
7.1	Pascal's Wager	149
7.2	Two-God MGO	150
7.3	Three-God MGO	151
7.4	Three-God MGO with Finite Utilities	155
8.1	Pascal's Wager	169
8.2	MGO	171
8.3	Trickster Deity	175
8.4	Universalism	177
8.5	Evil Deceiver	179
8.6	Our Judgment Flawed	180
9.1	Metaphysical Wager	189

9.2	Mundane Wager	190
9.3	Hybrid Wager	190
9.4	Tragedy of the Commons	201
9.5	Wagerer's Dilemma	202
9.6	Strategies Compared	206
11.1	Wager Assuming P-Theology	226
11.2	Wager Assuming X-Theology	226
11.3	Wager Assuming God Does Not Exist	226
11.4	Wager with No Constraints on Prudentialism	228
11.5	Winding Road	230
11.6	Conjunctive Wager	233
12.1	Pascal's Wager	237
12.2	Many Jealous Gods	240
12.3	Canonical Wager, Relative Decision Table	244
12.4	Many Jealous Gods, Relative Decision Table	245
12.5	The Pantheon	246
12.6	Jealous and Grouchy	246
12.7	Jealous and Universalist	249
12.8	Jealous and Nice Cartel	251
12.9	Finite Utilities	252
12.10	X-Theology	253
12.11	Grouchy Cartel	254
12.12	Jealous and Grouchy Cartel	255
12.13	Grouchy and Nice	255
12.14	Grouchy, Nice, and Jealous	256
12.15	Jealous and Virtuous	257
13.1	Pascal's Wager	261
13.2	Relative Utilities	267
13.3	Mixed Strategies	267
13.4	Jealous and Nice	269
13.5	Grouchy and Nice	269
13.6	Jealous, Nice, Indifferent, and Perverse	271
13.7	Jealous, Nice, Perverse, and a Cartel	272
13.8	Jealous and Virtuous	276
14.1	Pascal's Wager	279
15.1	Pascal's Wager	294

Contributors

Paul Bartha is Professor and Head of the Department of Philosophy, University of British Columbia.

Adam Buben is Assistant Professor of Comparative Philosophy at Leiden University in the Netherlands.

Craig Duncan is Associate Professor and Chair of the Department of Philosophy and Religion at Ithaca College.

James Franklin is Professor at the School of Mathematics and Statistics, University of New South Wales.

Joshua Golding is Professor in the Department of Philosophy at Bellarmine University.

Alan Hájek is Professor at the School of Philosophy, Australian National University.

Jeffrey Jordan is Professor of Philosophy at the University of Delaware.

Paul Moser is Professor of Philosophy at Loyola University Chicago.

Graham Oppy is Professor of Philosophy and Director of the Graduate Program in Philosophy at Monash University.

Lawrence Pasternack is Professor of Philosophy at Oklahoma State University.

Susanna Rinard is Assistant Professor of Philosophy at Harvard University.

Paul Saka is Professor at the University of Texas, Rio Grande.

Elliott Sober is Hans Reichenbach Professor and William F. Vilas Research Professor at the University of Wisconsin (Madison).

Sylvia Wenmackers is Research Professor at the Institute of Philosophy at KU Leuven.

William Wood is a Fellow and Tutor in Theology, and Associate Professor of Philosophical Theology at Oriel College, University of Oxford.

Acknowledgments

The co-editors would like to thank the contributors for providing a diverse and engaging set of perspectives on Pascal's Wager. We are immensely grateful to UBC PhD candidate Sina Fazelpour, who created the combined Bibliography and edited all fifteen chapters for consistency of citations. We thank our many colleagues and students who read and commented on draft material; they are acknowledged in the individual chapters. Finally, we would like to thank Hilary Gaskin for her patience, advice, and support.

Introduction

Paul Bartha and Lawrence Pasternack

1 Pascal's Wager: An Argument with Many Audiences

Pascal's Wager is one of the great classic arguments for belief in God. The other great theological arguments – ontological, cosmological, and teleological – aim to establish that God's existence is necessary or probable. Pascal's Wager, by contrast, is a prudential or "pragmatic" argument. The conclusion is a recommended action: it is in your interest to believe (or to strive to believe) in God, even if you insist that the probability of God's existence may be very low. In its most familiar contemporary formulation, the basic argument runs as follows: so long as the probability of an infinite afterlife reward for belief in God is greater than zero, "wagering for God" has infinite expected utility and is therefore superior to "wagering against God" (which has at best finite expected utility). In short, wagering is a good gamble.

The argument is embedded deep within the *Pensées*. That work was addressed to Pascal's first audience, worldly seventeenth-century Parisians for whom religion was largely an object of indifference. For Pascal, by contrast, nothing could be more serious. Influenced by Jansenist theology, Pascal believed that human nature was thoroughly corrupt. God's grace, freely accepted, could put us on the path to salvation.[1] In Pascal's libertine audience, however, a hard shell of skeptical hostility, reinforced by secular reasoning and habits, formed a barrier to divine grace. The Wager was just one in a series of manoeuvers designed by Pascal to chip away at that barrier. A distinctive feature of his argument is its appeal to practical reason (and gambling instincts), in recognition of the limitations of theoretical reason when it

---

[1] Here we set aside Pascal's belief in the existence of a predetermined class of *elect individuals* who will be saved. Several of the chapters in this collection (Wood, Moser, Franklin) address the complex theological background to the *Pensées* and its seeming conflict with an argument designed to promote belief in God.

comes to faith. Pascal explains that those who lack faith can only hope to achieve it through (practical) reasoning, "until God gives it by moving their heart" (L110/S142). The conclusion of the Wager is not that one should instantly believe in God – neither a realistic nor a possible act given Pascal's understanding of grace – but rather that one should *take steps* toward a "cure," measures that eliminate obstacles to faith. Pascal writes: by "taking holy water, having masses said, and so on," you can "diminish the passions which are your great obstacles" (L418/S680).

The argument has been vigorously debated ever since its earliest appearance. We might identify a second, distinctively philosophical audience for the Wager beginning with Diderot, and extending through Kant, Kierkegaard, Nietzsche, and William James.[2] This group of readers recognized both the virtues and the weaknesses of Pascal's argument, and in some cases developed descendant versions. Increasingly, the Wager came to be considered as a stand-alone argument, removed from the context of the *Pensées*.

This volume is dedicated to a third audience: contemporary philosophers and philosophy students. As is clear from its enduring influence and appeal, Pascal's Wager is a remarkable argument in many ways. It is remarkable for bringing together *big ideas*: infinity, God, salvation, and prudential and evidential reasoning. It is remarkable for its invention of a new style of argument, ultimately formalized as *decision theory*. Decision theory provides both insight into the original argument of the *Pensées* and the tools to develop elaborate contemporary variations. The Wager is also remarkable for its influence on later philosophers. Finally, it is remarkable for its enduring relevance to many areas of philosophy: philosophy of religion, decision theory, formal epistemology, and broader currents of thought. Pascal's Wager continues to have many audiences.

This collection is designed for contemporary students and philosophers. It includes chapters that explore the historical context of Pascal's argument and its influence on later philosophers. It includes discussion of some of the central objections and debates about the Wager. Additional chapters show how ideas in the philosophy of probability and decision theory – imprecise and infinitesimal credences, non-standard decision theory, infinite utility – shape current debates about the argument. These chapters also reveal the many ways in which Pascal's Wager, in turn, has had and continues to have an important influence on different areas of philosophy.

---

[2] Pascal's influence on these later philosophers is discussed in this collection's chapters by Buben and Jordan.

This introduction begins by setting out the technical background for contemporary discussions of Pascal's Wager. This consists of the basic elements of decision theory (§2) and assumptions needed for reasoning about infinite utility (§3). We then review (§4) three versions of Pascal's Wager that appear to be present in the *Pensées*, and provide a brief review of some of the classic objections to the most familiar version. Finally (§5), we provide an overview of the contributed chapters in this book.

2 Standard Decision Theory: Dominance, Expected Value, and Expected Utility

2.1 Dominance

Consider the following gamble. A fair coin will be tossed. On a result of *Heads* you win $100, while *Tails* pays you nothing. If you reject the gamble, you gain nothing. Assuming that you prefer $100 to $0, you should take the bet!

A *decision table* helps to illuminate the reasoning that leads to this conclusion. A decision table has one row for each possible *act* and one column for each possible *state*. Each act–state combination results in an *outcome*, identified here with a monetary sum.

One act *dominates* another if it does *at least as well* as the other act in every possible state, and *strictly better* in at least one state.[3] An act *strictly dominates* another if it does *strictly better* in every state. In Table I.1, *Bet* dominates *Don't bet* (although not with strict dominance). A good decision rule for this example is the *Dominance Principle*.

> *Dominance Principle*: Choose an act that dominates all other available actions, if such an act is available.

We have to be a bit careful with this rule, but it works well here.[4]

Table I.1 Dominance Reasoning

	Heads	Tails
Bet	$100	$0
Don't bet	$0	$0

---

[3] This is sometimes called *weak dominance*, but we will simply call it *dominance*.
[4] The main restriction is that the probabilities of the possible states should *not vary* depending on what action is selected; see Hájek's chapter in this volume for discussion.

Table I.2 EMV Reasoning

	(0.5) Heads	(0.5) Tails	
Bet	$100	$0	EMV(*Bet*) = (0.5)($100) + (0.5)($0) = $50
Don't bet	$40	$40	EMV(*Don't bet*) = $40

2.2 Expected Monetary Value (EMV)

The *expected monetary value* (EMV) if you take the bet is $50, since the probability of *Heads* is 0.5. This number represents your average winnings: you expect to win $100 half the time and $0 half the time.

Suppose you learn that the gamble is not free. The price is $40. No longer do we have a dominant act, but it still looks like we have a good bet (see Table I.2). We justify this conclusion with a decision table that is a slight variation from the previous one. Each outcome is now your *final amount of money* (assuming that you start with $40 and keep it if you *Don't bet*).

The EMV for each act is calculated as a weighted average of the possible outcomes: you multiply the monetary outcome of that act under each state by the probability of the state, and sum over all possible states. (We have indicated the probability above each state. In any decision table, the states must be mutually exclusive and the probabilities must sum to 1.) A good decision rule for this case is *EMV Maximization*.

> *EMV Maximization*: Choose an act that maximizes EMV.

In this case, that act is *Bet*. Notice that EMV maximization is consistent with the *Dominance Principle*, and would also recommend *Bet* in Table I.1.

2.3 Expected Utility (EU)

In many cases, we need a more general decision rule than either the *Dominance Principle* or *EMV Maximization*. That will certainly be true in decisions where there may be *non-monetary* outcomes: a trip to the beach, an enjoyable movie, or a miserable cold. Even with purely monetary outcomes, EMV maximization may not be your best guide. Although you prefer $100 to $40, suppose that you desperately need the $40 for a cab fare to get home. You don't wish to risk it in a gamble even though the odds are in your favor. You represent your preferences by replacing the dollar values in the decision table

Table I.3 EU Reasoning

	(0.5) Heads	(0.5) Tails	
Bet	10	1	EU(*Bet*) = (0.5)(10) + (0.5)(1) = 5.5
Don't bet	9	9	EU(*Don't bet*) = 9

with *utilities*: $u(\$100) = 10$ in place of $100, $u(\$0) = 1$ in place of $0, and $u(\$40) = 9$ in place of $40. (Don't worry too much about how we assign numerical values to the utilities; we'll get to this point shortly.) This yields the data laid out in Table I.3.

The *expected utility* (EU) of each act is calculated by multiplying the utility of that act under each state by the probability of the state,[5] and summing over all possible states. (This quantity is also referred to as the *expectation* of the act.) The decision rule for this case is *EU Maximization*.

> *EU Maximization*: Choose an act that maximizes EU.[6]

In this case, that act is *Don't bet*. Notice that *EU Maximization* can differ from *EMV Maximization*.

In this analysis, a *utility function* u represents your preferences. The function u assigns a numerical value (a real number) to each possible outcome. The most basic requirement for faithful representation is that higher numbers correspond to preferred outcomes: $u(\text{Outcome 1}) \geq u(\text{Outcome 2})$ exactly when Outcome 1 is as good as or better than Outcome 2.

EU Maximization is more generally useful than *EMV Maximization*, and it is the most fundamental principle in standard decision theory. For this decision rule to make sense, however, you have to be careful about how you represent your preferences. Suppose that you change the numbers in Table I.3 to give a different utility function u^*: $u^*(\$100) = 10$, $u^*(\$40) = 2$, and $u^*(\$0) = 1$. Then $EU^*(Bet) = 5.5$ as before, but $EU^*(Don't\ bet) = 2$. The new utility function u^* accurately represents your preference ordering, but now *EU Maximization* seems to recommend *Bet*! To prevent this type of

---

[5] EU calculations use subjective probabilities, also called *credences*. In this example, we assume that the subjective probabilities coincide with the objective probabilities of 0.5 for *Heads* or *Tails*.

[6] In decision problems where several acts are tied for maximum EU, there is no clear recommendation. One other subtlety: in many decision problems, the probabilities of one or more states may vary depending on the act. For these problems, the expected utility of each act is calculated using *conditional probabilities* rather than fixed probabilities for each possible state. In the case of *Pascal's Wager*, we can reasonably suppose that the agent's choice has no influence on the probability that God exists.

contradiction, the *intervals* between utilities must also reflect your preferences accurately. In Table I.3, the interval between having $40 and having $0 is eight times as large as the interval between having $100 and having $40. If this reflects the importance of keeping $40 for cab fare, then the second utility function u^* is not a faithful representation of your preferences. In order to be useful in expected utility calculations, the utility function u should represent not just the preference ordering but also the *structure* of your preferences.

Standard decision theory proves a remarkable result that justifies the rule of EU maximization and ensures that there will be no conflicting recommendations, but it requires some strong assumptions about rational agents. The theory assumes, first, that rational agents have a well-defined *preference ordering* among all possible outcomes. This means that for any two outcomes O_1 and O_2, the agent either prefers O_1 to O_2 (written $O_1 \succ O_2$), prefers O_2 to O_1 (written $O_2 \succ O_1$), or is indifferent between O_1 and O_2 (written $O_1 \sim O_2$).[7] Second, standard decision theory identifies a set of *preference axioms* which are supposed to be satisfied by rational agents. For instance, preferences should be *transitive*: if the agent prefers O_1 to O_2 and prefers O_2 to O_3, then the agent prefers O_1 to O_3.[8] Given these assumptions, the fundamental result of standard decision theory is the *Expected Utility Theorem*: there is a way to represent the agent's preferences with a utility function u so that the utility of any gamble is identical to its expected utility and the best act is always an act that maximizes expected utility. The function u that represents the fine structure of your preferences in this way is almost unique.[9]

Mixed strategies are also an important part of decision theory. In our decision tables, each row corresponds to a *pure act*, something that is within the agent's control. An example of a mixed strategy is to flip a fair coin and then *Bet* if the result is *Heads*, but *Don't bet* if the result is *Tails*. We can represent this as [0.5 *Bet*, 0.5 *Don't bet*], which indicates a probability 0.5 attached to each possible pure act. Of course, decision tables can have more than two rows. In general, a mixed strategy is *any* assignment of probabilities to a set of possible pure acts, so long as the probabilities sum to 1. The *Expected Utility Theorem* shows that the utility

---

[7] Standard decision theory assumes that you have these clear preferences even when the outcomes are gambles, rather than simple outcomes.
[8] For a full list of the axioms of standard decision theory, see Resnik (1987).
[9] If u and v are two utility functions that can be used in EU calculations, then there are constant numbers a and b with $a > 0$ such that $v = au + b$.

of any mixed strategy is its expected utility. For example, based on Table I.3,

$$EU([0.5\ Bet,\ 0.5\ Don't\ bet]) = 0.5\ EU(Bet) + 0.5\ EU(Don't\ bet)$$
$$= 0.5(5.5) + 0.5(9)$$
$$= 7.25$$

Since EU(*Don't bet*) = 9, you are better off with the pure act *Don't bet* than with the mixed strategy of the coin toss.

3 Infinite Utility

Pascal's Wager is commonly represented as a decision-theoretic argument. Pascal seems to suggest several different versions of the argument, but all of them share the same basic decision table (see Table I.4).[10]

The numbers f_1, f_2, and f_3 are all *finite* utilities. If God does not exist, then the rewards and penalties for wagering or not wagering are finite. Pascal seems to think that if God does exist, then the penalty for not wagering is also finite. In the top-left corner, however, we have the possibility of salvation, which Pascal characterizes as "an infinity of infinitely happy life." Since this is better than any finite reward, we represent the utility as ∞.[11]

Infinity in mathematics is an extremely useful concept for talking about certain kinds of limits. For instance:

$$1 + 2 + 4 + 8 + \ldots = \infty.$$

On the left, we have an *infinite series*. What the equation means is that the finite partial sums (first 10 terms, first 100 terms, and so forth) increase without bound. No matter what number we pick, we'll eventually pass it by adding enough terms in the series. If we replace the terms with –1, –2, and so on, then the finite partial sums decrease without bound.

Table I.4 Pascal's Wager

	God exists	God does not exist
Wager for God	∞	f_1
Wager against God	f_2	f_3

---

[10] Here we follow Hájek (2012b).
[11] Note that even if wagering only produces a *positive chance* of salvation if God exists, the correct value for this outcome is still ∞, as will be explained shortly.

In order to talk about such limits, it is convenient to add the two elements +∞ (usually written ∞) and −∞ to the real numbers. The new set is called the *extended real numbers*.[12] We extend the definition of *less-than*, <, by the assumption

(1) For all real numbers x: $-\infty < x < \infty$.

Note that real numbers are all finite; the new elements ∞ and −∞ are *not* real numbers. We extend the definition of basic arithmetical operations by making the following assumptions for operations where one argument is finite and one is infinite:

(2a) For all real numbers x: $x + \infty = \infty$ and $x - \infty = -\infty$.
(2b) For all real numbers x: $x \cdot \infty = \infty$ if $x > 0$ and $x \cdot \infty = -\infty$ if $x < 0$.
(2c) For all real numbers x: $x \cdot -\infty = -\infty$ if $x > 0$ and $x \cdot -\infty = \infty$ if $x < 0$.
(2d) $0 \cdot \infty = 0$ and $0 \cdot -\infty = 0$.

Finally, we make some assumptions about operations where both arguments are infinite:

(3a) $\infty + \infty = \infty$ and $-\infty + -\infty = -\infty$. Note that $\infty - \infty$ is *undefined*.
(3b) $\infty \cdot \infty = \infty$ and $\infty \cdot -\infty = -\infty$.
(3c) $-\infty \cdot \infty = -\infty$ and $-\infty \cdot -\infty = \infty$.

These assumptions correspond to results about limits. For example, (3a) corresponds to the fact that if we take two infinite series, each of which sums to ∞, and form a new series by adding them together term by term, the new series will also sum to ∞. But if we subtract them term by term, the new series might sum to ∞, −∞, or any finite number; hence, in general, $\infty - \infty$ is undefined.

Using the extended real number ∞ in the decision table is the easiest way to represent the infinite value of salvation. Our assumptions, especially (2a) and (2b), give us all the mathematics that we need to perform the expected utility calculations that we encounter in many discussions of Pascal's Wager. It is also plausible to argue that these assumptions faithfully represent Pascal's own statements about infinity (Hájek, 2003). But there is still a big problem. As McClennen (1994) notes, the introduction of ∞ as a possible utility value takes us beyond standard decision theory. Infinite utility is incompatible with one of the preference axioms needed in standard decision theory. Most discussions of Pascal's Wager simply ignore this problem and assume that

---

[12] The following assumptions are taken from Royden (1968).

decision theory, augmented with ∞ and $-\infty$ as possible utility values, works in almost exactly the same way as standard decision theory.[13] We refer to this non-rigorous approach as *naïve infinite decision theory*. One way to avoid this problem is to use a *very large but finite* value for the utility of salvation. This is a sensible option adopted in several of the chapters below, but it has significant consequences for Pascal's argument. Finally, there are a variety of rigorous approaches, unavailable in Pascal's day, which allow us to represent the value of salvation as infinite. We refer to them as *non-standard decision theory*, but we leave their discussion to individual chapters of the book.

4 Pascal's Wager

Hacking (1994 [1972]) has identified three distinct decision-theoretic arguments in *Pensées* L418/S680. In this section, we present the core text (Krailsheimer translation), along with the three decision-theoretic arguments.[14]

Pascal begins by characterizing the decision about belief in God as one that reason cannot decide (on the basis of evidence), but also as one that cannot be avoided. He notes that two things are at stake, knowledge and happiness, but (at least initially) it seems that knowledge is not to be had, so that the case for wagering must rest on happiness.

> Let us then examine this point, and let us say: "Either God is or he is not." But to which view shall we be inclined? Reason cannot decide this question. Infinite chaos separates us. At the far end of this infinite distance a coin is being spun which will come down heads or tails. How will you wager? Reason cannot make you choose either, reason cannot prove either wrong.
>
> Do not then condemn as wrong those who have made a choice, for you know nothing about it. "No, but I will condemn them not for having made this particular choice, but any choice, for, although the one who calls heads and the other one are equally at fault, the fact is that they are both at fault: the right thing is not to wager at all."
>
> Yes, but you must wager. There is no choice, you are already committed. Which will you choose then? Let us see: since a choice must be made, let us see which offers you the least interest. You have two things to lose: the true

---

[13] It won't be exactly the same because we might occasionally encounter a bet whose expected utility is undefined because it involves the calculation $\infty - \infty$.
[14] These three arguments, together with Hacking's analysis, receive detailed discussion in Hájek's chapter in this volume.

and the good; and two things to stake: your reason and your will, your knowledge and your happiness; and your nature has two things to avoid: error and wretchedness. Since you must necessarily choose, your reason is no more affronted by choosing one rather than the other. That is one point cleared up.

Pascal now proceeds to the first version of the Wager, which Hacking calls the *argument from dominance*.

4.1 Argument from Dominance: $f_1 \geq f_3$

The text continues:

> But your happiness? Let us weigh up the gain and the loss involved in calling heads that God exists. Let us assess the two cases: if you win you win everything, if you lose you lose nothing. Do not hesitate then; wager that he does exist.

Consider the decision table (Table I.5). Pascal states, "if you lose you lose nothing." Hacking writes: "if God is not, then both courses of action are pretty much on a par. You will live your life and have no bad effects either way from supernatural intervention." If we interpret this as $f_1 = f_3$ in Table I.5, we have the argument from dominance. *Wager for God* weakly dominates *Wager against God*; the *Dominance Principle* tells us to accept the Wager.

Much later in the text, Pascal suggests that we might actually have what amounts to an argument from *strict dominance*, i.e., $f_1 > f_3$. He offers the following reassurance for one who opts to wager that God exists:

> Now what harm will come to you from choosing this course? You will be faithful, honest, humble, grateful, full of good works, a sincere, true friend ... It is true you will not enjoy noxious pleasures, glory and good living, but will you not have others?
>
> I tell you that you will gain even in this life, and that at every step you take along this road you will see that your gain is so certain and your risk so

Table I.5 Pascal's Wager

	God exists	God does not exist
Wager for God	∞	f_1
Wager against God	f_2	f_3

negligible that in the end you will realize that you have wagered on something certain and infinite for which you have paid nothing.

Both the argument from weak dominance and the argument from strict dominance concentrate exclusively on the right-hand column of the decision table. If $f_1 > f_3$, then the decision to *Wager for God* is an easy one.[15]

Of course, Pascal recognizes that the religious life involves certain sacrifices ("glory and good living"). Although the mature believer may come to see these as "noxious pleasures," it is doubtful that Pascal would have considered the strict dominance argument as persuasive to an audience of hardened skeptics. The other two versions of the argument, accordingly, proceed on the assumption that *Wager against God* is the better option if God does not exist, i.e., $f_3 > f_1$.

4.2 Argument from Expectation

Let us go back to the earlier point in the text. Pascal continues by raising doubts about the Wager that only make sense if $f_3 > f_1$. Since the *Dominance Principle* is inapplicable, the issue has to be settled by an appeal to expected utility.

> [Do not hesitate then; wager that he does exist.] "That is wonderful. Yes, I must wager, but perhaps I am wagering too much." Let us see: since there is an equal chance of gain and loss, if you stood to win only two lives for one you could still wager, but supposing you stood to win three?
>
> You would have to play (since you must necessarily play) and it would be unwise of you, once you are obliged to play, not to risk your life in order to win three lives at a game in which there is an equal chance of losing and winning. But there is an eternity of life and happiness.

In this argument, Pascal returns to the coin-toss analogy: both outcomes, God exists and God does not exist, have equal probability. Even if we allow for a significant finite loss "–L" (the value of a life) in the case where God does not exist, Pascal imagines that the case where God does exist offers a gain two or three times as great, "+2 L" or "+3 L". Calculation shows that the expectation of *Wager for God* is equal to that of *Wager against God* if the prize is +2 L (a case of "double or nothing"):

---

[15] Perhaps not quite so easy as one might think. If there is a correlation between God's existence and your wagering against God, then dominance reasoning fails (see fn. 4) unless we suppose both $f_1 > f_2$ and $f_1 > f_3$. This problem does not arise under the assumption that there is no correlation between your wagering behavior and God's existence.

$$\text{EU}(\textit{Wager for God}) = 0.5 \ (+2 \ L) + 0.5 \ (0) = L$$
$$\text{EU}(\textit{Wager against}) = L$$

Since we have a tie, as Pascal notes "you could still wager" in this situation. Once the prize is +3 L, however, the argument for accepting the Wager is compelling:

$$\text{EU}(\textit{Wager for God}) = 0.5 \ (+3 \ L) + 0.5 \ (0) = 1.5 \ L$$

Consequently, Pascal notes that it would be "unwise" not to wager in this case. His final point is that the case is infinitely stronger than this, because the potential gain is infinite if God exists.

Of course, this argument requires what Hacking refers to as the "monstrous premiss" that God's existence and non-existence are equally probable. The final and most famous version of the argument dispenses with that premise. Hacking calls this the *Argument from Dominating Expectation*; following Jordan (2006), we also refer to it as the Canonical Wager.

4.3 Argument from Dominating Expectation: The Canonical Wager

Pascal continues:

> [But there is an eternity of life and happiness.] That being so, even though there were an infinite number of chances, of which only one were in your favor, you would still be right to wager one in order to win two; and you would be acting wrongly, being obliged to play, in refusing to stake one life against three in a game, where out of an infinite number of chances there is one in your favor, if there were an infinity of infinitely happy life to be won.

In this remarkable passage, Pascal appears to invent the concept of *infinitesimal probability* (one chance out of an infinity of presumably equal chances), and to argue that you should be willing to stake one finite life even for an infinitesimal chance of winning an infinity of infinitely happy life.[16] However, we need not dwell on the difficulties of the calculation, since Pascal immediately sets this possibility aside. In the very next sentence, he insists that although the prize is indeed infinite, the probability of winning (i.e., that

---

[16] Wenmackers's chapter in this volume discusses how Pascal's argument might accommodate infinitesimal probabilities.

God exists) is finite rather than infinitesimal, and that what we are staking is merely finite (the "glory and good living" referred to earlier):

> But here there is an infinity of infinitely happy life to be won, one chance of winning against a finite number of chances of losing, and what you are staking is finite. That leaves no choice; wherever there is infinity, and where there are not infinite chances of losing against that of winning, there is no room for hesitation, you must give everything.

Pascal goes on to dispel the misconception that the infinite distance in *utility*, between the value of salvation and the value of worldly existence, is somehow cancelled out by an infinite distance in *probability*, between the chance of salvation and the certainty of worldly existence, which would nullify his argument. That objection might succeed if the probability that God exists were infinitesimal, but Pascal insists that it is not. He concludes that the case for gambling with a finite stake is compelling when the probability of infinite gain is finite rather than infinitesimal:

> Thus our argument carries infinite weight, when the stakes are finite in a game where there are even chances of winning and losing and an infinite prize to be won. This is conclusive and if men are capable of any truth this is it.[17]

Here is the standard way to represent the Canonical Wager in decision-theoretic terms. Following (Hájek, 2012b), the argument has three key assumptions:

Premise 1: The decision is appropriately modeled by Table I.4.
Premise 2: The probability p that *God exists* is a positive (non-infinitesimal) real number, $p > 0$.
Premise 3: Rationality requires choosing the act with maximum expected utility (if there is one).

We replicate the decision table in Table I.6. As before, we assume that the utility of salvation is *infinite*, and that all other utilities in the table are finite.[18] We write the probabilities above each of the two columns. We then calculate the expected utility of each act:

---

[17] Although Pascal speaks of "even chances of winning and losing," it is clear from context that he means that the relevant probabilities are finite and comparable, rather than infinitesimal.

[18] We may concede that $f_3 > f_1$ so that we don't have a simple dominance argument, but this assumption plays no role in the argument.

Table I.6 Pascal's Wager with Probabilities

	(p) God exists	$(1-p)$ God does not exist
Wager for God	∞	f_1
Wager against God	f_2	f_3

$$\text{EU}(\textit{Wager for God}) = p \cdot \infty + (1-p) \cdot f_1 = \infty$$
$$\text{EU}(\textit{Wager against God}) = p \cdot f_2 + (1-p) \cdot f_3 = \textit{finite}$$

Since *Wager for God* has higher expected utility, that is the prescribed act. This is an argument from dominating expectations because even though p is unknown, *Wager for God* has higher expectation than *Wager against God* over the entire range of *acceptable* values ($p > 0$).

4.4 Review of Objections

We have tried to let Pascal speak for himself, with minimal help from decision theory. His argument has faced many objections, beginning immediately in his own time and extending over the centuries. Here is a brief review, concentrating on the Canonical Wager. Following (Hájek, 2012b), which offers an excellent summary and discussion of the objections, we organize them in terms of the above premises.

Objections to *Premise 1*

- Does infinite utility belong in the table? Can humans experience infinite utility?
- The many-gods objection: should there be more columns (and rows) in the table? If infinite expectation attaches to Pascal's Christian god, then it must also attach to any rival deity who offers an infinite reward and has positive probability of existing. Since there is more than one wager with infinite expected utility, there is no longer any clear recommendation.
- Does belief based on Pascal's Wager count as authentic religious belief? As William James (1956 [1896]) writes:

We feel that a faith in masses and holy water adopted willfully after such a mechanical calculation would lack the inner soul of faith's reality; and if we were ourselves in the place of the Deity, we should probably take particular pleasure in cutting off believers of this pattern from their infinite reward.

In terms of Table I.4, this is a challenge to the assumed value of ∞ in the top-left corner.

Objections to *Premise 2*

- What is the justification for the assumption $p > 0$? What happens if $p = 0$, or if p is infinitesimal, or if we have only a vague idea about the probability that God exists?

Objections to *Premise 3*

- Is it rationally permissible to modify *beliefs* on the basis of pragmatic considerations rather than evidence? *Premise 3* is a principle of *prudential rationality*: it concerns the rationality of *action*. There is a companion principle for *epistemic rationality*: one should always believe on the basis of the available evidence. Call this *evidentialism*. A particularly well-known formulation due to Clifford (1879) states: "It is wrong always, everywhere, and for anyone to believe anything on insufficient evidence." If we adhere to Clifford's Principle, then we have a counterexample to *Premise 3*: although *Wager for God* maximizes expected utility, it is nevertheless not rational to wager for God.

Objections to the validity of the argument (even if *Premises 1–3* are granted)

- Is it legitimate to appeal to naïve infinite decision theory in Pascal's Wager? The rule of *EU Maximization* is derived in standard decision theory; that is the appropriate domain of application for *Premise 3*. We have no analogous rule about maximizing EU for naïve infinite decision theory. Hence, the conclusion does not follow.
- Even if we assume that *Premise 3* can be legitimately extended to naïve infinite decision theory, does the conclusion follow? The *mixed-strategies* objection alleges that it does not. Consider the strategy of flipping a coin and taking the wager on a result of *Heads* but rejecting it on a result of *Tails*. This strategy has infinite (naïve) expected utility. So does any mixed strategy that offers a positive chance of taking the wager (provided all other outcomes have finite utility). While our

premises tell us to choose *one* of the strategies with infinite expected utility, they do not uniquely prescribe the pure act, *Wager for God*. Hence, the conclusion does not follow.[19]

Since the middle and latter sections of this book are devoted to a discussion of the above objections, and to how Pascal's Wager might be modified in order to address them, we will not offer an extended discussion.

5 Overview

This volume consists of fifteen new essays on Pascal's Wager. Reflecting the complex history and philosophical status of Pascal's Wager, the book is organized into three main sections. Part I: Historical Context and Influence provides two chapters that set the stage by exploring the philosophical and theological context of Pascal's Wager, followed by three chapters that explore how the argument shaped later philosophical work. Part II: Assessment consists of five chapters devoted to the exposition and appraisal of Pascal's argument. Part III: Extensions contains five chapters of a more technical nature, exploring what happens to the Wager and its objections when we introduce contemporary ideas about probability, utility, and belief dynamics.

Chapters in this volume can be read independently and there is no need to follow any particular order. However, anticipating the interests of our readers, we offer the following suggestions. For those interested in the *historical context* of Pascal's argument, Franklin and Wood provide natural starting points. In addition, the chapters by Buben, Moser, Jordan, and Pasternack all engage with historical themes. Golding and Saka explore contemporary issues in *philosophy of religion*, and Moser works through the background theology of the Wager in light of some contemporary issues in the philosophy of religion. Readers particularly interested in the many-gods objection may wish to start with Duncan's and Pasternack's chapters, though this famous objection receives attention in many of our chapters. For readers oriented toward *contemporary decision theory*, Hájek's chapter helps to set the stage for the analyses found in the *Extensions* portion of this volume. Within that part, the first three papers (Sober, Bartha, and Oppy) emphasize belief

---

[19] There might be *additional premises* that allow us to "break the tie." For instance, Schlesinger (1994) argues that when multiple acts have infinite expected utility, we should choose the act with the highest probability of achieving an outcome with infinite utility. Schlesinger's Principle is discussed in several of the chapters in this collection.

dynamics and the utilities in the Wager, while the latter two (Rinard and Wenmackers) examine how Pascal's argument fares when we allow the probability of God's existence to be either imprecise or infinitesimal. In general, we recommend Hájek's chapter, which closely examines the basic structure of Pascal's Wager, as an excellent all-round starting point for most readers. Those who are familiar with contemporary decision theory may choose to begin with that chapter; for those unfamiliar with decision theory, the first few sections of this introduction will serve as a primer.

Part I Historical Context and Influence

In Chapter 1, James Franklin provides valuable historical context for understanding Pascal's Wager through an appreciation of its intended audience. Beginning with early versions of the Wager, Franklin shows that the argument emerged from deeply ingrained ideas about religion as a practical way of reasoning about high-stakes situations. A very important point in understanding these early versions, and Pascal's formulation as well, is that the intended audience for the Wager was "real decision-makers" rather than philosophers. Real decision-makers, such as Pascal's "man of the world" of 1660, face a range of religious options they take to be serious, with fixed probabilities grounded in their evidence and with utilities that are fixed quantities in actual minds. The many ingenious objections to the Wager dreamed up by philosophers, with their limitless range of possible gods, abstract probabilities, and fanciful payoffs, are irrelevant for the intended audience. Using a sober "real-world" decision matrix, the Wager looks like a good bet. Furthermore, Franklin argues, the Wager is adaptable and still powerful even for a contemporary intellectual (though perhaps less so for a contemporary philosopher), although the range of options and actions is not the same as in Pascal's day and the conclusion is more modest: one should undertake *serious investigation* of religious options.

In Chapter 2, William Wood argues that Pascal's Wager was originally intended not as a stand-alone argument but as one manoeuver in Pascal's long apology for Christianity: the *Pensées*. Pascal's Jansenist theology at once provides context for this apology (and the Wager in particular) and a deep puzzle about Pascal's motivation. As a Jansenist, Pascal was convinced that humanity was sunk in sinful habits and desires, with redemption possible only through divine grace. The Wager ingeniously appeals to those very habits (calculating self-interest as exemplified by gambling) to bring the reader to a state receptive of grace. The deep puzzle, however, is why Pascal would bother to write an apology when, as a Jansenist, he believed that God had predestined the elect for salvation. Specifically: if humans can do nothing to draw themselves closer to God without a special infusion of grace, which is only offered to the elect and cannot be resisted, what is the point of the Wager argument? Wood suggests that Pascal's theological principles allow for a number of solutions, including the idea that the Wager itself could be an instrument that plays a part in individual redemption.

In Chapter 3, Paul Moser examines how Pascal's Wager relates to a number of contemporary issues in the philosophy of religion. In particular, he explores the place of divine hiddenness in Pascal's theology as well as how various norms for theological inquiry can generate concerns about that theology and its Jansenist framework. He develops a critique of the doctrine of Election as found in Jansenist theology, and how that doctrine carries over to the Wager. Moser recommends an alternative to the Jansenist God, emphasizing the moral necessity of postulating a God who makes salvation possible for all. Moser then considers the tensions between this theological alternative and the implied divine hiddenness of God in Pascal's Wager. He argues that while divine hiddenness lends toward fideism, and in turn allows for the more severe Jansenist theology, our moral concepts nonetheless ought to steer us away from Jansenism and toward a theology which preserves at least the possibility of universal salvation.

In Chapter 4, Adam Buben explores Pascal's influence on several eighteenth- and nineteenth-century philosophers, especially Immanuel Kant, Søren Kierkegaard, and Friedrich Nietzsche. Although Kant does not discuss Pascal beyond just a few passing comments, there are a number of conceptual affinities between the Wager and Kant's "Moral Argument" for belief in God and an afterlife. After a brief exploration of these, Buben turns to Kierkegaard's reading of Pascal. He examines in particular Kierkegaard's reservations about Pascal's apologetics, and whether or not the latter's conception of faith involves merely a "suspension" of reason or a more radical "dying to" it. Buben then turns to Nietzsche's various comments on Pascal amidst his infamous critique of Christianity, focusing in particular on the connections drawn between Pascal and Schopenhauer, their shared "pessimistic gloom," as well as their greatness as "higher, rarer men." Lastly, Buben considers to what extent we may extrapolate from Nietzsche's views on Pascal to what Nietzsche would have thought about Kierkegaard.

In Chapter 5, Jeff Jordan begins with an examination of the differences between Pascal's Wager and William James's "Will to Believe" argument. While these are both pragmatic arguments, the Wager, Jordan emphasizes, is a special case as it is not merely an argument for one particular action or belief among many, but is rather directed to what one will choose as the object of their "ultimate concern." Jordan then considers the objections to pragmatic belief raised by W. K. Clifford and what Jamesian principles are available to subdue these objections. In particular, he challenges Clifford's view that beliefs ought only be adopted based upon their evidential support, as well as the "Agnostic Rule" which maintains that we ought to withhold belief whenever the evidence is insufficient. Jordan then applies these principles to Pascal's Wager, using them as part of a defense against the many-gods objection. He argues that wagerers do not have to consider every logically possible scenario, and can instead limit their decision matrices to only those options which they regard as "live hypotheses."

Part II Assessment

In Chapter 6, Alan Hájek simultaneously addresses the questions of how best to formulate Pascal's Wager using decision theory and whether the argument is valid. Hájek begins with careful scrutiny of Pascal's text and of Hacking's influential reading that identifies three versions of the Wager (as explained in section 4). There are some surprising plot turns: Hájek finds that Hacking misrepresents Pascal's reasoning in a number of places, and he reaches the radical conclusion that *all three versions* of Pascal's argument are invalid. Hájek then turns to a review of twelve distinct reformulations of the Wager, obtained by modifying either the decision matrix or the decision rule. All of these reformulations are formally valid, but Hájek notes ways in which they depart significantly from Pascal's original reasoning.

In Chapter 7, Craig Duncan distinguishes between two broad versions of the many-gods objection: what he calls the "ambitious many-gods objection" and the "modest many-gods objection." The former presents us with a plurality of theological hypotheses each offering the possibility of infinite utility; hence, there is no singular choice with infinite expected utility. Duncan provides a rebuttal to this version of the many-gods objection by replacing the standard infinite utility of our afterlife salvation with an "arbitrarily large" finite value. Distinct theological hypotheses then have different expected utilities because they have distinct probabilities. Duncan defends this substitution against an objection raised by Alan Hájek, and then moves on to the "modest many-gods objection," addressing the relative plausibility of a traditional Christian god versus a god who judges us based upon our epistemic virtues. Duncan suggests that both deities remain plausible enough to remain in the decision matrix, but that we are in no position to justify any judgment as to which deity has the higher probability. Hence, he concludes that once we replace infinite utilities with "arbitrarily large" finite values, the Wager remains vulnerable to the "Modest" version of the many-gods objection.

In Chapter 8, Lawrence Pasternack challenges many of the most common strategies used to defend the Wager against the many-gods objection. He worries that these strategies either compromise the status of the Wager as a decision under ignorance or carry over beyond the "cooked-up" hypotheses and "philosopher's fictions," which are their intended targets, to atheism as well. As such, Pasternack wonders whether the standard rebuttals to the many-gods objection broadly undermine the Wager, for they may have the unintended consequence of not only blocking "cooked-up" hypotheses and "philosopher's fictions," but also removing atheism from the decision matrix. Pasternack then suggests an alternative approach which more adequately preserves the Wager as a decision under ignorance. Rather than following the most common strategies against the many-gods objection, which draw upon considerations independent of the Wager, Pasternack moves through a series of characteristics already built into the Wager's decision-theoretic structure, showing how they can be leveraged against an array of "cooked-up" hypotheses and "philosopher's fictions."

In Chapter 9, Paul Saka explores whether or not it is rational to cultivate religious belief by way of the Wager. He distinguishes between the standard "metaphysical wager" about our afterlife fate, and the "mundane wager" which instead argues for the value of religious belief in this life. In this regard, Saka's chapter may be seen as an inquiry into whether or not the overall decision matrix of the Wager should be one of weak/simple dominance (or even strict-dominance). More fully, Saka raises a series of questions about the merits of various models of rationality. He considers different norms for prudential evaluation, the context-dependence of the value of the mundane wager's outcomes, and whether the benefits of religious belief are best modeled atomistically or in terms of collective well-being. Crucial to his argument are empirical studies suggesting that religious belief and practice are positively correlated with well-being at the individual level, but negatively correlated at the societal level. Saka argues that these empirical studies suggest different implications for the mundane wager, depending on whether one embraces an individualistic or cooperative concept of rationality.

In Chapter 10, Joshua Golding explores the concept of authentic belief, and more specifically, what it means to have an authentic *religious* belief. Golding thereby raises a number of concerns about the sort of belief-states that would come about through the Wager, whether the Wager could engender an actual belief-state, whether it promotes something more akin to a "self-brainwashing," and whether its appeal to self-interest can be understood as consistent with the belief-states proper to religion. Golding suggests that at least some worries about the Wager can be overcome if, instead of belief, it is understood as an argument in support of a "pragmatic assumption," wherein one does not hold to the truth of a proposition, but rather uses it more as a maxim to guide one's actions. Golding then applies the distinction between belief and pragmatic assumption to Judaism, differentiating between aspects of Jewish religious life for which a "pragmatic assumption" is sufficient and aspects that require full-fledged and authentic belief.

Part III Extensions

The traditional "evidentialist" objection to Pascal's Wager, outlined at the end of §4 (above), holds that Pascal's argument illegitimately imports prudentialist norms into the context of belief, where epistemic norms alone should apply. In Chapter 11, Elliott Sober alleges that the Wager actually suffers from the opposite problem: once we are willing to embrace prudentialism about belief, we should acknowledge that Pascal does not go far enough! Pascal holds fixed certain facts about what God would have to be like – call these assumptions a "theology" – and invites the reader to opt for belief based on prudential considerations. Mougin and Sober (1994) have argued that you could instead hold fixed your evidence and beliefs about God's existence, and undertake prudential scrutiny of your background theology. In this chapter, Sober observes that on a more broadly prudentialist approach, both degree of belief in God and theological commitments should be up for grabs. He sketches a dynamic model (along the lines of Brian Skyrms's dynamics of rational deliberation) for how our beliefs, of both types, might evolve toward a stable equilibrium. Different starting points lead to different end points, raising new challenges for Pascal's Wager.

In Chapter 12, Paul Bartha develops a very different dynamical approach to the Wager. Bartha raises the question: given that the conclusion of Pascal's argument is that you should act in ways that are likely to promote your belief in God, what happens if doubt remains? If you keep taking the Wager, your subjective probabilities should continue to evolve until they reach some type of equilibrium. Bartha's dynamical approach is developed via analogy with Brian Skyrms's replicator dynamics. Wagers for distinct theological hypotheses compete for our adherence; the formalism is similar to that used in evolutionary game theory where population groups compete for survival. Bartha argues that this approach allows us to respond to some familiar difficulties with Pascal's Wager, such as some complex versions of the many-gods objection, but that it also opens up some new challenges.

In Chapter 13, Graham Oppy takes on the challenge of making sense of the infinite utility that plays a central role in the canonical version of Pascal's Wager. Oppy runs through a catalogue of difficulties. Can humans meaningfully assign infinite utility as the outcome of wagering for God? Can naïve infinite decision theory overcome the mixed-strategies objection (§4d, above)? Oppy then turns to non-standard mathematical representations of infinity in the Wager (as presented in Hájek, 2003). He argues that since these approaches countenance infinite utilities, they should also permit infinitesimal probabilities. The decision about whether to wager then turns on the precise combination of the probability and utility values. The last part of Oppy's chapter is a critique of the approach taken by Bartha (2007, 2012, and Chapter 12 of this collection). Bartha's "relative utilities" avoid many of the problems that plague naïve infinite decision theory. However, Oppy observes that Bartha's approach canvasses too limited a range of possible deities, requires a strong equilibrium requirement that has no independent justification, and cannot handle infinitesimal probabilities. He concludes that there is as yet *no* satisfactory treatment of infinite utility.

In Chapter 14, Susanna Rinard asks what happens to Pascal's Wager if we drop the usual assumption that a rational agent must have a sharp, real-valued credence $p > 0$ that God exists and suppose instead that it is reasonable to have an *imprecise* credence. For instance, your credence might be vague over the interval (0, 1/100) if you think that the probability that God exists is greater than 0 but less than 1/100. Rinard shows that decision theory can accommodate imprecise credences, and that Pascal's Wager still succeeds for an agent whose credence that God exists is vague over a range that does not include 0.

For an agnostic whose credence interval includes 0, Pascal's Wager fails. However, Rinard provides three distinct arguments that for any contingent proposition P, including the proposition that God exists, no rational agent has a credence interval for P that includes 0. Thus, she concludes that the move from precise to imprecise credences makes no difference to the success of Pascal's Wager for any rational agent.

In Chapter 15, Sylvia Wenmackers explores the implications for Pascal's Wager if we allow agents to have an infinitesimal probability that God exists. As noted above (§4c), Pascal anticipated and rejected infinitesimal probabilities, but his reasons are not entirely clear. A number of philosophers (e.g., Oppy, 1991) have argued that infinitesimal probabilities are not merely permissible but sometimes rationally required – for example, in a scenario in which an agent has non-zero credence for infinitely many, equally likely deities. It is natural, then, to ask whether Pascal's Wager succeeds for an agent who combines an infinitely small credence in the existence of God with infinite utility for salvation. This question only makes sense relative to some non-standard decision theory. Wenmackers argues that any such theory should satisfy a "harmony" condition that allows us to combine infinite utilities and infinitesimal probabilities, and she shows that hyperreal decision theory meets this criterion. Within hyperreal decision theory, infinite utility can be "neutralized" by infinitesimal probability, provided the infinitesimal is small enough. There is an interesting analogy here between hyperreal and finite-utility versions of the Wager: in both cases, success or failure depends upon the precise combination of utility and probability values. Wenmackers's chapter also reviews important concerns about whether the hyperreal formalization is sufficiently faithful to Pascal's original argument.

Part I

Historical Context and Influence

1 Pascal's Wager and the Origins of Decision Theory: Decision-Making by Real Decision-Makers

James Franklin[*]

Philosophers approach Pascal's Wager with an initial impression that it is "too good to be true" (Hájek, 2003). That gives them permission to indulge a professional disposition to invent "what-ifs" involving gods with strange motivations, arbitrary distributions of probabilities and payoffs, and calculations with hyperreals and infinitesimals.

Risk analysts involved in actual decision-making, such as safety engineers for nuclear power plants, cannot allow themselves such intellectual luxuries. Serious decision theory keeps to realistic ranges of possibilities and associated risks. As Pascal and his interlocutors knew, choice of religion is equally serious. The person faced with that choice has an obligation to avoid frivolity.

A survey of Wager-like thinking before Pascal is useful. It shows how ingrained into real religious thinking the decision-theoretic considerations of the Wager are. The idea of the Wager applies much more widely than to the particular choice of options and payoffs that Pascal describes. As we will see later, the same applies, though differently, to the present-day situation.

1 Pre-Pascalian Versions of the Wager

Arguments with a broad similarity to Pascal's Wager are traditional in both popular and learned religion (Ryan, 1945). No doubt the "worth a shot" or "insurance policy" aspect to religion has always been part of its appeal, as well as a source of the lukewarmness in faith that can so exasperate the zealous. The Christian apologist Arnobius, about 300 CE, is probably the first to express the argument in a form that recognizably involves the rationality of a decision

---

[*] I am grateful to Vlastimil Vohánka, Paul Bartha, and Lawrence Pasternack for helpful comments.

in a case of doubt. His reasoning involves doubt, rewards, and decision in response to the doubt and rewards, but not exactly probability:

> There can be no proof of things still in the future. Since, then, the nature of the future is such that it cannot be grasped and comprehended by any anticipation, is it not more rational, of two things uncertain and hanging in doubtful suspense, rather to believe that which carries with it some hopes, than that which brings none at all? For in the one case there is no danger, if that which is said to be at hand should prove vain and groundless; in the other there is the greatest loss, even the loss of salvation, if, when the time has come, it be shown that there was nothing false in what was declared. [Arnobius, 300 CE, 2.4; 1907–1911, p. 434]

That is more generic than Pascal's version, and so directs attention to the core of the argument, as well as showing its adaptability. As often noted, the reasoning has the same form as the "precautionary principle" applied to such cases as global warming. A twenty-first century "Arnobian" might argue: "there is no danger [in carbon reduction, even if some inconvenience] if that which has been said to be at hand [about the risks of climate change] should prove vain and groundless; in the other there is the greatest loss, even the loss of [much of the world's dry land], if, when the time has come, it can be shown that there was nothing false in what was declared." That suggests that reasoning of a Pascalian nature is commonly taken to be sound, in cases of genuine risks.

The generic nature of Arnobius's argument, expressing common hopes, should be kept in mind when we consider whether objections to the details of Pascal's argument carry over to more general versions.

The comparison of choosing religion to a bet and the infinite payoff of choosing religion, both ideas that are important in Pascal's version, were added by William Chillingworth in his 1638 book *The Religion of Protestants a Safe Way to Salvation*. Chillingworth also uses the language of probability, which Pascal avoids but which has been commonly used since. He writes:

> For who sees not that many millions in the world forego many times their present ease and pleasure, undergo great and toilsome labours, encounter great difficulties, adventure upon great dangers, and all this not upon any certain expectation, but upon a probable hope of some future gain and commodity, and that not infinite and eternal, but finite and temporal? Who sees not that many men abstain from many things

they exceedingly desire, not upon any certain assurance, but a probable fear, of danger that may come after? What man ever was there so madly in love with a present penny, but that he would willingly spend it upon any little hope, that by doing so he might gain a hundred thousand pounds? And I would fain know, what gay probabilities you could devise to dissuade him from this resolution. And if you can devise none, what reason then or sense is there, but that a probable hope of infinite and eternal happiness, provided for all those that obey Christ Jesus, and much more a firm faith, though not so certain, in some sort, as sense or science, may be able to sway our will to obedience, and encounter with all those temptations which flesh and blood can suggest to avert us from it? [Chillingworth, 1840 [1638], pp. 430–31; Franklin, 2001, p. 250]

For later reference, we should note Chillingworth's observation that Wager-like reasoning typically works with large finite payoffs. The infinity of religious payoffs is not essential.

The closest parallel to Pascal's Wager in earlier work occurs in the 1637 book *On the Immortality of the Soul*, by Antoine Sirmond, one of the Jesuits most severely attacked in Pascal's *Provincial Letters*. Sirmond's version contains very explicit balancing of risks and rewards. As in Arnobius, but unlike in Pascal, much play is made of the downside risk of eternal damnation:

> Without favouring one party or the other, let us suppose there were nothing decided in the matter, and that it were problematic and equally doubtful on either side ... There is no man of good sense who would not rather lose a day or an hour of his pleasures than risk an eternity of happiness, or who would not choose to endure in the present a pinprick for a quarter of an hour, rather than put himself in danger of a torment which would have neither moderation of its rigour, nor limit to its duration. Compare the goods of this life with those to be feared or hoped for in the next, if there is a next life, and you will find there is no more proportion between the terms of this comparison than there is between the stakes and the rewards of this choice. What will happen to the man of vice? He prefers to play for the present than to attend to the future. And if the future deals with him otherwise than he thinks, if his soul finds itself taken on leaving the body, and finds itself existing in the midst of the sufferings it will have deserved, what will be his condition? Truly to have inherited an eternal evil for a moment of the pleasures of

which this life has delivered the enjoyment; to have lost eternal happiness that could have been bought at the price of a little pain in the exercises of several virtues contrary to his humour. Will he not reason, and tell himself thereafter: if I die completely when I quit this earthly coil, my lot will be to have avoided the evils of this life, and to have embraced its goods, as far as I could. In addition, I have naught either to fear or to hope, beyond the experience of sixty or a hundred years or so, that will be the most that will roll past me. If, on the contrary, I were to find after death a land where one lives longer than one does here, I would see myself condemned to torments intolerable in their gravity and infinite in duration. I would feel myself excluded from a state happy in proportion, full of all sorts of goods, and assured for eternity. What then is to be done? [Sirmond, 1637, pp. 456–61, quoted in Blanchet, 1919, pp. 628–30; Franklin, 2001, pp. 251–52]

Sirmond considers the objection that what is in the present is certain, while the supposed future life is only a prospect. He replies: "It is true that certainty in the present is worth more than uncertainty in the future, as long as there is some proportion between the two. But when it is a matter of eternal life or death, how can there be any comparison with a temporal life or death?" There is a Latin edition of Sirmond's book, which generally agrees with the French edition but goes further in comparing the choice between religion and "the world" to a game of chance:

No-one is a man of truly sound mind, who with a not unequal partner, wants to play dice or ball or any kind of game, such that if he wins, he gains a penny, while if not, he loses a most flourishing and opulent and everlasting kingdom ... However long and happy the space of this life may be, while ever you place it in the other pan of the balance against a blessed and flourishing eternity, surely it will seem to you, if the weight of things be known, that the pan will rise on high even as if you were to weigh a penny against the weight of gold of the most splendid kingdom ... [Sirmond, 1635, pp. 390–92, quoted in Blanchet, 1919, pp. 633–34]

Examining these versions in conjunction with Pascal's thus allows us to extract certain features which might not be so clear if we focused solely on Pascal's text. They include the adaptability of the Wager to more generic spaces of choices, the importance of downside risks, and the applicability of the Wager when payoffs are large but finite.

2 Pascal and the Decision Theoretic Perspective

The strength of Pascal's own version of his Wager is its focus on the decisions of a real agent. It is not about probabilities and payoffs in an abstract Platonic space of pure reasons and disembodied probabilities, but about the decisions faced by a person in doubt in the Paris of 1660 (and by extension, the decisions faced by other real persons in different religious contexts).

That already explains why some of the simpler objections to the Wager are misconceived. It is no use arguing that belief is not a matter of decision so that one cannot decide to believe in God. Pascal is well aware of that and his conclusion is strictly about action: "taking holy water, having masses said, and so on." The point of those actions is not to gradually deceive oneself into belief, either. Pascal, the radical Jansenist skeptical of free will, has no time for free actions leading to belief. In his view, faith is a gratuitous and undeserved gift of God, and the most one can do is to remove the obstacles to the action of divine grace – even that is doubtfully compatible with the rigorous determinism of Jansenism. Pascal certainly does not believe in salvation through pretending to believe, or acting "as if" religion is true – that is exactly what he criticized Sirmond for, when Sirmond had argued that one might be saved with good works "as if" one passionately believed in God but without inwardly doing so (Elster, 2003).

Nor is it any use complaining about the moral tone of the Wager. The Wager has often been obscured by a caricature of itself: "Being base and greedy, we want lots of goodies in this life and, if possible, the next. So we are prepared to give up some pleasures now, on the off chance of a lot more later, if our eye to the main chance makes it look worth our while. Since the loot on offer is infinite, even a small chance of raking it in makes it worth a try to grovel to any deity that might do what we want" (Franklin, 1998). Voltaire makes a bid for the high moral ground in saying, "That article seems a bit indecent and childish; that notion of gambling, of losses and winnings, does not suit the gravity of the subject" (Voltaire, 1961 [1778], p. 123). Possibly for this reason, the Wager has been almost as unpopular in religious circles as it has been in atheist ones. It is all beside the point. The motives anyone has for adopting an argument's conclusion have no bearing on the logical validity of the argument. That is the point of classifying *ad hominem* arguments as fallacies. Pascal's strict decision-theoretic approach bypasses any such complaints. An agent – in his time or ours – faces a forced bet, and so is obliged to consider the payoffs. He may take advice from his psychoanalyst on his motives if he wishes, but that will not affect what he hears from his philosophical adviser.

Pascal has the answer. "You must play." Having admitted that, it is time to consider the options, the probabilities and the payoffs.

3 The Range of Options (of a Real Agent)

Proponents of a Pascalian wager argument, including Pascal, are inclined to represent the religious options as two: theirs versus the rest (or often, theirs versus atheism). They would say that, wouldn't they? Others may be skeptical. "An imam could say as much as Pascal," as Diderot (1875 [1746], p. 167) quite rightly said, and a modern perspective may well see a considerably wider range of religious options as having some initial credibility.

That has often been claimed as a serious weakness of the Wager. Flew says, "The central and fatal weakness of this argument as an argument is that Pascal assumes, and has to assume, that there are only two betting options" (Flew, 1976, p. 66).

That cannot be right, since decision theory applies to any spectrum of hypotheses, with any distribution of probabilities across them, and with any set of payoffs. Whether Pascal's reasoning applies in some wider setting than the one he assumes remains to be seen. Perhaps the differences when the spectrum of options is wider are inessential to the argument, perhaps not. The applicability of the Wager to different spaces of options cannot be ruled out beforehand.

So, for a real agent contemplating the choice of religion, what is the spectrum of options?

That is for the agent to say. As for any real decision problem, the agent has a (typically narrow) range of options that are genuine, or really on the table, or realistically under consideration. For one thing, a real agent has only limited cognitive power and so can only feasibly manage a few options. But even apart from that, any case requiring decision has to constrain attention to those options that are in some sense realistic. In risk analysis and criminal trials, for example, risks have to be initially divided into genuine and fanciful. In evaluating the guilt of the accused in court as "beyond reasonable doubt," scenarios where he might be innocent have to be sorted into those genuinely probable and those merely possible to think up. In Lord Denning's famous statement,

> Proof beyond reasonable doubt does not mean proof beyond a shadow of a doubt. The law would fail to protect the community if it permitted fanciful possibilities to deflect the course of justice. If the evidence is so

strong against a man as to leave only a remote possibility in his favour which can be dismissed with the sentence "Of course it is possible but not in the least probable," the case is proved beyond reasonable doubt; nothing short will suffice. [Denning, 1947]

That is what decision in real life is like. It should be the same for decision in choice of religion.

It is a stock example in Philosophy 101 that the hypothesis of there being an elephant in the next room is a logical possibility, but should have no effect on anyone's action as it is a bare possibility and has no reasons in its favor. Similar points are made endlessly when discussing demon or vat skepticism: it is usually said that though logical possibilities, they need not be taken seriously as there are no reasons for them. Mere possibilities that one is wrong, or that some theory one has not investigated or heard of is right, do not impel the reasonable person to any action.

It is a pity then that when philosophers come to consider Pascal's Wager, those wise strictures are thrown out the window. Philosophers are the very ones eager to invent merely possible hypotheses that are alleged to show that Pascal has not taken into account the correct space of hypotheses.

Thus Bartha explains the common reasoning, "If we are prepared to assign positive probability to one deity, why stop there? Suppose that each of these deities offers an infinite reward to believers" (Bartha, 2016). Oppy (1991) invites us to consider, for each positive integer n, the hypothesis that there are "n deities (all much like the Christian God) who reward all and only those people who believe that there are n deities who are much like the Christian God." Other versions of the "many gods" objection exist, no doubt infinitely many.

It seems one could just as well consider merely possible religions, not just actual ones. Could not any would-be prophet whip up a structure of hopes of infinite future rewards and punishments, and barter them for tithes in the present? And perhaps one should consider such hypotheses as that God punishes especially severely those who hypocritically assume the forms of religion as a cover for greed. An uncommitted skeptic, in particular, is one who will consider a larger than average range of possibilities, including ones that promise punishment for belief and/or rewards for intellectual honesty (Cargile, 1966).

None of that impacts Pascalian reasoning. One can consider, suppose, invent, or imagine all the hypotheses one likes. Those "philosopher's fictions"

(Jordan, 2006, p. 75) have no impact on the decisions of a real decision-maker and the range of options he has for serious consideration (argued at length in Jordan, 1994b).

It remains to be explained exactly what distinguishes solid and realistic hypotheses, which must go in the decision matrix, from fanciful and speculative ones that can be ignored. That is traditionally expressed by assigning non-zero "probability" to the realistic ones. That shifts the problem to the meaning of probability in decision theory, to which we turn in the next section.

What Pascal's Wager looks like in the context of a range of options that is more attuned to the epistemic situation of contemporary agents will be considered later.

A final question is whether Pascal fairly stated the range of options confronting the decision-maker to whom he was addressing his argument, the "man of the world" of 1660 – as Rescher puts it, "the ordinary, self-centred, 'man of the world' preoccupied with his own well-being and his own prudential interests ... the glib worldly cynic – the free-thinking *libertin* of his day, the sort of persons who populated the social circle in which Pascal himself moved prior to his conversion" (Rescher, 1985, pp. 26–27). That is of historical interest, but also of philosophical interest in raising the general question of honesty in stating the range of options.

The answer is no. Pascal's two stated options are the "Catholic" one and the atheist one. But the informed Parisian audience well knew – as much as anything through Pascal's vigorous polemic in the *Provincial Letters* – that there were really two Catholic options, Jansenist and Jesuit, which differed on a question directly relevant to the Wager. The Jesuits such as Sirmond and the moral theologian Antonio Escobar y Mendoza, very laxly in Pascal's view, did not believe that wagering against God necessarily resulted in the loss of salvation, since good works performed with a good intention might suffice for salvation. That threatened to give the action of wagering against God also an infinite payoff. Thus Pascal had to spend time ridiculing the Jesuits so that their position could be ruled out from the start: "Ridiculous to say that an eternal reward is offered for morals *à la* Escobar" (L692/S571). Not everyone in 1660 agreed, or should have agreed.

4 The Grounded-Subjective Nature of the Probabilities in the Wager

There is a dilemma. To be relevant to decision theory of an agent, the probabilities involved must be actually held by the agent, so in some sense subjective. On the other hand, if they are disconnected from what the evidence really implies, anyone can think what he likes and construct "probable" religious beliefs to suit himself. In particular, Pascal's Wager only has purchase if the probability of God is really (in some sense) greater than zero. So the probabilities involved must be in some way objective. The problem then is to say in what way the probabilities are subjective and in what way objective. (Or is the combination impossible?)

Saying that the relevant probabilities are subjective does not have to mean that anything goes. Three parallels will make it clear. According to Catholic moral theory, one has an absolute obligation to obey one's conscience, but also an obligation to inform that conscience (to conduct "due diligence," as is said in the legal world). So one must follow one's own (subjective) views, but one is at fault if those views are culpably ill-informed (Catholic Church, 1992, pars 1783–1793). Or again, the utilitarian who is required to calculate the probable future utility of his actions needs this probability to be subjective (since that is all that he knows) but informed (so that a self-serving but crazy distribution of probabilities does not serve as an excuse for wrong action) (Smart, 1986, p. 31). The same applies to more practical cases such as deciding which risks to a nuclear power plant justify spending money on precautions. The probabilities of those risks must be entertained in the mind of the planner and dependent on his beliefs, but also must be grounded in the evidence available. Risk committees in nuclear power plants do not employ philosophers to expatiate on the "many catastrophes objection" as to why no precautions are worthwhile – they get on with addressing those risks they know about which have, to their knowledge, a real chance of being realized.

The combination of subjectivity and objectivity of the probabilities in such cases is not hard to understand. It is just a special case of the general need of knowledge and opinion to be in a reasoner's mind yet grounded in the evidence available to the reasoner.

Action, like conscience, is serious. It requires the beliefs on which it is grounded to be serious too.

That explains why the probabilities of fanciful theories, such as those considered in the last section, can be rightly regarded as zero and hence need not appear in the decision-maker's payoff matrix. Mere possibilities dreamed up by inventive philosophers do not provide the decision-maker with any reasonable ground for believing them, and he cannot assign non-zero probability to them. Hence his grounded subjective probabilities for them can and ought to remain at zero. They could acquire a non-zero probability if someone found a half-way reasonable argument for one of them. But their role in the many-gods objection was exactly to appear as mere possibilities.

It is true that if the probabilities are subjective though grounded, a zero probability for God may in some circumstances be reasonable. If someone grows up with only atheist indoctrination, for example in the old Soviet Union, and has never thought to take seriously the possibility of God, they do not have, subjectively, any substantial reason actually available to assign the theistic hypothesis in general a non-zero probability. Oppy (1991) does defend that position as reasonable on more normal background knowledge. But a brief amount of thinking should call that into question. "Non-zero" is a very low bar, and the mere knowledge that debate continues among apparently intelligent people in philosophical journals about classical arguments for the existence of God ought to be sufficient to clear it. (The difference between that and demon skepticism is instructive: the latter is not defended as a real possibility, even by those impressed by the traditional skeptical arguments, while the former is. Attempts by some atheists to assert a symmetry between the Christian god and the Flying Spaghetti Monster and thus assign zero probability to each are no better; they are frivolous just because everyone knows at least second-hand the seriousness of standard theistic arguments.)

It is still problematic whether a non-zero probability should be assigned to the more specific theistic hypothesis that is needed for the argument of Pascal's Wager to gain traction, namely that there exists a God who offers an infinite reward dependent on the response of the decision-maker. Pascal reasonably regards that as obvious for his interlocutor, the "man of the world" of 1660. Again, what it means for a contemporary informed person will be considered later.

5 The Payoffs (for a Real Agent)

Payoffs, like options and their probabilities, must be in the mind of the decision-maker if they are to have any effect on decisions. That has not always been kept in mind in the discussion on Pascal's Wager. That applies especially to the meaning of "infinite" payoffs.

Pascal does not draw a payoff matrix, but the one that translates his text is laid out in Table 1.1.

Before discussing the meaning of "infinite gain," which is crucial to the Wager, we consider briefly some other aspects of the matrix.

Table 1.1 Pascal's Wager

		States of the world	
		God exists	God does not exist
Actions	Subjective probability	Non-zero p	$1 - p$
	Wager for God	Infinite gain	Inconvenience in life
	Wager against God	Misery	Worldly pleasures

On Pascal's religious views, neither of the payoffs in the first column are in fact certain, if one wagers as indicated. Far from it. Wagering for God gives a chance for God's grace to act, but one still may be mired in sin, or be outside the number of the elect to whom God freely chooses to grant salvation. On the other hand, if one wagers against God, there is still some chance that things will work out positively later and one will scrape home with a deathbed repentance. That may make the expected outcomes of each action harder to calculate and compare, but it depends on what is said about infinity.

The outcome "misery" was played up by earlier writers like Arnobius and Sirmond. They mean by it eternal punishment and plainly intend that the Wager be seen as a threat as much as a promise. Pascal avoids calling attention to that, for reasons that remain unclear. Diderot suggests (1875 [1746], p. 167) that "eternal," said of punishments, is a mistranslation of a Hebrew word that merely means "long-lasting," and it may be doubted whether the difference could be clear in the minds of ancient writers.

The outcomes "inconvenience in life" and "worldly pleasures" come from the fact that wagering for God will require some lifestyle adjustments, such as renouncing sinful pleasures and spending time in prayer, while "worldly pleasures" may continue to be enjoyed by one who decides not to wager on God. One should not forget also the cognitive inconvenience of having to seriously think oneself into a religious point of view, which many find painful – perhaps especially philosophers, trained as they are in a professional skepticism. Pascal allows it to be understood that "inconvenience" is a negative and "worldly pleasures" a positive, but that is a concession to the point of view of his interlocutor, the "man of the world" (again revealing how the Wager must be understood as aimed at a decision-maker with a particular point of view). Pascal himself does not really believe that, as he thinks that a life of virtue and prayer would be better absolutely speaking than one mired in sinful pleasures (even if it should prove that both ended with death). He argues:

> But what harm will come to you from taking this course [committing to God]? You will be faithful, honest, humble, grateful, doing good, a sincere, true friend. It is true you will not enjoy noxious pleasures, glory and good living. But will you not have others? I tell you that you will gain even in this life ... [L418/S680]

Similar reasoning plays an important role in the recent defenses of the Wager by Rota (2016a, pp. 58–63; 2016b), which are addressed to those who are already seriously considering Christian commitment. That is worth arguing, when a real decision-maker is addressed. It also simplifies the decision problem because the strategy of wagering for God is then "superdominant," as is said in decision theory: it is the best strategy in *every* state of the world.

However, it is widely agreed that the reasonableness or otherwise of Pascal's Wager hinges mainly on the "infinite gain" promised to one who wagers on God, in the case where God does exist.

Recent discussion has often been conducted by philosophers whose main interest in Pascal's Wager is in infinite decision theory, rather than its application to choice of religion. They have taken a rather crude and uncritical understanding of Pascal's "infinity" and identified it with the mathematicians' ∞, a number or quasi-number that satisfies such simple formulas as $\infty + \infty = \infty$ and $100 \times \infty = \infty$. That allows the deployment of colorful mathematical technologies like infinitesimals and extended reals and the mounting of extended objections involving, for example, "mixed strategies" (the strategy "Toss a fair coin: heads, you wager for God; tails, you wager against God" will have a payoff $\frac{1}{2} \times \infty = \infty$ and so will be as good as simply wagering on God) (Hájek, 2012b; Bartha, 2016).

That is plainly disconnected from what faces real decision-makers. If confronted by the choice of "99 percent chance of infinite gain" versus "1 percent chance of infinite gain," it would be excessively fundamentalist about infinitary arithmetic to conclude that $0.99 \times \infty = 0.01 \times \infty = \infty$, and hence that the bets are equally good.

Schlesinger (1994, p. 90, supported in Bartha, 2016) suggests, "In cases where the mathematical expectations are infinite, the criterion for choosing the outcome to bet on is its probability." That is reasonable as it is a straightforward extension to the infinite case of the answer for a fixed finite payoff. If infinitary arithmetic gets in the way of that, one might better conclude that the whole idea of modeling Pascal's "infinite gain" by the mathematical ∞ is misconceived.

The same conclusion can be reached by recalling the importance in modern decision theory of utility functions. To account for the fact that $100 is of much greater concern to a poor person than to a rich person, a "utility function" is commonly assumed which translates from actual money outcomes to "what it's really worth" to the decision-maker involved. The utility function discounts large monetary sums to an assumed psychological weight. The discounting is necessary to correctly model the weight of the outcome in the real considerations of the decision-maker. The utility alone plays a role in the decision matrix.

If we now ask what "infinite gain" means for a real decision-maker, with the usual human mind with finite computational resources, it is unclear whether it should be called literally infinite at all. Perhaps a finite mind can only include a finite utility of an actually infinite outcome. Whether that is true or not, the promised infinite gain is a long way off, spread over an infinite period, and its discounted present value may well be finite. Clergy are always complaining about the difficulty of extracting from the faithful finite sums of money in return for the infinite good in prospect, suggesting a widespread heavy discounting in psychological reality. That degree of discounting might be put down to irrationality through a lack of imagination. But many even after serious thought would agree that "I would not, for every chance no matter how small for eternal bliss, cut off my arm, or suffer stupefaction or madness," (or allow someone I care about to suffer serious harm) (Sobel, 1996, p. 42). At some point, the smallness of the probability of the payoff renders it near-fanciful, implying that the payoff is not really taken to be infinite.

We can take our lead from Chillingworth's observation above that the reasoning of the Wager works in ordinary life in cases of finite payoffs. A more realistic model of the utility of Pascal's "infinite gain" would be a finite but enormous payoff HUGE. It is a number very far beyond the utility of the ordinary goods of life, but is not infinite. Half of HUGE is not equal to HUGE (though HUGE + 1 is indistinguishable from HUGE – indistinguishability being the correct notion of equality for psychological entities – agreeing with Pascal's remark that "Unity added to infinity does not increase it at all ... by adding a unit it does not change its nature" (L418/S680). Since a finite mind is barely able to distinguish between HUGE and the really infinite, a payoff of HUGE is as motivating as a payoff of infinity. Again, it must be emphasized that a utility is a quantity in the mind of a decision-maker, where it can act to motivate. (Similar considerations lead to Bartha's (2007) approach of primitive relative utilities.)

Table 1.2 Pascal's Wager with HUGE Payoff

	God exists	God does not exist
Subjective probability	Non-zero p	$1-p$
Wager for God	HUGE	Small finite
Wager against God	Negative	Small finite

Hájek (2003) argues that a proposal of that kind would offend against a literal understanding of some of Pascal's sayings, such as "The finite is annihilated in the presence of the infinite, and becomes pure nothingness" and that the wagerer for God "gains all," with its implication of maximality. That may be so, but Pascal is a propagandist fond of hyperbolic expression. The real decision-maker will reasonably be left cold by such considerations as that two eternal lives might be better than one (or 2 × HUGE > HUGE). Just as $2 billion is no more motivating than $1 billion (for the ordinary person) although the sums differ in reality, that difference is not relevant to the utility in the mind of the decision-maker contemplating the Wager.

The payoff matrix of Pascal's Wager, in the case where "infinite gain" is modeled by HUGE is then (simplifying also the payoffs to their quantitative essentials) as laid out in Table 1.2.

In the bottom-left cell of Table 1.2, where one wagers against God and he exists, a negative payoff of some magnitude is reasonable as the rare case of a deathbed conversion will not balance the overall expected large negative: the situation will be something like "1 percent of HUGE after a deathbed conversion summed with 99 percent of negative HUGE otherwise." It can be seen in any case that even a small positive finite value – or even a moderate positive one, less than HUGE – would not affect the conclusion; perhaps that goes some way toward explaining Pascal's avoiding discussing the matter.

With that matrix, Pascal's reasoning is still sound. The Wager is a good bet. It is true that one might wonder whether HUGE is really big enough to balance a sufficiently small probability for the hypothesis "God exists." It may well not be enough to balance a strictly infinitesimal probability, but what about a very small finite one? Since we are envisaging real decision-makers, that can be regarded as a genuine possibility and left for the decision-maker to say. Is his grounded subjective probability for "God exists" really small enough to render nugatory the product of it with HUGE, when all due allowance is made for the cognitive difficulty of comprehending HUGE? In the end, that is for the decision-maker to decide, honestly.

Philosophers may find that a less exciting conclusion than the "all or nothing" version of the Wager with real infinities that they are accustomed to, but real decision-makers may be less concerned.

6 The Adaptability of the Wager to Different Spectra of Options

Convenient as it may be to quarantine Pascal's disturbing thought to the Paris of 1660, that is not correct. Pascal is pursuing decision theory, so his rhetoric is addressed to real agents, namely "men of the world" in the Paris of 1660. The editors of the Port-Royal edition of the *Pensées* added on his behalf: "This chapter addresses only a certain kind of person ... The author claims merely to show that by their own principles and by the pure light of reason they should judge it to their advantage to believe" (Thirouin, 1991, p. 168). For those persons, as for Pascal, there were just those two options – that is, the spectrum of religious theories to which they attached grounded subjective non-zero probability consisted of just Catholicism and atheism (give or take, as we have seen, some different interpretations of Catholicism).

For other people, such as Muslims and contemporary post-Christian intellectuals, the range of credible options is different. That does not mean that the Wager loses its force. It just means it has to adapt to the range of subjective reasonable non-zero probabilities that different agents actually have.

If someone in the early twenty-first century faces a different range of options from the Parisian of 1660, that does not make the Pascalian game-theoretic perspective irrelevant. On the contrary, the richer the choice of options considered reasonable, the more the need for careful calculation. (Just as with nuclear power plants, the more risks there are with some chance of being realized, the more work needed to calculate where money should be spent on precautions.) Perhaps, like the Jesuits, we take more seriously than Pascal the idea that if God has all the goodness claimed for him, then "he's a good fellow, and 'twill all be well." Or perhaps a kind of lowest-common-denominator of religions attracts us more than any particular faith. Or perhaps in the chaos of the post-modern global village, there are as many reasonable distributions of opinions as there are people. It doesn't matter. The essential point is that decision theory applies in the first instance to the reasonable subjective probabilities of any real agent. Pascal got under the skin of the worldly, by understanding what options were serious ones for them;

Table 1.3 Contemporary Wager

	Materialist atheism	Sect of parents	Amorphous spirituality	...	Advaita Vedanta	Mormon	...
Subjective probability	0.7	0.1	0.1	...	0.01	0.005	...
Wager for	Worldly pleasures	HUGE	Oneness with nature?		Nothingness	HUGE	
Wager against	Unknown until investigation	Negative	Pleasures/ suffering?		Many cares, few pleasures	Suboptimal afterlife	

to that extent, but only to that extent, his argument is *ad hominem* and does not survive the cultural context of Louis XIV's France. Whatever options are serious ones for us, Pascal's approach applies to them – unless we can attribute zero probability to the sum of all theistic and perhaps pantheistic options (which as explained above, is not easy).

The payoff matrix for a contemporary Western educated person might have many columns, looking something like that laid out in Table 1.3. The initial subjective probabilities will depend on the evidence the person happens to have grown up with; that is inevitable.

In the modern context, the result of accepting the Wager would again be action, but not necessarily the action that Pascal envisages. It would more likely be a serious investigation of the claims of religion in general.

7 Information Foraging Decisions

Pascal imagined at least in outline not only decision theory itself, but also one of the more subtle aspects of it, namely, how to decide on action in case of uncertainty as to how much more information one should acquire.

Pascal is masterful on the folly of not examining religion because its claims are uncertain on present evidence. In a remark leading up to the Wager, he says "An heir finds the deeds of his house. Will he say, perhaps, that they are false, and not bother to examine them?" (L823/S664) The action indicated in case of uncertain information about a possible large reward is: investigate the evidence to see if it firms up.

An animal foraging for food provides a well-understood analogy. An animal is faced with a choice of committing to a search for food in some direction, based on current knowledge, or postponing commitment in the hope that gathering further information instead will lead to a better decision (Nishimura, 1992; Lawes and Perrin, 1995). There is a trade-off between exploitation and exploration. In cases where the current alternatives have little evidence for them – they lack "weight of evidence" (in the sense of Keynes, 1921, ch. 6), the probabilities of the alternatives are not robust – it is likely that a little further evidence will drastically change them. So if such evidence can be gathered at low cost, it is worthwhile to do so.

These considerations apply equally well to intellectual foraging, in such paradigm cases as computer programs to play chess (Frey, 1983). The problem there is to use some quick-and-dirty heuristics to decide which of the huge number of possible sequences of moves and counter-moves are worth calculating further, before deciding which move to actually make. Similar issues arise with finding relevant information on the internet; for example, finding an optimal hotel in an unfamiliar city using a hotel booking site which gives information on price, star-rating, location, and so on. We all have intuitive strategies on how to home in on likely candidates without wasting too much time on dead ends, strategies which can be formalized (Pirolli, 2007).

It is the same with foraging in the space of worldviews, in the light of the very partial information I have on them. Given the wealth of alternatives, I must decide on the basis of current knowledge not so much which to commit to, as which to investigate further. How much evidence must be collected is itself something that must be judiciously appraised, in the light of the time to be wasted in blind alleys, the prospects of success somewhere else, and my estimated capacity to understand and critically evaluate theories. And the payoffs, if the theory being investigated were to turn out to be true.

The process would of course be a dynamic one. Having chosen a religious position to investigate and reached a conclusion, I would update my (now better founded) subjective probability of it and hence of all the other options. Standard Bayesian methods of updating probabilities may be appropriate, but there is more to it than simply collecting new facts as in evaluating a doubtful historical thesis. As with a change in philosophical views, new understandings are more important than new facts. If some religion looks worth investigating, it will require intellectual efforts to understand the point of it, discussions with believers, perhaps prayer. Bartha (2012) suggests updating on the basis of relative utilities, iterating until equilibrium is reached. The question of how to proceed is a complex one; the considerations here suggest it is a problem worth intensive study.

For a contemporary seeker after truth with normal rationality, curiosity, and easy access to information on almost any aspect of any worldview, information foraging just is the decision that a modern version of Pascal's Wager would recommend.

2 The Wager and Pascal's Theology

William Wood

Pascal's Wager has taken on a life of its own in contemporary philosophy. Like Anselm's ontological argument, the Wager has spawned a vast secondary literature that often seems quite remote from Pascal's own concerns. Of course, there is nothing wrong with this outcome. Like the ontological argument, the Wager is of independent and ongoing philosophical interest. We really do want to know whether there are good pragmatic arguments for religious belief. If thinking along with Pascal aids contemporary philosophical inquiry, then it hardly matters that what philosophers call "Pascal's Wager" names an argument that may bear only a tenuous relationship to anything that Pascal actually wrote. Sometimes, however, we want to understand an argument in its original context, and to grasp the intentions of the one who formulated it. That is my task in this chapter.

To the best of our knowledge, Pascal intended the Wager as an early move in his broader, unfinished apology for the Catholic Christian religion. The apology itself was – unsurprisingly – aimed squarely at the unbelievers of Pascal's own society: worldly skeptics and libertines in seventeenth-century France. The Wager was probably not intended as an independent philosophical argument, nor even as the culmination of the apology. Pascal planned to use the Wager to induce his readers into taking the rest of his apology seriously, by chipping away at their initial resistance to Christianity.[1] If we want to understand the Wager as a historical artifact, these contextual details matter. What matters even more are the complex, interrelated views that comprise Pascal's "theology." Pascal was not a theologian but he did have a theology. It was a very distinctive theology, and it affected the way he thought about everything else, including the Wager.

---

[1] To the extent that there is anything like a scholarly consensus about the place of the Wager in Pascal's apology, this is it. Important statements of this consensus include Wetsel (1994), Thirouin (1991), and Sellier (1999). More recently, Cantillon (2014) has argued that even establishing the underlying text of the "Wager fragment" is more problematic than many scholars have assumed.

Pascal advocated a very traditional, austere form of Augustinian Christianity. He was associated with the followers of the Dutch theologian Cornelius Jansen (hence, the "Jansenists") who thought that the contemporary Catholic Church had departed too far from traditional orthodoxy and become decadent. The only philosophical or theological writings that Pascal actually published in his lifetime, *The Provincial Letters*, are a defense of Jansenist views on grace (Pascal: Oeuvres Completes, tome 1 (OC 1), pp. 579–818).[2] Pascal was thus a very active controversialist in one of the most heated theological disputes of his day, between the followers of Jansen and the followers of the sixteenth-century Spanish Jesuit theologian Luis de Molina. Molina developed an elaborate theory of divine foreknowledge in order to reconcile the traditional Catholic teaching on grace with a libertarian account of human freedom. Pascal thought that the Molinist view of human nature was far too optimistic. It did not take seriously enough the reality of human sin and depravity.

When considered in the light of his theology and the broader aims of his apologetic strategy, Pascal's Wager can be seen as an ingenious attempt to subvert our sinful habits and desires and bend them toward God's saving purposes. At the same time, however, the more we understand Pascal's theology, the more we find ourselves confronted with difficult questions about why he would attempt to write any kind of apology at all, no matter how ingenious. As a committed Augustinian, Pascal held that God predestines the elect for salvation, and that human beings can do nothing whatsoever to draw themselves closer to God without a special infusion of grace, which is not offered to all and is never resisted.[3] Given these views, why would Pascal write an apology? I consider several possible responses to this question, and argue that even on his own terms, Pascal could still coherently write a Christian apology that includes the Wager.

1 The Mystery of the Redeemer and the Mystery of the Fall

Pascal's theology and personal piety were both strongly centered on Christ. In a sense, all Christians have a Christocentric theology. But Pascal insists that unless we embrace "the mystery of the Redeemer," we cannot understand anything important about ourselves or the world. For Pascal, only Christ, the God-man, who is both fully human and fully divine, can answer the otherwise

---

[2] For Krailsheimer's English translation, see Pascal (1982).
[3] I discuss below why Pascal would say "is never resisted" instead of "cannot be resisted."

inexplicable questions posed by the human condition: "Not only do we only know God through Jesus Christ, but we only know ourselves through Jesus Christ; we only know life and death through Jesus Christ. Apart from Jesus Christ, we cannot know the meaning of our life or our death, of God or of ourselves" (L417/S36). This Christocentric theme sounds throughout the *Pensées*:

> [T]hey take occasion to blaspheme against the Christian religion, because they know so little about it. They imagine that it consists in worshipping a God considered to be great and mighty and eternal, which is properly speaking deism, almost as remote from the Christian religion as atheism... But let them conclude what they like against deism, their conclusions will not apply to Christianity, which properly consists in the mystery of the Redeemer, who uniting in himself the two natures, human and divine, saved men from the corruption of sin in order to reconcile them with God in his divine person. [L449/S690]

For Pascal, only the specifically Christian story of the Fall can unlock the mystery of the human condition, and yet we must have faith in Christ in order to accept that story. Without faith in Christ, the notion of the Fall and original sin seem thoroughly repugnant (L131/S164, L695/S574). Yet only when we see ourselves as fallen can we really see ourselves aright.

Pascal accepted as historical the Genesis account of the Fall of Adam and Eve, and he believed that the entire human race receives the punishment for Adam's sin. As a result of the Fall, every human being deserves damnation; moreover, the consequences of the Fall extend to human nature itself, especially our faculties of reason and will. Following Augustine, Pascal drew a sharp distinction between human nature before and after the Fall. Before the Fall, Adam was "just, healthy, and strong," and possessed a will that was libertarianly free ("equally flexible between good and evil") but still naturally inclined toward what was best for his happiness (Pascal: Oeuvres Completes, tome 2 (OC 2), p. 287).[4] The highest and objectively most desirable good for Adam, as for all of us, is the good of uninterrupted communion with God. Yet even in the unfallen state, a human being is a natural creature, and uninterrupted communion with God is a supernatural end. Because no natural creature can pursue a supernatural end with natural powers alone, even Adam

---

[4] All translations from works other than the *Pensées* are my own. These quotations are from Pascal's unpublished treatise "Writings of Grace." An excerpt from this treatise may be found in Levi (1995).

required the supernatural gift of God's grace. Before the Fall, Adam was free to cooperate with God's grace, and thereby to merit the reward of eternal life, or to reject it. Adam rejected God. As a result, he lost the supernatural assistance of grace, and even his natural faculties became weaker. His mind, originally "very strong, very just very enlightened," became "cloudy and ignorant," and his will, previously oriented toward his own genuine happiness, now "voluntarily, very freely and with joy" embraces evil as if it were good. As with Adam, so with us. Every human being bears the legacy of Adam's disastrous choice: "This sin having passed from Adam to all his descendants, who were corrupted in him as a fruit issuing from a rotten seed, all the men who descended from Adam are born in ignorance, in concupiscence, guilty of the sin of Adam, and worthy of eternal death" (OC 2, p. 289).

Only this traditional, Augustinian account of the Fall can explain the paradoxical dialectic of greatness and wretchedness that Pascal sees as the fundamental datum about the human condition (L199/S230, L122/S155). We are great because we are made in the image of God, and so we are capable of being happy and loving the truth. At the same time, we are wretched because we have fallen, and so those capacities are corrupt: in fact, we are not happy, and we hate the truth. "We are incapable of not desiring truth and happiness and incapable of either certainty or happiness. We have been left with this desire as much as a punishment as to make us feel how far we have fallen," he writes (L401/S20). According to Pascal, the Christian doctrine of the Fall and original sin is the best explanation for this seemingly paradoxical, attraction-yet-aversion to truth and happiness.

2 The Wager and the Consequences of the Fall

This understanding of the Fall colors everything about Pascal's apology, including the Wager. Pascal was convinced that the Fall had disastrously marred the human reason and will. In the wake of the Fall, our reason is "corrupted" (L60/S94, L600/S497) and "impotent" (L131/S164); as he sometimes puts it, reason is not reasonable (L76/S111). Similarly, the fallen will is "depraved" because of its systematic bias toward the self and self-oriented goods (L421/S680). According to Pascal, we typically treat ourselves as the absolute center and source of all value (L668/S547). We therefore seek to dominate others: "The self is hateful ... it is unjust in itself for making itself the center of everything: it is a nuisance to others in that it tries to subjugate them, for each self is the enemy of all the others and would like to tyrannize them" (L597/S494). In the sharpest possible contrast to our excessive self-regard, the true religion – Christianity – teaches us that we are sinful and broken, and that we must humble ourselves before God. These are deeply unwelcome lessons. As a result, "Men despise religion. They hate it and are afraid it may be true" (L12/S46). Indeed, the proofs of the Christian religion "lie before their eyes, but they refuse to look" (L428/S682).

This, in brief, is Pascal's account of the consequences of the Fall. With this account before us, we will be able to see why Pascal regarded the Wager as a valuable tool in his apologetic arsenal, and what he hoped to accomplish with it. Given our fallen condition, the primary barriers to Christian belief are not evidential but emotional and affective. The problem is not that Christianity lacks the support of good evidence and arguments, but that people are too sinful and self-absorbed to appreciate them. We cannot appreciate their force because, as a result of the Fall, we always reason in biased ways, under the influence of sinful self-love, which systematically leads us away from God.[5]

Although most commentators ignore the evidentialist strain of Pascal's thought, in fact Pascal did hold that Christianity was very well-supported by evidence, and he intended to present that evidence forcefully in his apology (Fouke, 1989; Wetsel, 1994). It is easy to ignore this element of the *Pensées*, because most of what Pascal considered to be the best evidence for Christianity no longer seems like evidence to us, and so we do not even notice that he is making evidential arguments. (It does not help that most of those arguments occur in the parts of the *Pensées* that are tedious and unpleasant to read, and therefore little read.) By modern lights, Pascal's evidential arguments all seem excessively *theological*. For example, as noted above, he regarded the explanatory power of the doctrine of the Fall as an indirect proof of the truth of Christianity. This inference to the best explanation would have been a major structuring principle of the apology. He also planned to make extensive appeals to historical miracle reports, which he assumed to be reliable, and to Old Testament prophecies, which he assumed were fulfilled by Christ. He even had rudimentary plans to make comparisons with the other religions that he knew about – especially pagan religions, Judaism, and Islam, but also religions of Asia and the Americas – in order to show that they were not serious rivals to Christianity.[6]

---

[5] Throughout the *Pensées*, Pascal makes the point (using the terminology of his day) that we often engage in motivated, biased reasoning (e.g., L539/S458).

[6] For a comprehensive review of Pascal's fragments on non-Christian religions, see Wetsel (1994).

All of this suggests a very specific way to understand Pascal's dilemma as an apologist. How could he get fallen human beings to take seriously the evidence for Christianity, given their selfish biases and their aversion to genuine self-knowledge? Rational arguments that God exists will not persuade, because we recoil at the bare possibility that they may be true. Pascal is clear that the traditional proofs are not unsound, but we nevertheless find their premises and conclusions so threatening and unwelcome that we are unwilling to appreciate their force (L190/S223, L135/S167, L781/S644). The same goes for the moral, historical, and biblical evidence in favor of Christianity that Pascal regarded as so persuasive.

In this context, the Wager can be seen as an attempt to "bootstrap" conversion by appealing to the unbeliever's shallow self-interest. The apologist – being a faithful Augustinian Christian – knows that the unbeliever, like every fallen human being, is caught in the grips of prideful self-love. The apologist therefore offers the unbeliever a choice that assumes that the unbeliever wants only to maximize his own happiness. After the Fall, human reason no longer functions properly, and so we might expect that the unbeliever lacks the rational capacity to weigh his options well, even when his happiness is at stake. But the fallen intellect and will malfunction in the direction of self-interest, and the apologist already factors this tendency into the terms of the Wager. The Wager works entirely with the grain of the unbeliever's biases, and asks him only to be maximally self-interested.

The Wager also appeals to the unbeliever's love of empty diversions (*divertissement*). Pascal thought that people wallow in mindless, trivial activities like hunting and gambling as a way of distracting themselves from considering whether they are really happy. If they were instead to focus on their own deep sense of existential unrest, they might consider turning to Christ as the remedy for that unrest. Some of the most eloquent and persuasive fragments of the *Pensées* make this very point (L136/S168). Pascal's theological critique of diversion also sheds light on the Wager. In the Wager fragment, the apologist initially presents the Wager as a pleasantly diverting thought experiment. ("At the far end of this infinite distance a coin is being spun which will come down heads or tails. How will you wager? Reason cannot make you choose either, reason cannot prove either wrong . . ." (L418/S680.) From this innocent beginning the Wager builds, steadily, toward an argument which "carries infinite weight" that it is rational to become a Christian. Once again, the apologist takes an aspect of the unbeliever's fallen nature (in this case, his ingrained love of diversion) and uses it against his unbelief.

Few commentators notice that in the text, the Wager fails. The unbeliever admits the force of the Wager ("I confess, I admit it ...") but still resists its conclusion ("I am so made that I cannot believe"). When the apologist offers his well-known remedy for this resistance (act as if you already believe, take holy water, go to mass, "diminish the passions," and become "more docile"), the unbeliever never explicitly accepts the remedy, although he does exclaim that the apologist's words fill him with "rapture and delight." The fact that the unbeliever in the text never explicitly agrees to the terms of the Wager is further evidence that Pascal did not intend it as a standalone argument or as the culmination of his apology. The response of the unbeliever in the text makes more sense if Pascal intended the Wager only as an initial salvo meant to soften the resistance of a potential convert so that he could better appreciate the rest of Pascal's arguments.

So far so good. If our story ended here, we would have a tidy account of why Pascal might believe that the Wager could succeed where traditional apologetic arguments fail. Given his understanding of sin and the Fall, such a position makes eminent sense. Unfortunately, however, the story does not end here. Pascal also held a very strong view of predestination and grace, one that calls into question the value of any kind of apologetic argument, however novel.

3 God's Unconditional Predestination of the Elect

Every orthodox Catholic Christian affirms that fallen human beings cannot save themselves with their natural capacities alone, and that we all require the supernatural aid of God's grace in order to love God as we ought and keep his commandments.[7] Catholic Christians are also committed to the further thesis that we require supernatural grace even to ask for God's help. That is, we cannot simply "do the best we can" with our unaided natural capacities and thereby merit grace as a reward.[8] Beyond these points of general agreement, matters quickly become divisive, complicated, and murky. Seventeenth-century France was wracked by heated disputes over the relationship between grace and freedom. These disputes are now opaque and often quite technical. We can see our way through them by focusing on two central questions:

---

[7] The denial of this position constitutes the Pelagian heresy, condemned at the Council of Carthage in 418, and again by the Council of Trent in 1547.
[8] The denial of that position constitutes the "semi-Pelagian" heresy, condemned at the Council of Orange in 529, and again by the Council of Trent in 1547.

Is grace genuinely resistible, such that we can freely choose to accept it or reject it? And does God offer grace to everyone, or just to the elect? The answers to these two questions will tell us what room there is for human effort in coming to salvation: whether people are simply saved or damned by divine fiat, or whether there is room to say that they freely cooperate with grace.

According to Pascal every human being deserves eternal punishment, and so God could justly damn everyone. Instead, purely out of mercy, God elects a small number of people from the corrupt mass of human beings and infallibly predestines them to salvation. The others – the majority of the human race – he justifiably damns. God's motives for saving some and damning others are entirely inscrutable. Pascal is emphatic that prior to creation, God makes an absolute, unconditional choice to save his elect. In other words, God does not grant eternal salvation conditionally, as a reward for those who live good lives or otherwise do their best. Contra Molinism, as discussed above, God does not predestine the elect as a result of his foreknowledge that they will accept his offer of salvation. On the Molinist view, God wants to save everyone, and gives everyone grace sufficient to restore their ability to choose between good and evil. He then gives saving grace to the elect as a result of his foreknowledge that they will cooperate with initial grace and thereby merit salvation. Pascal regarded the Molinist view as little more than a sophisticated form of the Pelagian heresy. Even though God does foreknow who will cooperate with grace, God's gift of grace is not in response to any foreknown human action:

> As all human beings are in this corrupt mass and are equally worthy of eternal death and of the anger of God: God could justly abandon them all without mercy to damnation. Nevertheless, it pleased God to choose, elect and discern from within this corrupt mass, in which he saw only evil, a number of people of all sexes, ages, conditions, and complexions, of all countries, and indeed, of all kinds ... God distinguished his elect from the others for reasons unknown to men and angels through pure mercy and without any merit ... [OC 2, p. 289]

By contrast, God does not absolutely will the damnation of the reprobate, but wills their damnation only conditionally, as a result of his foreknowledge that they will not die in a state of grace (OC 2, p. 261). Pascal was keen to maintain this distinction between God's absolute choice to save the elect and his conditional choice to damn the reprobate. Something like this distinction is required to avoid the "Calvinist heresy" of double-predestination, which had been condemned by the Council of Trent.[9] In any case, Pascal thought it monstrous to hold that the God of love absolutely predestines anyone to damnation. He held this view even though he himself agreed that God could have predestined the reprobate to salvation, but did not. No doubt many readers will find Pascal's account of grace unsatisfying, even repulsive. If God could have given saving grace to all, why didn't he? Moreover, his appeal to foreknowledge seems question-begging. The future would have unfolded differently if God had acted differently. So why didn't he?[10] Pascal has no further answers to these questions: the fact that God saves anyone at all is an act of unfathomable mercy.

The doctrine of the absolute predestination of the elect can seem dark, and Pascal affirmed it in an especially strong form. He even thought that "God sometimes abandons the justified before they have abandoned him" (OC 2, p. 217). In other words, God gives saving grace to some people, but then withholds it, so that without the continuing help of grace, the person inevitably turns away from God again and justly deserves damnation. This choice, too, springs from God's unknowable will, and is not a response to any prior human action. In summary, for unknowable reasons, God absolutely predestines some people to salvation and gives them the grace that they need to love God properly and keep the commandments throughout their lives. The others he justly damns, because – without the help of grace – that is what they deserve.

However difficult it may be for us to accept, Pascal regarded the doctrine of predestination as a source of comfort. If our salvation is entirely in God's hands, rather than our own, then we do not need to worry about whether we are good enough to deserve salvation. Given the severity of the Fall, the suggestion that we must achieve salvation through our own merits seemed both ludicrous and terrifying to Pascal. In the *Pensées*, in a long meditation that has come to be called "The Mystery of Jesus," Pascal writes, in the voice of Christ: "Take comfort; you would not seek me, if you had not found me" (L919/S751; L929/S756, tm). A bit later, in the same fragment, he writes in the same voice: "Your conversion is my affair. Do not be afraid, and pray with confidence as though for me." The key point, succinctly put: our conversion is God's affair, and never ours alone.

---

[9] Canon 17 of the Sixth Session, concerning justification (Leith, 1982, p. 408).
[10] I thank Lawrence Pasternack and Paul Bartha for pressing these points.

4 Can We Resist Grace?

Pascal and his Jansenist colleagues faced an especially thorny question: do human beings have the power to reject grace? The logic of their own position implies that the answer is no. God gives grace only to those whom he wills to save, and surely God's sovereign will cannot be thwarted by mere creatures, so it makes no sense to suggest that human beings could reject grace. Furthermore, if the human will is free to accept or reject grace, then the beginning of our salvation is up to us, and not to God, which just is (so they thought) the semi-Pelagian heresy. Plausible readings of Augustine, especially the later Augustine, seem to concur that grace is irresistible.[11] Yet even though their view had a venerable pedigree, the Jansenist account of grace and predestination seemed uncomfortably close to that of "heretical" Protestant reformers like Luther and Calvin. It was also at odds with the relatively lax moral theology of their own day. Worst of all, it seemed directly to contradict the Council of Trent, which all Catholics regarded as authoritative.

In response to Reformation-era controversies, the Council of Trent took the view that grace can be resisted. Indeed, Trent's fourth canon concerning justification anathematizes the opposite view:

> If anyone says that man's free will, moved and aroused by God, by assenting to God's call and action, in no way cooperates toward disposing and preparing itself to obtain the grace of justification, that it cannot refuse its assent if it wishes, but that, as something inanimate, it does nothing whatever and is merely passive, let him be anathema. [Leith, 1982, p. 430]

Pascal and the Jansenists had no wish to dissent from the teachings of an ecumenical council. Yet their views on predestination and grace put them at odds with some of Trent's teachings. To address this tension, they developed an account of the human will on which the will retains some bare capacity to resist grace, even though God alone decides who is saved, and even though God's grace is always triumphant.

5 The Mechanisms of Free Choice, Sin, and Grace

The key to this account lies in the Augustinian notion that grace is experienced as a kind of pleasure, a new feeling of delight at the prospect of loving God and keeping his commandments. This delight – called a "victorious

---

[11] For a way into this enormously complicated issue, see Burns (1999), Kolokowski (1995).

delectation" (*delectatio victrix*) by Augustine – is stronger than the sinful attraction to evil that we all feel as a result of the Fall.[12] Although the term "desire" comes closest to capturing Pascal's idea that grace causes a newfound delight in the law of God, it is important to emphasize that there is also a cognitive component to the reception of grace. The elect not only have God-oriented desires, they also come to believe that they will find greater satisfaction in following God. Indeed, the gift of grace begins a process that reorients everything about the sinner's life (see Wood, 2013, pp. 19–50, 212–26).

Given the fragmentary state of his writings, it is not easy to reconstruct Pascal's exact account of free choice. But he seems to hold the broadly compatibilist view that an action counts as free as long as it is caused by the agent's own desires. More formally, the choice to A counts as free when it is sufficiently caused by one's desire to A. According to Pascal, explanations that advert to "the will" are tautologous because they resolve into the claim that one always does what one most wants to do. To speak of "willing" some goal is merely to assert that we want what we want. He writes: "For what is more clear than this proposition, that one does always what delights one the most? Since this is nothing other than saying that one always does what pleases one most, that is, one always wants what pleases one, that is, one always wants what one wants" (OC 2, p. 272). The implication is that by definition, we always act to satisfy our all-things-considered strongest desires.

Pascal's account of grace as pleasure, combined with his account of free choice as doing what one most wants to do, gives him a straightforward way to understand how God infallibly brings about the salvation of the elect. When the elect receive grace, they are given a new set of God-oriented desires. These holy desires make it the case that they now find sinning comparatively unattractive; instead, they want to obey God and draw closer to him. Their (graced) choice to obey God is genuinely free, according to Pascal, because their wills are not coerced by anything external. When the elect reject evil and turn toward the good, they merely follow their own strongest desires. The fact that God caused them to desire good and reject evil does not mean that the resulting desires are alien. They are now the authentic desires of the graced will itself, according to Pascal, and so when the elect act as a result of these desires, their actions count as free. Under the influence of grace, the elect do what they genuinely want to do.

---

[12] See Augustine, *De peccatorum meritis et remissione* 2.19.33 and *De spiritu et littera* 1.29.5. See also Moriarty (2003, p. 148).

One might wonder whether Pascal is really entitled to say that the elect count as properly free – even on his own terms – since their new, grace-based desires are implanted in them by God. This is certainly a fair question. In other circumstances, we would regard acting on the basis of externally implanted desires as a kind of compulsion. If Harry gives Ron a magic potion that causes him to want to dance all night, it no longer seems correct to say that Ron himself genuinely wants to dance all night, or that the desire to dance all night is genuinely his own.

Pascal does not raise this question with respect to the causal power of grace, and so we do not know exactly how he would answer it. But given his Catholic Christian commitments, it is not difficult to imagine what he would say. God's actions are not relevantly similar to the actions of a human agent who manipulates someone else's will. (God is not like Harry in the scenario above.) God is the creator and sustainer of all things, and everything other than God himself exists only because God continues to create it. In this sense, *everyone's* desires are *always* caused by God. Pascal would insist that the new, graced desires that God gives to the elect are as much their own as any of their previous desires were, or could be.

6 The Free Will Always Can Resist Grace but Never Does Resist Grace

According to Pascal, this account of grace and freedom shows that humans are free to resist grace, as the Council of Trent requires, while also ensuring that God's absolute will to save the elect is never thwarted. On this account, the capacity to resist grace is just the capacity to do something that, all things considered, one has absolutely no desire to do. We may have such a capacity, according to Pascal, but it is little more than notional. Pascal is therefore comfortable saying that God's grace "infallibly" causes us to choose the good:

> [Grace is] nothing other than sweetness [*suavité*] and delight [*délectation*] in God's law poured into the heart by the Holy Spirit, which, not only equaling but even surpassing the concupiscence of the flesh, fills the will with a greater delight in good than concupiscence offers in evil; and so free will, charmed more by the sweetness and pleasures inspired in it by the Holy Spirit than by the attractions of sin, infallibly chooses God's law for the sole reason that it finds greater satisfaction there ... So that those to whom it pleases God to give this grace, of themselves, by their own free will, infallibly prefer God to created things ... [OC 2, pp. 289–90]

God changes man's heart by pouring into it a celestial sweetness ... and [man] finding his greatest joy in the God who charms him, infallibly inclines towards God of his own accord by a movement that is wholly free, wholly voluntary, and wholly loving; so that it would be a pain and a punishment to be cut off from him. It is not that he is not always able to turn away [from God], and would actually turn away if he wanted to; but how could he want to, since the will invariably inclines towards what gives it greatest pleasure, and nothing pleases it as much as this unique good, which encompasses all other goods ... ? [OC 1, pp. 800–1]

In summary, "the free will always can resist grace, but never wishes to do so," according to Pascal (OC 1, p. 801). Considered in isolation, the claim that the free will always can resist grace might suggest a libertarian interpretation, on which the will retains the liberty of indifference to accept or reject grace. But when we recall that Pascal is committed to a strong form of predestination, and to a general account of willing on which human actions are caused by our all-things-considered strongest desires, then the libertarian interpretation seems untenable.

Nevertheless, Pascal was convinced that this account of grace cohered with the Council of Trent's canons on justification.[13] He can agree with Trent that the human free will is able to "refuse its assent [to grace] if it wishes" and that the will is active, not passive, in the process of justification. The will can indeed refuse assent *if it wishes*, but never does so wish; and the will is indeed active in the process of salvation, even though its movements are caused by divinely given, efficacious desires. Because the will is active and free, on Pascal's account, he can even agree that the human will "co-operates toward disposing and preparing itself to obtain the grace of justification," as Trent requires.

Before turning to the question of how Pascal can reconcile this account of grace with his Christian apology in general and with the Wager in particular, it is useful to summarize his views. Pascal is committed to the following propositions:

No Pelagianism: Only those who receive the aid of divine grace can be saved.

No Semi-Pelagianism: Without grace, we cannot perform any action that merits grace. Divine grace is required in order to repent of one's sins, turn toward God, and begin the process of salvation.

---

[13] The following quotations are from the 4th canon of Trent Sixth Session as in Leith (1982, p. 420).

No universal salvation: Not everyone is saved. In fact, God justly condemns most of the human race to damnation.

Absolute predestination to salvation: God alone decides who is saved. His reasons for choosing to save the elect are inscrutable. God's choice is unconditional and in every respect prior to any foreknowledge of what the elect will do.

Grace as desire and delight: When the elect receive grace, they begin to love God and desire to please him, and they find obeying his commandments pleasurable. By doing what they want to do, they also do what God wants them to do, and thereby cooperate with God's decision to save them.

Grace is never resisted: Even if it is possible to resist God's grace, God's grace is in fact never resisted.

7 Grace, Freedom, and the Wager

Pascal's understanding of grace and freedom threatens to render the Wager entirely otiose. Even though people do make choices about whether to accept or reject God, those choices are ultimately outside their own control. It is hard to see why someone with such views would write a Christian apology, still less one that includes the Wager. As Jan Elster has noted, the Wager argument would make much more sense if it had been offered by one of Pascal's theological opponents (2003, pp. 70–71). On rival views, the Wager seems like a plausible way to convince some people to begin a process that could eventually lead to their own salvation. But Pascal holds that whether someone is saved or not ultimately depends on whether God has numbered that person among the elect. In which case, why does Pascal commend the Wager? We can ask a parallel question from the point of view of an unbeliever who undertakes the Wager. If my conversion is even partly under my control, then the Wager might convince me to turn to God. But on Pascal's account, God infallibly brings about my salvation if I am among the elect, and if I am not among the elect, then I am damned no matter what I do. So why bother to wager?

It is, of course, possible that Pascal decided that he could not reconcile his views on grace with his plan to write an apology. He left the apology unfinished, and we do not know whether he abandoned it, or simply found himself unable to complete it before his death. Pascal says very little to help us. The most pertinent statement comes in his unpublished "Writings on Grace":

all persons on earth are obliged ... to believe that they are among the small number of the elect whom Jesus Christ wanted to save, and to judge the same with respect to everyone else on earth, however wicked and impious they may be, as long as they have a moment of life, leaving the discernment of the elect and the reprobate to God's impenetrable secrecy. [OC 2, p. 262]

Pascal here counsels us not to speculate about who is saved. We lack the epistemic resources to distinguish the elect from the damned, because we have no access to the "impenetrable secrecy" of the divine intellect and will. In the face of such epistemic uncertainty, we must think and act on the basis of what we do know, from Scripture and Church tradition: that God has commanded everyone to love and serve him, and has further commanded Christians to bring the gospel to unbelievers. We cannot refuse to obey these commands just because we are unsure about who is among the elect. God's commands bind everyone – both the elect and the non-elect alike. It follows that in practice, we must treat everyone, including ourselves, as if they are among the elect, because that is the only way to carry out the command to obey God and spread the gospel.

This passage also suggests an obvious reason why Pascal, as apologist, might commend the Wager. God can use the Wager to bring about the conversion of someone predestined to salvation, and he can also use Pascal as his instrument to offer it. Pascal can agree that God is the cause of every conversion, and consistently maintain that when he himself offers the Wager to an unbeliever, he merely carries out God's will. The fact that God's plan to save the elect is shrouded in "impenetrable secrecy" entails that we do not know how God carries it out in specific circumstances. For all anyone knows, those who are convinced by the Wager are convinced by the power of God's grace.

Even if Pascal's account of grace is not logically incompatible with the Wager, we might wonder whether the two are psychologically incompatible. Can someone who shares Pascal's views about grace (for whatever reason) find the Wager, and the post-Wager process of habituation, plausible from her own, first-person, point of view? The answer is yes. Someone who provisionally accepts absolute predestination still has a motive to wager. In effect, the unbeliever would say to Pascal, "If Christianity is true, then God alone brings about the salvation of the elect using inscrutable methods known only to him. You have convinced me that it is rational to wager that God exists, and for all I know, this is precisely the way that God is bringing about my salvation, so I choose to wager."

If pressed to reconcile his views on grace with the argument of the Wager, I suspect that the point in the fragment above would be Pascal's final word on the matter. He would appeal to the inscrutable divine will and the universal obligation to obey God and preach the Gospel to unbelievers. Yet even if Pascal himself would not say more, we can draw on his writings to construct a more extensive reply on his behalf. I offer three additional, mutually reinforcing lines of argument to which he could appeal. They are not (quite) arguments that Pascal makes in his own voice, but they are arguments that he could make, because they are consistent with all that he does say.

7.1 Appeal to Altruism

Although the point is sometimes overlooked by commentators, Pascal thought that unbelievers can only experience inferior forms of happiness and pleasure. Yet even in this life, higher forms of happiness and pleasure are available to Christians. It follows (to Pascal) that the downside of the Wager is no real downside at all. Even if God does not exist – or, closer to the matter at hand – even if we wager but are not among the elect, we will still be objectively better off if we endeavor to live as faithful Christians. In the Wager fragment itself, Pascal writes that someone who gambles on God "will gain even in this life," and will be "faithful, honest, humble, grateful, full of good works, a sincere, true friend." True, such a person will "not enjoy noxious pleasures, glory, and good living" but to Pascal, these are not reliable goods that lead to genuine happiness anyway. Throughout the *Pensées*, he argues that worldly goods are unsatisfying and actually make us unhappy.[14] So one way that Pascal can reconcile the Wager with his views on grace is with a straightforward appeal to altruism. He can say that anyone persuaded to live as a Christian will find that they are happier, better people – whether or not they are among the elect.

7.2 Appeal to the Sweetness of Grace

Recall the way Pascal describes grace as "sweetness," "delight," and "that which pleases the will the most" (OC 2, p. 289; OC 1, pp. 800–1). This understanding of grace has important implications for how we should understand what is happening in any apologetic scenario, including the Wager. Suppose Pascal presents the Wager argument to his worldly, skeptical friend,

---

[14] For discussion and the relevant fragments see Wood (2013, pp. 34–50).

Damien Mitton. By hypothesis, God causes Pascal first to formulate and then to offer the Wager; subsequently, God also causes Mitton to accept it. With respect to their motives, Pascal and Mitton cannot distinguish between God's causation and their own first-order desires – Pascal wants to offer the Wager, and Mitton wants to accept it. Similarly, with respect to their beliefs, they cannot distinguish between God's causation and their own internal sense that the Wager argument is credible.

The picture is not significantly different if Mitton is not among the elect. In that case, God does not give Mitton a newfound delight in the law of God, and a sense that the Wager argument is credible, so Mitton listens with interest but ultimately declines to wager. In neither case does God force or compel Mitton. In fact, in both scenarios, Mitton simply does what he most wants to do. "The way of God, who disposes all things with gentleness, is to instill religion into our minds with reasoned arguments and into our hearts with grace, but attempting to instill it into hearts and minds with force and threats is to instill not religion but terror. *Terror rather than religion*," writes Pascal (L172/S203; L460/S699). So the second way that Pascal can reconcile the Wager with his views on grace is by pointing out that on his account, no one ever has to do anything other than what they most want to do. Some people will want to wager and others will not. In either case, God has chosen to save his elect in the most gentle, non-coercive way possible: by allowing everyone to think and do exactly as they wish.

7.3 Appeal to Divine Accommodation

The third line of argument available to Pascal is an additional specification of the second. On this line, Pascal's views on grace do not work against the force of the Wager, but support it. Grant that God's grace takes the form of desire and delight, such that one turns to God (or not) merely by doing what one most wants to do. It then seems straightforward to say that God accommodates his gift of grace to those whom he wishes to save by ensuring that grace builds on their existing desires. No doubt God could bring it about that Mitton, the worldly gambler – all-at-once and out of nowhere – suddenly finds himself with an overwhelming desire to go out into the desert and live in solitude as a religious recluse. In such a case, Mitton might feel obliged to satisfy this overwhelming desire, but he would likely experience the desire itself as a kind of alien compulsion, something external, since it would seem completely unconnected to his existing desires and his deeper sense of personal identity. (Recalling the fragment above, such a compulsive desire would produce "terror rather than religion.") Far better, then, for God to lead Mitton to conversion by means of the very thing that he loves most – gambling.

When we bear in mind Pascal's understanding of grace, one of the most common philosophical objections to the Wager loses its power. Since Voltaire, at least, critics have argued that Pascalian wagering is too rooted in narrow self-interest to count as a genuine expression of religious piety. Yet this objection cannot succeed once we stipulate that God himself brings it about that some people are converted by the Wager. (Plausibly, any course of conversion that God infallibly brings about will count as a genuine expression of piety.) Moreover, like Pascal, we can even marvel at the generosity God displays in meeting self-interested sinners exactly where they are. God offers them a path to salvation that begins from their own corrupt desires for domination and diversion. For such people, the Wager might be exactly what grace looks like.

I have tried to show that Pascal's account of grace is compatible with his endorsement of the Wager. In his contribution to this volume (Chapter 3), Paul Moser agrees that I have succeeded at this task: he admits that it is "undeniable" that God can use the Wager to convert the elect. But he worries that I have not really established the "key issue": that "a Jansenist use of Pascal's Wager is indispensable in salvation" and can accommodate a God who is "worthy of worship and hence morally perfect." Moser is correct. I have not established these things. I have not attempted to establish them. I have not argued that the Wager is indispensible for salvation (quite the opposite), that absolute predestination is just, or that the Jansenist God is morally perfect. Moser's "non-Jansenist God" is indeed more palatable to our moral sensibilities – exactly as in Pascal's own day. But Pascal would utterly reject any attempt to judge God according to human moral standards (L131/S164). In short, Moser's real quarrel is with Pascal.[15]

I have defended only the claim that Pascal's views on grace and the Wager are consistent. I have tried to stick as closely as I can to arguments that Pascal himself could plausibly make. Much of my defense implicitly relies on the notion of perspective: on how matters seem from a first-person point of view, in comparison with the way they seem from a third-person point of

---

[15] Moser in effect argues that Pascal's Jansenist views on grace are incompatible with the Wager, *given some of Moser's own additional assumptions*, drawn from analytic perfect being theology and appealing to broadly libertarian intuitions about free will. However cogent, they are not Pascal's assumptions, and so I have not taken them on board in this chapter.

view. According to Pascal, we cannot take an external, third-person perspective on whether we are among the elect. Such a perspective is not available to us, because it belongs to God alone. From an external, third-person perspective, all we can know is that for inscrutable reasons God has chosen to save some, when all deserve damnation.

Yet from one's own first-person perspective, the effects of God's eternal decree to save the elect are indistinguishable from one's sense of selfhood and personal agency. It may be the case that God's grace causes us to act as we do, but from our own perspectives, we act freely and deliberately, respond to reasons and incentives, and carry out plans and desires that are authentically our own. Even if we agree that our actions are determined, we still have to live in the world as we find it, and we have no alternative but to live from the first-person perspective, as free, rational human agents.

8 Conclusion: The Wager as a Theological Argument

I leave it to others, better versed in decision theory, to determine whether Pascal's Wager succeeds as a standalone philosophical argument that it is rational to believe in God. I do think that the more one agrees with Pascal's theological premises, the more appropriate his Wager will seem as an apologetic tool. Pascal's own Wager is really a theological argument, one that depends on a network of controversial theological premises about human nature, sin, and redemption. Ironically, someone who accepts Pascal's full-blown account of predestination and grace actually has the least reason of all to recoil at the Wager. Surely God, in his absolute mystery, can bring the elect to glory by means of the Wager. Who could ever be in a position to say otherwise?

As a historical matter, Pascal almost certainly did not intend the Wager as a standalone argument. He likely intended it as a propaedeutic argument that would leverage our excessive self-love – itself a result of the Fall – and then deploy it against our unbelief. On this account, the Wager is meant to lead the unbeliever to recognize that it is his sinful passions, and not primarily his reason, that prevent him from taking Christianity seriously. Once the hypothetical unbeliever comes to that realization, he is more likely to engage with the rest of Pascal's apology in a sympathetic way. More importantly, he might decide to live as a faithful Christian, which would in turn further diminish his passions, better equip him to appreciate the truth of Christianity, and, in general, make him a happier, better person.

3 Pascal's Wager and the Ethics for Inquiry about God

Paul Moser*

1 Introduction

How does Pascal's Wager fit, if it does, with the ethics for inquiry about God's reality, in particular, with *conceptually* and *evidentially responsible* inquiry about God's reality? Such responsible inquiry about God's reality must be agreeable to a fitting *notion* of God and to relevant *evidence* about God's reality, in a manner needing specification. In that case, we shall see, a wager based solely on prudential or pragmatic considerations will face a problem. It then will have no necessary connection to the relevant evidence of God's reality. This chapter will ask whether Pascal's Wager falls short on this front, and, if so, whether there is any way to rehabilitate the Wager from the perspective of conceptually and evidentially responsible inquiry about God's reality.

We shall consider, in connection with Pascal's Wager, how evidential factors figure in responsible inquiry about God's reality when prudential or pragmatic factors are available to motivate belief that God exists. We also shall attend to how divine elusiveness or hiding toward some humans at times bears on evidentially responsible inquiry about God in connection with Pascal's Wager. We shall see that Pascal's Wager fails as a result of its reliance on Jansenist theology (stemming from Augustine) regarding God's dominating selective will relative to humans.

2 Pascal's Wager and Jansenism

Whatever else Pascal's Wager is, it is a reflection on human inquiry about God's reality. Pascal is clear on this matter:

> "Either God is or he is not." But to which view shall we be inclined? Reason cannot decide this question. Infinite chaos separates us. At the far end of

---

* Thanks to Lawrence Pasternack and Paul Bartha for helpful comments.

this infinite distance a coin is being spun which will come down heads or tails. How will you wager? Reason cannot make you choose either, reason cannot prove either wrong. [L418/S680]

Pascal has in mind a distinctive kind of inquiry about *God's* existence. He contrasts one's being "inclined" to, or one's choosing, the "view" that God exists with one's being inclined to, or one's choosing, the view that God does not exist. He asks what can incline us to choose one view rather than the other, and he denies that reason can answer his question.

If we are left to our "natural lights" in inquiry, Pascal confesses to being skeptical about our acquiring knowledge about God. He writes: "If there is a God, he is infinitely beyond our comprehension, since, being indivisible and without limits, he bears no relation to us. We are therefore incapable of knowing either what he is or whether he is" (L418/S680). Human reason by itself, according to Pascal, cannot bridge the chasm between humans and God that blocks "natural" human knowledge of God. In his view, God would have to self-reveal to humans, beyond the resources of human reason.

One might move immediately to the practical consequences for humans of the two options: inclination to the view that God exists and inclination to the view that God does not exist. This strategy, however, would jump the gun, and many commentators make this mistake. We shall see that Pascal himself makes this mistake in his Wager argument, in failing to give due consideration to a certain conception of God as perfectly good. We thus need to begin with careful attention to what Pascal has in mind with the slippery term "God." Fortunately, we have clear evidence of Pascal's conception of God, and we can recruit it to bear on an assessment of his Wager.

Human inquiry about God, like human inquiry about anything else, should attend to the specific nature of the object or subject-matter of the inquiry. In doing so, we can be conceptually responsible in inquiry and avoid working with concepts that mislead or otherwise obstruct our inquiry. In particular, we can identify concepts and corresponding assumptions that do not fit with conceptually responsible inquiry about a particular object or subject-matter. For instance, if we start with the assumption that a perfectly good God would have eliminated all evil, we will distort inquiry about God, given that the assumption is false. A perfectly good God may have a good purpose for not having eliminated all evil. So, one should not assume otherwise in inquiry about a perfectly good God.

Attending to the specific nature of the object or subject-matter of an inquiry can contribute to a needed understanding of relevant *evidence* in that inquiry. If God inherently has a morally perfect character, owing to being worthy of worship, then salient evidence of God's reality would reflect this character. It would reflect, in particular, a redemptive effort by God that somehow invites people to share in the kind of moral character distinctive to God, for their own good. An absence of such evidence would be an absence of evidence for a perfectly good God, because it would leave no indication of a perfectly redemptive God, who seeks to benefit people in what they truly need.

A perfectly redemptive God would try to give people the perfect goodness they need, in keeping with their being genuine agents. At the center of such goodness would be human reconciliation to God whereby humans resolve to cooperate with God's goodness for all agents concerned. This reconciliation, reflecting God's perfect moral character, would require unselfish love from humans toward all other people, including enemies. Such love would require that humans will what is good, all things considered, for those people. Failure on this front would entail falling short of the perfect goodness required of a God worthy of worship and of humans who reflect that God.

Pascal follows Augustine in a conception of divine goodness that entails a distinctive relation between divine mercy and divine justice. David Wetsel has identified the relation as follows: "In condemning some and saving others, God does not act arbitrarily or contrary to mercy and justice. By right, all human beings, who fell in Adam, justly merit eternal damnation. God simply *leaves*, abandons, the unjust to a fate that all humanity has always deserved" (2003, p. 165). In Pascal's words: "God, through an absolute and irrevocable will, wanted to save his elect, in a purely gratuitous act of goodness, and . . . he abandoned the others to the evil desires to which he could have justly abandoned all men" (1995 [1656], p. 222). Pascal thus proposes that "men are saved or damned according to whether it pleased God to choose them to be given this grace amongst the corrupt mass of men in which he could justly abandon them all" (1995 [1656], p. 223).

In Pascal's Jansenist story from Augustine, the decisive consideration is that "[God] wished to blind some and enlighten others" (L232/S264). As a result, "there is enough obscurity to blind the reprobate and enough light to condemn them and deprive them of excuse" (L236/S268). Pascal implicates the ancient Hebrew prophets and Jesus in this story: "This is one of the formal intentions of the prophets. *Shut their eyes* (Isa. 6:10) . . .

Jesus came to blind those who have clear sight ..." (L228/S260, L235/S267). This view stems from Pascal's position that "as all humans in this corrupt mass are equally worthy of eternal death and of God's anger, God could justly abandon them without mercy to damnation" (1995 [1656], p. 222).

According to some Pelagians (Pascal's term) who would dissent from Augustine and Pascal, "God would have been unjust if he had not wanted to secure [for redemption] all human beings (within the corrupted mass), and if he had not given to all sufficient help to save themselves" (1995 [1656], p. 224). In opposition to such Pelagians, Pascal invokes a contrary notion of (severe) justice that stresses the following consideration: "All those to whom [divine] grace is not given ... merit eternal death, since they chose evil through their own free will" (1995 [1656], p. 223). This position differs from a Calvinist view in acknowledging a role for free will, at least in the rebellion of Adam against God. Even so, questions of mere justice aside, we need to ask whether God would be *perfectly merciful, gracious, or loving* in not wanting to redeem all human beings and in abandoning some humans "without mercy to damnation."

Pascal's Jansenist story portrays God as "inspiring" some people, but not all people, to receive divine grace and salvation from damnation, even in keeping with human "free will." The story acknowledges the following kind of divine grace and inspiration:

> the grace of Jesus Christ which is nothing other than complaisance and delectation in God's law diffused into the heart by the Holy Ghost, which ... fills the will with a greater delight in good than concupiscence offers in evil; and so free will, entranced by the sweetness and pleasures which the Holy Ghost inspires in it, more than the attractions of sin, infallibly chooses God's law for the simple reason that it finds greater satisfaction there. [1995 [1656], p. 223]

God's Spirit, in this story, inspires or fills a human free will with a delight in God's law, and this inspiration leads to an infallible choice of God's law over contrary human ways.

Part of Pascal's story is: "The first thing that God inspires in the soul that he deigns to truly touch, is a knowledge of and an extraordinary insight by which the soul considers things and itself in a completely new way. This new light gives her [the soul] fear and brings her a troubled [spirit] that pierces the tranquility she found in the things that gave her pleasure" (1963 [1653], pp. 290–91). This kind of inspiration from God's Spirit, according to Pascal, saves some people, but not all people, from God's abandoning them without mercy to damnation. This selective inspiration comes ultimately from God's selective redemptive will, and not from human willing.

In the *Pensées,* Pascal acknowledges a central role for the divine inspiration of humans, beyond human reason and habit:

> There are three ways to believe: reason, habit, inspiration. Christianity...does not admit as its true children those who believe without inspiration. It is not that it excludes reason and habit, quite the contrary, but we must open our mind to the proofs, confirm ourselves in it through habit, while offering ourselves through humiliations to inspiration, which alone can produce the real and salutary effect. *Lest the Cross of Christ be made of none effect.* [1 Cor. 1:17, (L808/S655)]

Such divine inspiration of humans produces human desire, in accordance with free will, to yield to God and God's law.

Pascal denies that people must believe in God on the basis of argument in all cases. He remarks: "Do not be astonished to see simple people believing without argument. *God makes them* love him and hate themselves. He inclines their hearts to believe. We shall never believe, with an effective belief and faith, unless God inclines out hearts, and we shall believe as soon as he does so" (L380/S412, italics added). The motivational key for human salvation, in Pascal's account, is *God's* inclining or inspiring human hearts to obey God's will, and God does not await human compliance in advance of divine inspiration. God's intervention comes first for those, and only those, people who thereby become the elect rather than the damned. So, God selectively inspires some humans, abandoning all others to damnation without mercy, even though God easily could have inspired all humans and saved them from ultimate destruction. In Pascal's Jansenist story, only God can inspire humans in this way; the Church cannot (1919 [1659], sec. 881).

Faith in God, according to Pascal, is a gift from God received by the human "heart," which includes the human will: "It is the heart which perceives God and not the reason. That is what faith is: God perceived by the heart, not by the reason" (L424/S680). This view of faith includes a distinctive view of reasons of the heart: "The heart has its reasons of which reason knows nothing" (L423/S680). In Pascal's epistemology, God gives the reasons of the heart, and thereby gives faith in God, via direct inspiration from God, as a redemptive gift. In doing so, God fills the human will (a central part of the "heart") with inclination toward God's perfect will, and this originates from God's will ultimately, and not a human will.

Pascal opposes Luis de Molina (a Spanish Jesuit) and his followers, who portray salvation as depending ultimately on human free will rather than God's dominant will. The Molinist position, he claims, "flatters [human common sense] and, making men masters of their salvation or damnation, excludes from God any absolute will, and makes salvation and damnation derive from human will, whereas in Calvin's view both derive from divine will" (1995 [1656], pp. 216–17). An "absolute" and "irrevocable" divine will in the salvation of selected humans is a non-negotiable in Pascal's Jansenist theology. This absolute will accounts for the divine inspiration of selected humans that yields redemptive knowledge of God and salvation for them.

We plausibly can ask whether the Jansenist God defended by Pascal falls short of worthiness of worship in not being *perfectly* or *fully* merciful, gracious, and loving toward *all* people. Contrary to a previous quotation from Wetsel (2003, p. 165) regarding God's not acting in conflict with mercy, Pascal's position concerning God does violate perfect mercy, grace, and love toward some people, and settles instead for partial, imperfect mercy toward humans. That is, it settles for divine mercy toward some people, and rejects such mercy for other people. (See Pascal's aforementioned talk of God's abandoning some people to damnation without mercy; 1995 [1656], p. 222.) This is a failure of perfect mercy in the Jansenist God, and hence a failure of perfect grace and love in this God. It thus entails a moral imperfection in Pascal's God.

Marvin O'Connell has commented on the "harshness" of Pascal's Jansenism:

> Human choice plays a qualified part in the economy of salvation, but the divine will assumes full sovereignty, because, unless impelled by the "singular" and "efficacious" force exercised by the Creator, the creature's will remains ultimately impotent. Yet one might legitimately wonder what scope for human freedom was left in such disproportion, in such strain put upon apparently simple terms like "absolute" and "sufficient." It is difficult, in other words, to see how, given the final verdict, the *Écrits sur la grace* [by Pascal] have awarded anything more than a nominal role in the moral sphere to free choice, and difficult therefore to see how the Jansenist view differed in substance from an extreme version of Calvinist predestination. Its harshness in any case needs no comment. [1997, p. 164]

The perceived "harshness" stems from the selective mercy of God in Pascal's Jansenism that excludes some people from salvation. (Similarly, Graeme Hunter (2013, p. 153) speaks of "Pascal's harsh Jansenist universe," in connection with "supernatural blindness brought about by God, a species of anti-grace that is a punishment for sin.") God chooses, solely from the divine sovereign will, to forgo the inspiring and saving of many humans, even though God could have done otherwise without loss. God could have chosen to inspire and to save all people, with no loss of goodness for God or others. In failing to do so, the Jansenist God exhibits a kind of harshness incompatible with perfect mercy, grace, and love toward all people. Such a moral defect blocks worthiness of worship, and thereby precludes divinity for the Jansenist God.

We now can identify an alternative to the Jansenist God, an alternative that omits divine harshness of a morally defective sort. This non-Jansenist God supplies general or prevenient grace to endow *all* people with the power to decide, at some point, either in favor of God or against God. This God does not leave or abandon anyone to a divinely willed fate of condemnation and destruction. Instead, this God enables all people with the power to submit to God rather than to reject God's offer of reconciliation, even though they still have the freedom to reject God. God thus would preserve justice as fairness, while offering perfect, *non*-exclusive mercy, grace, and love to all humans, with no one omitted by God from an opportunity of reconciled life with God. We thus should contrast perfect, non-exclusive mercy, grace, and love with the exclusive mercy, grace, and love proposed by Pascal's Jansenism.

We are not left with a "Pelagian" account entailing that humans somehow earn their salvation from God. Human choice in favor of God can be free of human earning or meriting from God. God could expect humans to choose in favor of God without demanding that they earn their status before God (by obligating God to approve of them); so, divine grace would not be at risk here. If, however, one widens talk of a "Pelagian" account to involve any exercise of human will in salvation (as in what some call a "semi-Pelagian" account), we should be suspicious of *non*-Pelagian accounts for their depersonalizing human agents in salvation and making God morally defective with imperfect mercy.

If God does not do all that God can to attract humans as free agents to reconciled life with God, then God will be less than morally perfect in the divine redemptive effort. This God thus will fail to be worthy of worship. Here we have a serious problem for the kind of Jansenist God offered by Pascal in the wake of Augustine. It challenges the coherence of Pascal's idea of God, if God (by the perfectionist title "God") must be worthy of worship and hence morally perfect. If God is morally defective, then God is not truly God, but is instead a counterfeit god. Contrary to Pascal's suggestion above (1995 [1656], pp. 216–17), the proposed non-Jansenist God does not flatter human common sense or diminish the importance of God's role in salvation. Instead, this God offers mercy to all people in the hope that they will cooperate with God's morally perfect character and will (see Isa. 65:2, Rom. 10:21). This moral feature of unmerited prevenient grace runs afoul of a common-sense assumption that humans must earn or merit their standing with God. So, Pascal's concern is misplaced.

Pascal claims that his Christian God is a God of love:

> The God of Abraham, the God of Isaac, the God of Jacob, the God of the Christians is a God of love and consolation: he is a God who fills the soul and heart of those whom he possesses; he is a God who makes them inwardly aware of their wretchedness and his infinite mercy: ... who fills [their soul] with humility, joy, confidence, and love: who makes them incapable of having any other end but him. [L449/S690]

Contrary to Pascal's suggestion, however, his God is not a God of "infinite mercy," because, as noted, his God chooses *selective* mercy toward humans and thus excludes some humans from mercy and salvation. The non-elect do not get the kind of divine mercy extended to the elect, and therefore divine mercy has definite, divinely chosen limits and exclusions. God's redemptive love toward humans, then, is selective and exclusive at best, just for the elect. This fits with Pascal's remark that "God inclines the hearts of those whom he loves," a remark to which Pascal adds: "the Spirit of God is upon them and it is not upon the others" (L382/S414). So, Pascal's God is not a God of perfect "love and consolation" for all people. Instead, this God withholds redemptive love and consolation from many people, and thus falls short of perfect love. As a result, Pascal does not offer a morally perfect God worthy of worship.

3 Jansenism and Divine Hiding

According to Pascal's Jansenism, God may hide from a person as God chooses, for God's purposes, and this can include a divine decision not to give a person evidence of God's reality. Pascal remarks: "Wishing to appear openly to those who seek him with all their heart and hidden from those who shun him with all their heart, [God] has qualified knowledge of him by giving signs which can be seen by all who seek him and not by those who do not" (L149/S182). We have seen that, in Pascal's Jansenism, *God* inspires some people and thus makes them seek God, while abandoning other people to damnation without mercy. So, any human seeking is only courtesy of God's active will. Pascal adds: "God being thus hidden, any religion that does not say that God is hidden is not true, and any religion which does not explain why does not instruct ... Verily thou art a God that hidest thyself (Isa. 45:15)" (L242/S275; cf. L427/S681). The Jansenist explanation is ultimately in God's will, as God chooses selective hiding and redeeming.

God's hiding, in Pascal's Jansenist perspective, aims to make certain people stumble, rather than advance, in relation to God. Pascal comments: "What do the prophets say about Jesus Christ? That he will plainly be God? No, but that he is truly a hidden God, that he will not be recognized, that people will not believe that it is he, that he will be a stumbling-block on which many will fall, etc." (L228/S260). He adds: "As [Jesus] came *for a sanctuary and a stone of stumbling,* as Isaiah says (Isa. 8:14), we cannot convince unbelievers and they cannot convince us. But we do convince them by that very fact, since we say that his whole behavior proves nothing convincing either way" (L237/S269). Jesus has a special role in Pascal's story of knowing God, because Jesus offers the proper balance in showing us God and our wretchedness before God. In doing so, Jesus offers an alternative to pride and despair: "Knowing God without knowing our own wretchedness makes for pride. Knowing our own wretchedness without knowing God makes for despair" (L192/S225; cf. L190/S223).

The hiddenness of God, according to Pascal, bears on available evidence of God for humans, and blocks any familiar appeal to natural theology to convince unbelievers of God's existence. He remarks:

> I marvel at the boldness with which [some] people presume to speak of God. In addressing their arguments to unbelievers, their first chapter is the proof of the existence of God from the works of nature ... But ... people deprived of faith and grace, examining with such light as they have everything they see in nature, ... find only obscurity and darkness ... This is giving them cause to think that the proofs of our religion are indeed feeble, and reason and experience tell me that nothing is more likely to bring it into contempt

in their eyes. This is not how Scripture speaks, with its better knowledge of the things of God. On the contrary it says that God is a hidden God, and that since nature was corrupted he has left men to their blindness ... [L781/S644]

Pascal offers a sharp contrast between the biblical writers and the proponents of natural theology: "It is a remarkable fact that no canonical author has ever used nature to prove God. They all try to make people believe in him ... They must have been cleverer than the cleverest of their successors, all of whom have used proofs from nature" (L463/S702). Typical natural theology, in Pascal's view, neglects the role of divine hiddenness in human knowledge of God.

Pascal relates his account of knowing God to "the arrogance of the philosophers":

> [Christianity] teaches ... these two [points] alike: that there is a God, of whom men are capable, and that there is a corruption in nature which makes them unworthy ... It is equally dangerous for man to know God without knowing his own wretchedness as to know his own wretchedness without knowing the Redeemer who can cure him. Knowing only one of these points leads either to the arrogance of the philosophers, who have known God but not their own wretchedness, or to the despair of the atheists, who know their own wretchedness without knowing their Redeemer. [L449/S690]

The deficiency of the philosophers, according to Pascal, extends beyond their "arrogance" to their supposed God. He contrasts their God with the God of his Jewish-Christian theism: "'God of Abraham, God of Isaac, God of Jacob', not of philosophers and scholars" ((L913/S742, "The Memorial"). He also contrasts knowing God with loving God: "What a long way it is between knowing God and loving him" (L377/S409). Philosophical approaches to God typically neglect the divine desire for human love of God and others, and this neglect, according to Pascal, distorts an understanding of knowledge of the Christian God. This God is redemptive in requiring human love of God as a good lacking from humans but needed by them for their flourishing. Knowledge of God that omits such love will fall short of what God seeks from humans.

Philosophers, according to Pascal, do not know the true good for humans or the true state of humans. He adds: "If they gave you God for [an] object it was only to exercise your pride; they made you think that you were like him and of a similar nature" (L149/S182). This contrasts with what Pascal calls "the right way": "The right way is to want what God wants" (L140/S173). He remarks: "If their only inclination is to win men's esteem, if their only perfection lies in persuading men ... that there is happiness in loving them, then I say such perfection is horrible" (L142/S175). So, Pascal finds that a philosophical approach omits a desire for what God values; it allegedly omits the truth of Christian theism with its redemptive God who exposes and cures human wretchedness as something incompatible with God's moral character.

Pascal is clear about the importance for Christianity of what humans want and fail to want. He opens the *Pensées* with: "*Order*. Men despise religion. They hate it and are afraid it may be true. The cure for this is first to show that religion is not contrary to reason, but worthy of reverence and respect. Next make it attractive, make good men wish it were true, and then show that it is" (L12/S46). We shall see that an attempt to "make it attractive," beyond a case for its truth, lies behind Pascal's Wager, but that Jansenism intrudes with a debilitating vengeance. Pascal's effort fits with David Wetsel's remark about Pascal and the unbeliever: "The apologist's role is to awaken in [the unbeliever] a desire to seek by convincing him that his dominant nostalgia for a never-found happiness is none other than an unrecognized yearning for a lost God" (1994, p. 25). The obstacle, however, is that people cannot seek without an intervention from God's active will.

God, in Pascal's Jansenist view, takes the initiative in assigning hope to some people, via supplying divine inspiration to them, but despair to others, via hiding divine inspiration and evidence from them. Pascal finds this two-part strategy to be just, given human rebellion against God's will, and he finds that it fits with the mixed evidence humans have regarding God's reality. So, he recommends the Christian story about God as fitting our evidence, even in a manner better than its competitors (L242/S275).

Pascal offers the Christian God as exclusively good for humans: "The Christians' God is a God who makes the soul aware that he is its sole good; that in him alone can it find peace ... This God makes the soul aware of [its] underlying self-love which is destroying it, and which he alone can cure" (L460/S699). Even so, the same God chooses not to redeem some people or even to give them salient evidence of God's reality, when God could do so without loss (cf. L149/S182). The latter people have no genuine opportunity to know God or to be redeemed. So, Pascal's Jansenist story of divine hiddenness fails to meet any normal standard of mercy toward humans. (The role of Jansenism in this story gets inadequate attention from many commentators, including Nemoianu, 2015.)

4 Jansenism against Pascal's Wager

We now can approach the argument of Pascal's Wager in the light of his Jansenist epistemology and theology. The argument stems from Pascal's aforementioned aim regarding Christianity: "Make good men wish it were true" (L12/S46). This aim is not reducible to an aim to show that Christianity is true.

Pascal denies that reason can show either that God exists or that God does not exist (see L418/S680). His following rhetorical question includes a strong assumption in this connection: "Who then will condemn Christians for being unable to give rational grounds for their belief, professing as they do a religion for which they cannot give rational grounds? They declare that it is a folly, in expounding it to the world, and then you complain that they do not prove it" (L418/S680). Pascal here speaks of "giving rational grounds" for Christian belief, and of "expounding it to the world." The most plausible interpretation is that Christians cannot give convincing rational grounds for Christian belief *to those who desire it not to be true*. God must change that desire, according to Pascal's Jansenism, by a dominant divine will that inspires a person with faith in God. Reason has no effective power here without God's controlling, dominant will in human desire and faith.

According to Pascal, "since you must necessarily choose [between the view that God exists and the view that God does not exist], reason is no more affronted by choosing one rather than the other" (L418/S680). (We now can ignore, if only for the sake of argument, an option of withholding judgment on the matter.) Since a choice must be made, in Pascal's opinion, he asks which one offers a person the least "interest." He finds that one's happiness is at risk, and proposes that we assess the gain and the loss in happiness involved in wagering that God exists. This proposal leads Graeme Hunter to remark that "Pascal's apodictic philosophy ... is rooted in the objective and almost plebeian self-interest that attracts us to a good bet" (2013, p. 209). Arguments in favor of Pascal's theological teachings and their benefits, in this story, are not "metaphysical proofs," but are "inducements" that will yield, if received, happiness in this life and the next (see Hunter, 2013, pp. 221–22).

Pascal's Wager aims to induce human *desires* for theological beliefs of the sort offered by Pascal, without pretending to demonstrate the truth of those beliefs. Those desires can lead to actions which, in turn, can lead to the theological beliefs in question. (For more on inducements in contrast with proofs in Pascal, including the Wager as seeking to induce desires, see Jones, 1998.) The inducements of the Wager, however, are variable in their results among humans; in the Jansenist perspective, they will be met with actual desires for theological beliefs and faith in God only if *God* exercises a dominant will in inspiring (selected) people. Given human sin as alienation from God, the practical inducements of the Wager will not by themselves move humans to desire what God desires, in Pascal's Jansenist story. God's dominating, inspiring power is needed to effect human cooperation with God in desire and belief. This stark lesson lies behind divine hiding and redeeming toward humans.

Pascal identifies the relevant benefit and corresponding downside raised by the Wager: "Let us weigh up the gain and the loss in calling heads that God exists. Let us assess the two cases: if you win, you win everything; if you lose, you lose nothing. Do not hesitate then; wager that he does exist" (L418/S680). He adds: "Since you are obliged to play, you must be renouncing reason if you hoard your life rather than risk it for an infinite gain, just as likely to occur as loss amounting to nothing" (L418/S680). So, the Wager is about an "infinite gain" that is no less likely than a complete loss; it enables one, if God exists, to "win everything."

A problem arises, by Pascal's lights, for actual wagering under the influence of human passions or desires:

> If you are unable to believe, it is because of your passions ... Concentrate then not on convincing yourself by multiplying proofs of God's existence but by diminishing your passions ... Learn from those who were once bound like you and now wager all they have ... They behaved just as if they did believe, taking holy water, having masses said, and so on. That will make you believe quite naturally, and will make you more docile. [L418/S680]

Pascal thus assumes that specified behavior can diminish one's passions in a way that contributes to one's being able to believe the theological truths offered by Pascal. It is, of course, an empirical matter whether this is so for a person, and the answer is not obvious on our actual empirical evidence. We seem to have mixed empirical evidence on the matter. Even so, the real work in convincing is done by *God's* dominating will, according to Pascal, and not by human efforts: "It is God himself who inclines them to believe and thus they are most effectively convinced" (L382/S414). The latter Jansenist thesis colors Pascal's whole theology and philosophy, despite its neglect by many commentators.

Pascal allows for a distinctive kind of certainty that results from the Wager: "I tell you that you will gain even in this life; ... you will see that your gain is so certain [certitude] and your risk so negligible that in the end you will realize that you have wagered on something certain and infinite for which you have paid nothing" (L418/S680). Here, too, we seem to face an empirical matter whether this is so for a person, and the answer seems not to be obvious on our actual empirical evidence. Again, we seem to have mixed empirical evidence on the matter. Hunter offers the following reading: "What the inquirer learns at last by experience is that the faith he has set out to find is indeed supported by many reasons worthy of belief, even though it is not nailed down by irrefragable demonstrations ... It remains a road of inquiry ... because there is no prospect of decisive rational proof, ... lest grace should be thought unnecessary" (2013, p. 197).

A role for "grace" will be crucial in Christian theism, but Pascal's Jansenist understanding of grace threatens the viability of his Wager and his theology in general. Jansenist grace does not include God's offering and enabling a genuine opportunity to all people for reconciliation to God. Instead, it is a matter of God's using a dominating divine will to select some people for salvation and to abandon others to condemnation and destruction without mercy. As a result, a simple but fatal question arises: Why bother? If God inspires selected humans by an "absolute," "irrevocable," and "effectively convincing" divine will (L382/S414), there is no *real* need, practical or otherwise, for human wagering and seeking toward God.

God gives, by an absolute, effective, and irrevocable divine will, a human want for God's ways to selected humans, and not to others. Those omitted by God from having such a want will not turn to God under any circumstances; they are abandoned by God, irrevocably, to despair and destruction, without mercy. So, Pascal's Wager will be ineffective and dispensable, owing to the dominating role of God's will, and humans need not bother with any such wager. As noted, God alone inspires and convinces selected humans in Pascal's Jansenist story (L382/S414). A human calculation of benefit, then, is at best a disengaged wheel in that harsh story. What matters is a divine inspiration of a human to want what God wants; and *only* God, in Pascal's view, can supply that inspiration and the resulting want. We should not pretend otherwise, even with the option of a wager. Pascal's Wager can be, and is, misleading in that regard.

William Wood (Chapter 2 in this volume) has offered Pascal's Wager as being logically consistent with his Jansenism. In Pascal's compatibilist view of freedom, he suggests, "humans are *free to resist grace*, as the Council of Trent requires, while also ensuring that God's absolute will to save the elect is never thwarted. On this account, the capacity to resist grace is just the capacity to do something that, all things considered, one has absolutely no desire to do. We may have such a capacity, according to Pascal, but it is little more than notional" (p. 56; italics added). This is unconvincing, however, if it grants a capacity entailing an elect person's being *free to resist grace*. Pascal's God, as noted, uses his "absolute and irrevocable will" (L382/S414) to give the elect a desire to receive grace, and that divine will, being absolute, irrevocable, and effective, *precludes* in the elect a desire to resist divine grace. So, the proposed "freedom" to resist grace amounts to something empty, because there is no genuine capacity in the elect to resist. Everything depends here on what causally lies behind the desire of the elect to receive grace. Given that it is God's "absolute and irrevocable will" to save them, there is no room for a capacity in the elect to resist. Being free, then, is not just a matter of doing what one desires to do; it matters what causally bears on one's desires. So, the suggestion that, in Pascal's Jansenism, the elect are *free to resist grace* is misleading.

Wood adds: "God can use the Wager to bring about the conversion of someone predestined to salvation, and he can also use Pascal as his instrument to offer it. Pascal can agree that God is the cause of every conversion, and consistently maintain that when he himself offers the Wager to an unbeliever, he merely carries out God's will." (see p. 59 of this volume). It's undeniable that "God can use the Wager" to convert the elect, but that is not central to the pressing issue. Instead, the key issue is whether a Jansenist use of Pascal's Wager is indispensable (in any manner) in salvation, and can accommodate a role for a God set on genuine grace and love toward all humans, that is, for a God worthy of worship and hence morally perfect. This issue remains open even after granting that "God can use the Wager" to convert the elect.

Wood proposes that "Pascal can reconcile the Wager with his views on grace ... by pointing out that on his account, no one ever has to do anything other than what they most want to do. Some people will want to wager and others will not. In either case, God has chosen to save his elect in the most gentle, non-coercive way possible: by allowing everyone to think and do exactly as they wish." (see p. 61 of this volume). This is too quick to convince. It would be more "gentle," on God's part, if the divine saving of the elect were conjoined with the divine saving of all other humans, too. It would be more gentle in virtue of being more merciful and loving toward humans; that is, toward all humans. Something "most gentle" *toward the elect* does not generalize to what is "most gentle," particularly if we have the divine exclusion of people who could have been saved by God. In addition, we should consider that God would be more "gentle" toward the elect if they were allowed to form their own desires toward receiving divine grace, without "absolute and irrevocable" divine causation behind their desires. That seems to be a plausible option for a more "gentle" approach on God's part.

"Like Pascal," according to Wood, "we can even marvel at the generosity God displays in meeting self-interested sinners exactly where they are. God offers them a path to salvation that begins from their own corrupt desires for domination and diversion. For such people, the Wager might be exactly what grace looks like" (see p. 62 of this volume). Such marveling would be premature, however, for the reason just suggested. The "grace" and "generosity" in question are morally imperfect, owing to their excluding, by divine intent, many humans who could have been included by God among the elect. Being thus imperfect, they have no role in moral perfection or in worthiness of worship. They therefore leave a moral defect in God's moral character, quite aside from the issue of whether God's absolute will toward the elect harms, and even precludes, their being genuine agents with freedom to resist.

It is hard to see how Pascal's Jansenist grace is truly "generous," "sweet," "gracious," or "loving" for all concerned if God could have predestined the non-elect to salvation but nonetheless chose, by an "absolute and irrevocable will," to let them go to eternal damnation. The fact that God did not have to push them to nonbelief hardly makes God's approach "generous" or "sweet" for all concerned. So, Pascal's God fails as a model for perfect grace and love for all humans. As indicated, such a God could have humans use a wager, but that prospect would not redeem God's dubious moral character toward the non-elect; nor would it make a wager indispensable to salvation (in any manner) on Jansenist terms.

Did Pascal come to see that his Jansenist position undermines the need and the effectiveness of not only his Wager but also the *Pensées* in general? Perhaps, but our evidence is inconclusive, if only because Pascal did not comment on the matter. Anthony Levi notes: "It now seems clear that the project to write an apologetic was not abandoned for reasons of health, as is still often assumed, and even that, on Pascal's own premises, the intended apologetic could have served no purpose..." (1995, p. ix). The premises in question are his Jansenist assumptions about God's dominating will in human salvation and condemnation. Levi thus asks rhetorically: "If salvation was God's gratuitous gift to a minority of chosen human souls, how could any moral act, and in particular any freely chosen commitment of belief or behavior, affect the individual's eternal destiny?" (1995, p. ix). He adds: "As a Christian Pascal had to combine belief in a loving and lovable God with a revelation which he acknowledged to contain the truth that the majority of the human race is created by that God as a *massa damnata* – that is, in circumstances which admit only of eternal damnation" (1995, p. xxxvi).

The Jansenist "loving God" in question, I have indicated, is not perfectly loving, owing to the exclusive redemptive circumstances just noted. As a result, I have suggested that Pascal's Wager is not compelling, given God's suspect moral character that falls short of moral perfection. God's lack of perfect mercy, grace, and love blocks us from deeming, with adequately grounded confidence, the outcome of the Wager as supplying unmixed happiness for us. It leaves open the real prospect of the divine abandonment without mercy of more people, including people who have wagered for God. All bets are off, in the end, if God is morally imperfect regarding mercy in the way entailed by Pascal's Jansenism.

5 Jansenism and Inquiry about God

Graeme Hunter sums up Pascal on inquiry:

> The grandeur of [Pascal's] thinking lies... in his manner of inquiry. Pascal holds the secular modern (or postmodern) reader to account in a personal manner that recalls the manner and power of Socratic investigations. He mocks our pieties, challenges our certainties, and provokes us with the thought that we have settled for an impoverished life when infinite riches are within our grasp. Pascal's philosophy is not on trial. We, his readers, are. [2013, p. 222]

Perhaps we humans do need to be put on trial before a morally perfect God. It is doubtful, however, that Pascal's Jansenist line of inquiry about God, including the Wager, puts us on trial in a compelling manner.

The Jansenist Pascal must face the problem that conceptually and evidentially responsible inquiry about *God*, at least in the major monotheistic traditions, should be inquiry about an agent who is *worthy* of worship and hence morally perfect. So, a wager regarding this God must be a wager for an agent characterized by perfect mercy, grace, and love toward all humans, instead of imperfect, partial mercy, grace, and love toward (selected) humans. The Jansenist God who chooses to inspire and convince some humans, but not others, by a dominating divine will could choose by the same will, at any time, to terminate their being inspired by God. This God would have no acknowledged constraint by an obligation or a moral character to sustain perfect mercy, grace, or love toward all people or even the elect. So, Pascal's Wager for "infinite gain" could fail with that God, owing to the lack of an acknowledged constraint on that God by perfect mercy, grace, or love. As a result, the alleged "infinite riches within our grasp" are less than certain and at best doubtful.

The success of Pascal's Wager would be as much about God's moral character as about ours, and the Jansenist God has displayed moral imperfection in relating to humans selectively and exclusively with mercy. Even if *some* kind of (severe) "justice" is preserved, owing to human sin, that justice will not demand that God sustain the inspiration of the elect. The same is true of perfect mercy, grace, and love; they will not solve the problem, because those moral perfections do not constrain or guide the Jansenist God. So, Pascal's Wager becomes indeterminate as a result of the moral indeterminacy or unpredictability of the Jansenist God's moral character. The Wager and the Jansenist God's character fail to determine that the outcome will be "infinite gain," given the exclusive, morally suspect will of that God.

Conceptually and evidentially responsible inquiry about God is central to the ethics for inquiry about God. It constrains inquiry about God in a manner suited to conceptually and evidentially trustworthy inquiry about the relevant subject-matter. Pascal settles too quickly and easily for inquiry and wagering about a morally imperfect, exclusive Jansenist God. A conceptually responsible inquirer would need to look more carefully for a God of perfect prevenient mercy, grace, and love for *all* people. Given such a God, human wagering could be stable relative to God's moral character of perfect goodness toward all people. That stability would stand in sharp contrast to the situation of the Jansenist God.

Under the shadow of Augustine's theology, Pascal dismisses quickly a role for human volitional cooperation in salvation, as mere "flattery" of human common sense and as making humans "masters" of their salvation. This is not a conceptually responsible treatment of the complexity of the matter, and it leaves us with an exclusive, dominating God unworthy of worship. Conceptually responsible inquiry about a God worthy of worship, as suggested, would have to give more careful attention to perfect prevenient mercy by God toward all people. Otherwise, God's moral character will be left suspect at best, and any positive projection from a wager about God will be dubious. (For an approach to redemption that preserves a role for human freedom and cooperation with God, see Moser, 2013, 2017.)

Pascal takes evidence of divine hiddenness to support his Jansenism, including his view that, courtesy of the dominating divine will, God abandons some people to condemnation and destruction without mercy. Pascal's Wager in favor of God, however, cannot be evidentially responsible for those people to whom God has not given evidence of God. Those people would face the arbitrariness of fideism, and such arbitrariness would put at risk well-grounded belief in God. Even if Pascal's Wager concerns what we desire, our actionable desires regarding God will be suspect if they neglect evidence regarding God. A more defensible perspective would not propose a wager in the absence of well-grounded support. It would propose instead that God hides at times because God knows that some people are not ready to decide freely in favor of God or even to give the matter attention with due care and candor. As a result, a perfectly good God would respect the need of such people for more time and guiding experience, and would not force or otherwise promote a suspect decision on the basis of inadequate evidence. (For elaboration on such an approach to divine hiddenness, and an alternative to fideism, see Moser, 2008, 2017.)

For the sake of conceptually and evidentially responsible inquiry about God, Pascal should have attended more carefully to the Christian message of *universal* divine love, instead of the harsh theology of the later Augustine. Here he would have found a message of God's loving "the world" (John 3:16), including God's mercy and love toward God's enemies (Luke 6:32–36; Matthew 5:43–48). He would have found the message that God demonstrates divine love for sinners and even enemies of God without preferential treatment (Romans 5:8, 10). In addition, he would have found the message that God wishes that none should perish, but that *all* people should come to salvation, in reconciled life with God (2 Peter 3:9, 1 Timothy 2:4, Romans 10:21). The dominant New

Testament message of God's perfect mercy toward all people runs afoul of Pascal's Jansenist story, but Pascal neglects that message in a manner at odds with conceptually responsible inquiry about God. His Wager suffers accordingly; it leaves one in need of an alternative approach to responsible inquiry and decision-making about a God worthy of worship.

4 Pascal and His Wager in the Eighteenth and Nineteenth Centuries

Adam Buben

By the twentieth century, Pascal's Wager had become one of the most famous arguments in the philosophy of religion, and perhaps in the entire history of philosophy. But how was Pascal's thought received in the centuries following the first publication of the *Pensées* in the late seventeenth? This chapter will focus on examining the reception of Pascal and his Wager in the writings of several prominent eighteenth- and nineteenth- century philosophers, especially Immanuel Kant, Søren Kierkegaard, and Friedrich Nietzsche. While the comparison with Kierkegaard is probably the most striking, there are fascinating conceptual affinities to be found in Kant's views on religion, and a surprisingly sympathetic critique in the midst of Nietzsche's most venomous assault on Christianity. With some exceptions, what most of the commentary from this period has in common is an expression of broad admiration for Pascal and his intellect, coupled with a healthy suspicion about the reasoning involved in the Wager itself and apologetic strategies in general. Ultimately, it is this concern about Pascal's understanding of the relationship between faith and reason that will prove fertile ground for the most interesting conclusions.

1 Explicit French Criticism and Possible German Affinity

Voltaire and Denis Diderot were among the most notable critics of Pascal's Wager in the first century after the Port-Royal edition of *Pensées* was published in 1670. While they both praise Pascal as a profound and unique thinker (e.g., Diderot, 2009, p. 32; Voltaire 1763, p. 3), they also raise some serious objections to the Wager. Diderot (1875 [1756], para. 59), for example, suggests an early version of the many-gods objection when he, after briefly summarizing the Wager, retorts, "an imam can say the same as Pascal." In his *Philosophical Dictionary*, Voltaire (1856, p. 385) seems to make a similar point, also with reference to Islam, but his more explicit assault on the

Wager comes in his fairly extensive "Remarks on Mr. Pascal's Thoughts."[1] Here he lists several concerns about the Wager that mostly seem to miss the point of Pascal's thought experiment. For instance, in saying that "it is a very false assertion, that the not laying a wager that God exists, is laying that he does not exist," Voltaire (1763, p. 10) fails to acknowledge that when only an affirmative response to the question of God's existence would make one eligible for certain benefits, the neutrality of the agnostic, no less than the outright rejection of the atheist, rules out these benefits. His most interesting argument concerns the issue of predestination – if only a very small number of elect have any hope of realizing the advantages of God actually existing, it seems to make sense for the vast majority of people to hope that God does not in fact exist (Voltaire, 1763, pp. 10–11). Interesting as this idea may be, I am not sure that it would bother Pascal much, given that the uncertainty about one's place among the elect would seem to function in the same way as the uncertainty of God's existence within the logic of the Wager. As long as the possibility of an infinite reward cannot be definitively denied, the risk of any finite loss (however likely) seems to pale in comparison.

While his fellow Frenchmen offer fairly straightforward criticisms of Pascal's Wager in the middle of the eighteenth century, it will require a bit more effort to piece together the opinions of later thinkers. One complicating issue is that it seems to have taken some additional time for the Wager to catch the attention of significant philosophers outside of France. Kant, for example, has very little to say about Pascal, and provides no evidence that he was even aware of the Wager.[2] He does mention Pascal in passing on a number of occasions, but these statements mostly just touch on minor details of Pascal's biography (e.g., his precociousness as a child) or call attention to his work outside of philosophy and religion (e.g., in geometry). The only references to Pascal of any real relevance to the present

---

[1] Given more pressing interests, there simply is not space in the present chapter for a thorough consideration of all that Voltaire has to say about Pascal in this text. However, to characterize his general concern briefly, Voltaire believes that Pascal is far too harsh in his depiction of the depravity of humankind. Beyond this overriding worry, his remarks involve a rather random assortment of Pascal quotations followed in each case by his rebuttal. Voltaire (e.g., 1763, pp. 50, 54) is aware that *Pensées* was not a finished work, and claims that Pascal would have corrected or omitted some of his ideas if given the opportunity to complete the book as intended. With this in mind, Voltaire's remarks tend to read a bit more like incredulous (and occasionally somewhat rash and superficial) editorial suggestions than meticulously developed criticisms.

[2] It is not clear how familiar Kant was with the *Pensées*, but Pascal was apparently not found in his personal library (see Warda, 1922). Kant would have had access to one of the same German translations that Kierkegaard owned (Pascal, 1777), but he also read French so it is possible that he was familiar with earlier (and as we will see, incomplete and altered) editions. For more on Kant's library and his reading habits, see Frierson (2012, pp. 71–72).

endeavor concern his use of "fanatically frightening" ideas[3] (e.g., 7:162),[4] but even in these cases, there simply is not enough to go on in coming to a solid conclusion about Kant's view of Pascal. Nonetheless, there is certainly sufficient ground for an interesting comparison of their thoughts on belief in the existence of God.

What Kant and Pascal have in common is that they both consider the possibility that there are practical reasons for, or benefits in, believing in God even if there are no compelling theoretical demonstrations that such a being exists. It remains unclear whether or not Kant would approve of the Wager, but there is a certain kinship between Pascal's famous argument and Kant's account of the Highest Good and the Practical Postulates. The key connection is that they both seem to recommend the subjugation of one's own immediate and finite interests for the sake of some infinite or absolute good. In Kant's *Critique of Pure Reason*, this good is largely wrapped up with the rewards of the afterlife. At this relatively early stage in his work, punishment and reward function as the primary motivation for moral behavior. Why would one refrain from following one's immediate inclinations if not for the hope of something better to come as long as one does so? Since God is the being that is capable of meting out rewards and punishments based on our behavior, and this does not seem to happen in "the sensible world," we have a practical need to "assume" the existence of "God and a future life" (A 811/B 839). Putting all of this together, Kant says, "without a God and a world that is now not visible to us but is hoped for, the majestic ideas of morality are, to be sure, objects of approbation and admiration but not incentives for resolve and realization" (A 813/B 841). In other words, even if we cannot prove that God exists, we need God and the justice he brings to the universe in order to motivate moral behavior (cf. Pasternack, 2011, pp. 305–6). Not only does belief in God bring with it the possibility of infinite personal gain in a future world, it also makes possible the absolute value of living in a moral universe.

---

[3] There is really no way to be sure of which ideas Kant is referring to here, but Pascal is known to use frequent thoughts of impending death and damnation to motivate his arguments. For example, he says, "Let us ... judge on that score those who live without a thought for the final end of life, drifting wherever their inclinations and pleasures may take them, without reflection or anxiety ... eternity exists, and death, which must begin it and which threatens at every moment, must infallibly face them with the inescapable and appalling alternative of being either eternally annihilated or wretched, without their knowing which of these two forms of eternity stands ready to meet them forever" (L428/S682). For more on Pascal's use of death as a motivator, see Buben (2011, pp. 67–68).

[4] References to the *Critique of Pure Reason* will be to the standard A/B edition pagination, while references to Kant's other work will be to the Akademie-Ausgabe volume and page numbers. All quotations will come from the Cambridge Edition of the *Works of Immanuel Kant*.

After further developing his moral theory in subsequent work, most notably the *Groundwork of the Metaphysics of Morals* and the *Critique of Practical Reason*, Kant provides an argument for belief in God that does not rely so heavily on the desired personal outcome of moral behavior provided by the Highest Good, but rather on the nature and demands of morality itself. We do not need to believe in God in the hope of some future reward, but because God's existence helps us understand how morality works. Without getting into all of the specifics of his more mature view of moral matters, the main difference is that, in this later work, morality is more intimately bound up with the Highest Good; the latter is necessary in order for the former to make sense (5:114). Morality would be undermined if people did not get what they deserve in the long run. Despite its somewhat different role here, as we see in the first *Critique*, the Highest Good is simply moral worth dictating the final distribution of happiness.[5] And following a similar line of reasoning, God and the afterlife remain necessary for making this distribution possible.

Since pure practical reason establishes the moral law in accordance with the Highest Good, humans are justified, practically speaking, in postulating what is necessary to make it happen (cf. Pasternack, 2012, p. 172 n. 43). In Kant's words, "practical reason inexorably requires the existence of [God and immortality] for the possibility of its practically and absolutely necessary object, the highest good" (5:134). Only God is capable of justly pairing reward with desert, so belief in the existence of such a being is warranted. As Lawrence Pasternack (2011, p. 309) puts it: "In the case of God ... the perfect distribution of happiness to moral worth requires that there is some being capable of evaluating worthiness and distributing rewards accordingly." However, since "such a distribution does not seem to happen during life," it is quite reasonable to hold that "there must be an afterlife" during which all wrongs will be righted, so to speak (Pasternack, 2011, p. 309).

While Kant continues on to offer further variations on this theme in the third *Critique* and beyond,[6] it should be easy enough to see at this point what his discussions of the Highest Good and the Practical Postulates have in common with Pascal's Wager. Both Pascal and Kant (at least early on, in the latter case) contend that belief in God makes sense when considering the

---

[5] This definition of the Highest Good seems to hold throughout Kant's Critical Period, even while its role shifts (cf. A 810/B 838).

[6] Looking past the situation of particular individuals, Kant (see, e.g., 6:98–99) also argues that God is necessary if there is to be a truly ethical community aimed at the Highest Good. For a helpful discussion of such developments, see Courtney Fugate (2014, esp. pp. 152–53).

unlimited potential afterlife benefits, even if it means giving up something of more defined value in the here and now. In addition to this obvious similarity, Kant goes on to argue that belief in God and the afterlife is justifiable also when considering the absolute value of morality itself. Although Kant does not put either of these arguments in terms of gambling, he joins Pascal in suggesting that there are good reasons for such belief that have nothing to do with providing irrefutable evidence or theoretical proofs. But this is not the end of the links between them – their respective defenses of belief both offer the sort of practical justification for religion that Kierkegaard is likely to find objectionable.

2 Kierkegaard's Appreciation and Apprehension

Compared to Kant, Kierkegaard actually has a great deal to say about Pascal. According to the auction record after his death, Kierkegaard was in possession of three different German editions of the *Pensées* (Rohde, 1967, pp. 48–49), and he shows strong signs of familiarity with their contents.[7] Almost all of his comments about Pascal can be found in his journals and notebooks, and they come with increasing frequency in his later years.[8] It is in these passages that one can get a clear sense of Kierkegaard's deep respect for the man and his convictions (e.g., JP 1:69/ SKS 24:518–19), while also coming to understand his serious concerns about Pascal's apologetic tendencies. On the positive side, Kierkegaard seems to see a parallel between his own critique of the Danish Lutheran Church and Pascal's worries about the Jesuits and their scholastic foundations (e.g., JP 3:421–22/SKS 24:115–16).[9] As an example of Pascal's

---

[7] In addition to *Pensées*, Kierkegaard also had some knowledge of the *Provincial Letters*, and read biographical texts about Pascal (JP 3:421/SKS 24:113, 115). Here are the key texts and translations he made use of: Neander (1847); Pascal (1777, 1840); and Reuchlin (1839, 1840). References to (and quotations from) Kierkegaard's published work will be to the standard abbreviated title and page numbers from the Princeton edition of *Kierkegaard's Writings*. References to (and quotations from) Kierkegaard's *Nachlass* will be to the volume and page numbers from the Indiana edition of *Søren Kierkegaard's Journals and Papers* (abbreviated: JP). It has also become common practice to include reference to the volume and page numbers from the new Danish fourth edition of Kierkegaard's works (abbreviated: SKS) because the English editions listed above only provide a concordance with older Danish editions.

[8] Kierkegaard only mentions Pascal once, briefly, in his published writings (SLW 460/SKS 6:424).

[9] He also believes that there is a similarity in the mistreatment they each endured because of their views. The Jansenist movement, for which Pascal had some affinity, was openly condemned by the dominant Jesuits, at least in part, for its Protestant sympathies. Kierkegaard, on the other hand, suffered a bit of mockery and social isolation in his later years due to his criticisms of the lifestyle of ordinary Danish Christians and the teachings of the Church that contributed to it. One other point of connection Kierkegaard notices concerns their long battles with physical ailments (and one might

concerns about contemporary Christianity lining up with his, Kierkegaard claims that "there is much truth and pertinence in what Pascal says, that later Christianity with the help of some sacraments excuses itself from loving God" (JP 1:222/SKS 25:256). This kinship may have something to do with the shared Augustinian views of the Jansenist-leaning Pascal and the heavily Luther-influenced Kierkegaard, and they both end up advocating a version of Christianity that emphasizes the struggle involved in personally relating to Christ and the need for a divine helping hand, rather than reliance on tradition and ritual (cf. Maia Neto, 1991, p. 164).[10]

Despite Kierkegaard's apparent appreciation for and knowledge of Pascal, as in the case of Kant, there is no firm evidence suggesting familiarity with the Wager. The most famous formulation of the Wager (the one from the "Discourse on the Machine" section) was not included in the earliest editions (and translations) of *Pensées*, and while some version of it does appear as far back as the Port-Royal edition, it is not clear that this formulation caught Kierkegaard's attention.[11] Nonetheless, given his views on related issues and all that he does say about Pascal, it should still be possible to surmise what his opinion of the Wager would have been.

The purpose of the Wager seems to be to convince people that being a Christian or wanting to be a Christian without proof of Christianity's core tenets is not the blameworthy foolishness detractors often say it is (L418/S680). It is not the case, as is sometimes popularly believed, that the Wager is intended to lead directly to faith; Pascal recognizes that having faith requires divine assistance/intervention and is not something that one can just rationally decide to do. In any case, Kierkegaard is almost entirely uninterested in

also point out their similarly short lives). For more on these issues, see Maia Neto (1991, pp. 163–64).

[10] On the topic of the place of reason within this relationship of faith, I discuss in greater detail elsewhere the proximity of Kierkegaard and Pascal to Augustine, Luther, and scholastic luminaries such as Anselm and Aquinas (Buben, 2016, pp. 71–76). Briefly, while Pascal offers a corrective to the scholastic rational exaggeration (perpetuated by the Jesuits in his own day) by staying closer to Augustine's fairly moderate suspicion of reason – and saying things like "Two excesses: to exclude reason, to admit nothing but reason" (L183/S214) – Kierkegaard follows Luther in extending this suspicion as far as it can go. Although they learned some similar lessons from Augustine's legacy, Pascal was immersed in a largely Catholic context while Kierkegaard grew up in a thoroughly Lutheran one.

[11] In 1850, Kierkegaard himself actually learns that "not until just recently has Pascal's Pensees [sic] been published in its complete and original form by Prosper Faugère, 1844, that the older editions had omitted and altered portions, that similarly Anton Arnauld (Port Royal), who published them, took the liberty of making changes" (JP 3:419/SKS 24:98). However, even this relatively late claim of completeness may not be entirely accurate.

convincing people that Christianity is not a crazy and meritless proposition. This is, of course, largely because he is less concerned about converting unbelievers than he is in setting straight Danish Christians who have grown complacent and taken its alleged merits for granted. As for something like the Wager itself, consider Kierkegaard's early worry about Socrates's approach to the afterlife:

> If one also bears in mind that he still does not really know what the shape of the next life will be or whether there will be a next life, if amid this poetry we hear the prosaic calculating that it can never do any harm to assume another life…then one sees that the persuasive power of this argument is considerably limited. [CI 68/SKS 1:126][12]

While the uncertainty about what comes next is an important part of our faithful relationship with the afterlife, and the God responsible for it, Kierkegaard seems to believe that this sort of calculative bet-hedging ruins the relationship. Some might suggest that it is precisely this kind of strategy that is found at the core of the Wager, but even if this characterization is not quite accurate, Kierkegaard might still see in the Wager a troubling Socratic resemblance.

Pascal famously argues that since "you are embarked" (L418/S680, tm) – i.e., you already exist – you have to take a position for or against the existence of God and the afterlife. Practically speaking, to refuse to take a position is to be against since such a refusal means rejecting the potential benefits of God and the afterlife. Death is coming (and sooner than you might like) regardless of which way you go, so your life is staked no matter what, but because only one option (however unlikely) comes with the hope of "infinite gain," it makes sense to want to believe. As I stated above, this little thought experiment is not sufficient to generate faith, according to Pascal, because faith in God and the promise of an afterlife that comes with him is at least as much about how one lives out a relationship with him as it is about giving one's assent to the proposition "God exists"; and living out such a relationship requires a life-long commitment – only possible by the grace of God – which is fraught with difficulty that goes far beyond any momentary struggle with decision-making.

---

[12] While Kierkegaard seems to have the Socrates of Plato's *Apology* in mind (29b, 40c–41b), consider also that, in the *Phaedo*, Socrates's friends discover that he, while awaiting his execution, has been writing poetry for the first time in order to cover all his bases and appease his inner (often described as divine) voice (60d–61b). Socrates, it would appear, is a master of hedging bets. In later writings, Kierkegaard drops this criticism of Socrates, who is often instead portrayed as the paragon of pagan faithfulness and consistency.

Nonetheless, because it uses reason to suggest the attractiveness of Christianity, Kierkegaard would likely worry about the Wager's potential for corrupting the faithful relationship with the divine.

In order to understand why Kierkegaard and Pascal part company on the issue of apologetics, it will first be helpful to look a little closer at what they have in common. I have argued in greater detail elsewhere that one of the important things Pascal and Kierkegaard agree upon is that proper Christianity requires renunciation of our ordinary worldly tendencies, including our tendency to over-rely on rational justification for our behavior and beliefs (Buben, 2011). One of the main causes for concern about offering this sort of justification might be that it looks less and less effective given the increasingly rational scientific worldview that religion must contend with in the modern era.[13] Reason seems to be on the side of those who actively doubt the likelihood of virgin births, resurrected god-men, and inherited sinfulness.[14] But this is not news to Christianity, which has a long tradition of recommending that one cut off (in some cases literally) anything that becomes an obstacle to faith (see, e.g., Matt. 5:29–30, 18:8–9; Mark 9:43–47; Luke 14:26). In the case of reason, Paul even acknowledges that Christianity will appear foolish to those lacking faith (1 Cor. 1:23). Because reason, no less than other elements of our worldly experience, can become an obstacle to the acceptance of Christian doctrine, Pascal advises caution so that it can be used appropriately.

He claims that it is not "through the proud activity of our reason but through its simple submission" that we are able to "know ourselves" and take part in the relationship with the divine (L131/S164). An important component of his view of the diminished role of reason is his notion that the heart (*le coeur*), the instrument of faith, allows for a more intuitive, emotional, and spiritual kind of knowledge that is only possible when reason steps aside. Just after discussing his Wager, Pascal (2005, pp. 215–16) proclaims, "It is the heart that experiences God, and not reason. Here, then, is faith: God felt by the heart, not by reason. The heart has its reasons, which reason does not know" (L423-24/S680).[15] Now this is not to say that

---

[13] Pascal does express concern about those who "probe science too deeply" (L552/S461).

[14] On the latter issue, Pascal says, "Without doubt nothing is more shocking to our reason than to say that the sin of the first man has implicated in its guilt men so far from the original sin that they seem incapable of sharing it" (L131/S164).

[15] Once again, it is unlikely that Kierkegaard knew this exact passage, but he does seem to be familiar with its substance: "Pascal merely insists upon the practical and 'finds it ridiculous for reason to demand from the heart proofs for its first principles, just as it would be for the heart to demand that

Christianity has no use for reason, or that the heart is sufficient for getting by in either the secular or spiritual realms. In fact, one might argue that Pascal is doing something a bit sneaky here, playing loosely with the term "reason" and leaving open the possibility, and maybe even the necessity, of a kind of rational justification for faith once God has taken up residence within an individual. While faith (especially its initial formation) might require the suspension of reason at appropriate moments, this temporary suspension does not rule out the use of a sort of insider rationality for the preservation and defense of faith at other moments. Indeed, Pascal points out, "If we submit everything to reason our religion will be left with nothing mysterious or supernatural. If we offend the principles of reason our religion will be absurd and ridiculous" (L173/S204).[16] Since Christian faith must be allowed to offer reasons in its defense, Pascal finds a great deal of value in pointing to miracles and fulfilled prophecy as evidence for the truth of his beliefs (cf. L323–48/S354–80, L483–99/S718–36), even asserting that "It would have been no sin not to have believed in Jesus Christ without miracles" (L184/S215).

Despite any initial agreement between Kierkegaard and Pascal on the suspicion of reason, claims like this will raise Kierkegaard's ire. His position on the relationship between faith and reason, especially in his later years when he seems to have had a thorough engagement with Pascal's ideas, is a bit more extreme.[17] Kierkegaard argues that Christianity demands renunciation of "every human confidence in their own powers or in human assistance" (FSE 77/SKS 13:99). Chief among the powers and forms of assistance that must be surrendered for the sake of cultivating Christian faith is reason. This is because, according to Kierkegaard, "the way is narrow – it is ... impassable, blocked, impossible, insane [*afsindig*]! ... To walk this way is immediately, at the beginning, akin to dying! ... along this way sagacity [*Klogskab*] and common sense [*Forstand*] never walk – 'that would indeed be madness [*Galskab*]'" (FSE 61–62/SKS 13:84). Those same rational stumbling blocks of Christian doctrine listed above, coupled

the reason should feel all the propositions it proves in order to embrace them'" (JP 3:420/SKS 24:99).

[16] Following Paul, he does allow that Christianity is foolish, in a sense, but foolishness is evidently not irrational in the way that ridiculousness is for Pascal (L291/S323, L418/S680).

[17] I address the possible shifts in Kierkegaard's views on faith and reason across his authorship elsewhere (Buben, 2016, p. 85 n. 31). In the present chapter, I rely heavily on what Kierkegaard says in *For Self-Examination* (1851), not only because the writing of this short book lines up chronologically with the bulk of his comments about Pascal in his journals and notebooks, but also because it is not pseudonymous and might represent the mature Kierkegaard's more considered views.

with the dangers of misunderstanding, mockery, persecution, and martyrdom, signal to Kierkegaard that Christianity demands something of individuals that no rational person should be willing to give (FSE 60–62, 82–85/SKS 13:83–84, 103–5). Even though Kierkegaard can follow Pascal for a while when it comes to the rejection of reason for the sake of faith, ultimately the two diverge at the point where the latter recommends a mere suspension or humbling of reason (e.g., L110/S142), and the former demands a more radical "dying to" it.

Kierkegaard, describing no similar role for "the heart" as the source of intuitive knowledge, holds out little hope that reason of any sort can ever really contribute to or aid faith. Rather, he says things like, "faith is against understanding [*Forstand*]; faith is on the other side of death. And when you died or died to yourself, to the world, then you also died to all immediacy in yourself, also to your understanding" (FSE 82/SKS 13:103).[18] Given the apparent lack of cooperation between faith and reason, according to Kierkegaard, he is less disturbed than Pascal by the idea that Christianity might be necessarily absurd and unattractive; and this explains why he is also far less interested in seeking out the comfort of supporting evidence, such as miracles. Kierkegaard drives this point home in one especially amusing passage:

> Some ... sought to refute doubt with reasons [*Grunde*] ... they tried to demonstrate the truth of Christianity with reasons ... these reasons fostered doubt and doubt became the stronger. The demonstration of Christianity really lies in *imitation*. This was taken away. Then the need for "reasons" was felt, but these reasons, or that there are reasons, are already a kind of doubt ... thus doubt arose and lived on reasons ... the more reasons one advances, the more one nourishes doubt and the stronger it becomes ... offering doubt reasons in order to kill it is just like offering the tasty food it likes best of all to a hungry monster one wishes to eliminate. No, one must not offer reasons to doubt – at least not if one's intention is to kill it – but one must do as Luther did, order it to shut its mouth, and to that end keep quiet and offer no reasons. [FSE 68/SKS 13:90]

Although Pascal does not believe that the truth of Christianity can or should be absolutely demonstrated, his Wager and his discussion of

---

[18] The Danish *Forstand*, which is often translated as "understanding," is the term Kierkegaard uses most frequently in this context. Although some might wonder if there is any significance to his less frequent use of *Fornuft*, which is often rendered as "reason," I argue elsewhere that they are practically synonymous for Kierkegaard (Buben, 2011, p. 74 n. 22).

miracles seem like prime examples of feeding the monster insofar as they both attempt to assuage the anxiety of doubt by means of rational appeal. While Pascal and Kierkegaard agree that one cannot simply reason one's way into Christianity, they seem to disagree when it comes to the appropriateness of providing justificatory, or even merely comforting, reasons. For Kierkegaard, the provision of reasons is a reliance on one's own powers and a symptom of a foundering faith; silence suggests a more vigorous trust in God.

If one considers that Pascal's explicit intention was to produce an *Apology for the Christian Religion*,[19] it is not surprising that Kierkegaard ends up opposing him. In Pascal's own words, he means to argue that those living a Christian life are "reasonable and happy" and that those who reject this life are "foolish and unhappy" (L160/S192). Although Kierkegaard might agree that Christianity promises an eternal happiness beyond this life, he has a deeply rooted mistrust of apologetics, and an equally powerful disdain for any apparent softening of what is involved in Christianity in the here and now. Because the target of his polemic is a complacent and diluted contemporary Christianity that takes for granted the benefits of automatic membership in the Church (simply by virtue of having been born in nineteenth-century Denmark), his work is primarily aimed at providing a more accurate, however harsh, account of the potential misery that lies in store for anyone who genuinely takes up the cross. Such misery may be indicative of a spiritual blessedness, but there is nothing particularly happy or reasonable about it in a worldly sense.

Kierkegaard brings his worries about making Christianity appear more attractive to bear on Pascal specifically in one fascinating passage from 1850:

> If I had to find a beautiful expression for the Mynsterian approach and one which would please him, I would quote a passage from Pascal's *Pensees* [sic], where he speaks of how one should approach those who repudiate religion or are ill-disposed toward it: "One should begin with proofs, showing that religion does not quarrel with reason; next, show that it is venerable and try to inspire respect for it; then make it pleasant and appealing (ingratiate it) and awaken the desire in them for it to be true, something one shall then drive home with irrefutable proofs; but it mainly depends on making it pleasant and appealing in their eyes." [JP 3: 423–24/SKS 24:119]

---

[19] Although this was not his chosen title, it was the project he "outlined at a meeting at Port-Royal," and the notes for this project would become *Pensées* after his death (Ariew, 2005, p. xi).

Despite any parallel he acknowledges between his own critique of contemporary Christianity and Pascal's, Kierkegaard's well-known distaste for the "Mynsterian approach" of comforting rationalization, and his association of Pascal with the Danish bishop (Jacob Peter Mynster), suggests his ultimately negative verdict on Pascal's tendency toward apologetics.[20] It is worth noting that it is not just "proofs," but even more subtle forms of ingratiation (which might include the Wager) that Kierkegaard objects to. Even though Pascal has his concerns about reason overstepping its bounds in matters of faith, Kierkegaard clearly has a much more restrictive boundary in mind.[21]

In order to make sure that the difficult and often unpleasant demands of Christianity are not misunderstood or overlooked, Kierkegaard forgoes the common rationalizing strategy of using its potential rewards to offer comfort. It would certainly not be appropriate to say that Pascal routinely shies away from the more painful elements of the Christian experience, but in one 1854 passage, Kierkegaard does suggest that Pascal would rather "coddle" himself than make all of the necessary sacrifices, which might include "martyrdom, a bloody martyrdom" (JP 2:367–68/SKS 25:482). Four years earlier, Kierkegaard admits to his own similar shortcomings (JP 6:313–14/SKS 23: 271–72), but while his self-awareness about these moments of mitigating the sacrifices involved in true Christianity hints at the possibility of further empathy with Pascal, it is precisely this self-awareness that prevents him from adopting Pascal's verve for indulging in them. Carefully examining his own internal motivations, Kierkegaard recognizes them as moments of weakness.

3 Nietzsche on Pascal as Victim of Christianity

While the religious Kierkegaard obviously did not know about Nietzsche, and the irreligious Nietzsche may only have known a little about Kierkegaard,[22]

---

[20] However, as in the case of Pascal, Kierkegaard does harbor a great deal of affection for Mynster himself, whom he knew personally.

[21] For Kierkegaard, just because one could defend Christianity, it does not mean that one should. Elsewhere, I argue that the contrast with Pascal suggests that Kierkegaard is an anti-rationalist about faith (Buben, 2013). He refuses to rationalize, not because Christianity is necessarily irrational, but because rationalization weakens faith.

[22] Thomas Miles (2011) points out that although Nietzsche never read Kierkegaard directly, he did read Georg Brandes (1842–1927) and Kierkegaard's old adversary Hans Lassen Martensen (1808–1884), who discuss and quote from Kierkegaard's work in great detail. Miles (2011, pp. 274–78) even argues that some of Nietzsche's more positive depiction of primitive Christianity's "life-affirming" elements (as opposed to the more pessimistic doctrines that developed later within the Church) is influenced by his late second-hand encounter with Kierkegaard.

the apparent similarity in their so-called "existentialist" views has long been a source of fascination. I believe that their occasionally supportive, yet still predominantly critical, comments about Pascal's approach to faith can help to illuminate the relationship between them. But before considering how Pascal fits into this relationship, it will be instructive to examine the connections Nietzsche sees between his most important early influence, Arthur Schopenhauer, and Pascal. For his part, Schopenhauer (1958, p. 615) associates Pascal's personal asceticism with the sort of "denial of the will-to-live" that he praises as the primary insight of all religious traditions – even if the Indian traditions of Hinduism and Buddhism do a better job of explicitly acknowledging it.[23] Once Nietzsche abandons his youthful adherence to Schopenhauerian pessimism about life, he is left with a similar criticism of Pascal's (cf. L378/S410) self-abnegating tendencies to go along with a similar level of respect for the quality of the thinker.[24]

Comparing these two thinkers, Nietzsche (1968, p. 52) states, "In an important sense, Schopenhauer is the first to take up again the movement of Pascal: *un monstre et un chaos*, consequently something to be negated. – History, nature, man himself." And elsewhere he adds, "Schopenhauer ... involutarily [*sic*] steps back into the seventeenth century – he is a modern Pascal, with Pascalian value judgments *without* Christianity. Schopenhauer was not strong enough for a new Yes" (Nietzsche, 1968, p. 525). On the more approving side of the comparison, what Nietzsche loves about Pascal and Schopenhauer is their shared disdain for the achievements and capabilities of ordinary humans left to their own devices (cf. Nietzsche, 1997, p. 38; 2003, p. 90; also see Birault, 1988, p. 283). Humans are too fragile, limited, and temporary (and for Pascal, corrupted by sin) to be worth much. Nietzsche sees both Pascal and Schopenhauer as "higher, rarer men," who are painfully aware of these shortcomings, while the ordinary members of the herd continue to take a thoughtless pride in their prescribed and pathetic accomplishments. He says, "The *strong points* of [such] men are the causes of their pessimistic gloom: the mediocre are, like the herd, little troubled with questions and conscience – cheerful. (On the gloominess of the strong: Pascal, Schopenhauer.)" (Nietzsche, 1968, p. 157). The difference between them, of

---

[23] Schopenhauer actually refers to the two works by Reuchlin that Kierkegaard relies upon in coming to his understanding of Pascal.

[24] Paying no attention to the unfinished nature of *Pensées*, Nietzsche (1968, p. 229) also expresses some appreciation for Pascal's writing style: "The profoundest and least exhausted books will probably always have something of the aphoristic and unexpected character of Pascal's *Pensées*." This line is probably delivered with a hint of irony, given Nietzsche's own aphoristic approach.

course, is that while Pascal believes that Christianity offers some hope for the transformation and salvation of humans, Schopenhauer does not. However, because such transformation is only possible beyond this world, Pascal's view of life in the here and now is at least as bleak as Schopenhauer's.

Ultimately rejecting such bleakness, Nietzsche describes their outlook in these terms:

> our world is imperfect, evil and guilt are actual and determined and absolutely inherent in its nature; in which case it cannot be the *real* world: in which case knowledge is only the way to a denial of it, for the world is an error which can be known to be an error. This is the opinion of Schopenhauer on the basis of Kantian presuppositions. Pascal is even more desperate: he comprehended that, in that case, even knowledge must be corrupt and falsified – that *revelation* was needed even to understand that the world ought to be denied. [1968, p. 222; cf. 1968, p. 310; 1990, p. 49]

It will be necessary to consider this issue of the corruption of knowledge in greater detail below, but it should first be made clear that Nietzsche finds this world-denying tendency generally unacceptable regardless of its origins. Realizing that Pascal's Christian pessimism, like Schopenhauer's more Indian version, prevents him from affirming (i.e., finding valuable) any sort of existence within the world, Nietzsche (1990, pp. 83–84; 1992, p. 27; 2003, pp. 209–10) laments the damage Christianity has done to this great thinker.[25] His own approach, of course, involves a more open and optimistic "Yes-saying" – a willful appropriation of even what some might consider the crueler or more disappointing aspects of life. What Nietzsche finds disturbing about Christianity, and any similar ideologies, is the resentful attitude of the weak that pouts, gets lost in imaginary constructions of what life should involve, and refuses to accept any value in what life actually offers. In *The Anti-Christ* he states:

> If one shifts the centre of gravity of life *out* of life into the "Beyond" – into *nothingness* – one has deprived life as such of its centre of gravity. The great lie of personal immortality destroys all rationality, all naturalness of instinct – all that is salutary, all that is life-furthering ... *So* to live that there is no longer any *meaning* in living: *that* now becomes the "meaning" of life ... Christianity has waged a war to the death against every feeling of reverence and distance between man and man ... against everything noble,

---

[25] Also see Diderot (2009, p. 32), who seems to have anticipated this lamentation.

joyful, high-spirited on earth, against our happiness on earth. [Nietzsche, 1990, pp. 167–68; cf. 2003, pp. 240–41]

Pascal may well be the best Christianity has to offer, according to Nietzsche (e.g., 1968, p. 142; 1997, p. 113; 2002, p. 44; 2003, pp. 195–97), but insofar as he is caught up in its self/life/world-disparaging, he is the voice of a decadent "slave-morality" that seeks to undermine all that is natural, healthy, and strong within us.[26]

So what glorious natural qualities in particular does Pascal denigrate? Unsurprisingly, it is his belittling of rationality and the human capacity for knowledge that Nietzsche finds especially worrisome. He claims that Christianity:

> has depraved the reason even of the intellectually strongest natures by teaching men to feel the supreme values of intellectuality as sinful, as misleading, as *temptations*. The most deplorable example: the depraving of Pascal, who believed his reason had been depraved by original sin while it had only been depraved by his Christianity. [Nietzsche, 1990, p. 129]

Despite this critique of the "self-mutilation" involved in his "*sacrifizio dell'intelletto*," and his general disapproval of Christianity, Nietzsche (2002, p. 121) does prefer Pascal's more ascetic and terrifying version to the blatantly comforting and degenerate imitation of Christianity commonly encountered in nineteenth-century Europe. It is this preference that leads him to tolerate, or at least not explicitly criticize, Pascal's Wager.

Nietzsche (2003, pp. 89–90) states, "Even if Christian belief could not be disproved, Pascal, in view of a *dreadful* possibility that it might yet be true, considered it prudent in the highest sense to be a Christian," and he contrasts Pascal's "dreadful" Christianity with "an opiate Christianity ... which chiefly aims to soothe sick nerves [and] has absolutely *no need* of that dreadful solution, a 'God on the cross.'"[27] Nietzsche is right that Pascal (see, e.g.,

---

[26] Along these lines, Henri Birault (1988, p. 278) says that "it is perhaps because [Nietzsche] is thinking of this Christian strength which is *already* a weakness, that he does not confound the Pascalian strength with the more general and freer strength of searching, of fighting, of daring, of wanting to be alone." As we move onto Nietzsche's view of the Wager, also consider Gilles Deleuze's (1983, p. 37) related claim that "The whole alternative is governed by the ascetic ideal and the depreciation of life. Nietzsche is right to oppose his own game to Pascal's wager ... Nietzsche means that we have managed to discover another game, another way of playing ... we have managed to make chaos an object of affirmation instead of positing it as something to be denied. And each time we compare Nietzsche and Pascal (or Kierkegaard ...) the same conclusion is forced upon us – the comparison is only valid up to a certain point."

[27] Besides this more obvious reference to the Wager, Nietzsche (1968, pp. 64, 491) also seems to make a couple of more subtle allusions.

L428/S682), and the Wager specifically, are largely motivated by mortal terror, but what he does not acknowledge here (in his eagerness to ridicule something even worse) is the fact that the Wager, in establishing what is "prudent," is a lingering bit of rationalization that is at least in part meant to offer a kind of comfort to the aspiring or prospective Christian. Elsewhere, Nietzsche (2002, p. 44) does claim that – in contrast with the much-reviled Luther, who is all too eager to sacrifice reason for the sake of faith[28] – Pascal's rejection of reason is neither complete nor immediate; rather, he suffers from "a protracted suicide of reason." But given his obvious distaste for a Christianity that "soothes," I wonder what Nietzsche would make of Kierkegaard's more relentless attack on comforting Christianity, which does not even spare the likes of Pascal.

Thomas Miles believes that Nietzsche would see Kierkegaard as he sees Pascal. In particular, he thinks that Nietzsche would express a similar concern about how Kierkegaard was also led into sacrificing reason by Christianity (Miles, 2011, pp. 279–80). For the most part, I am inclined to agree with this piece of speculation given Kierkegaard's own views on Pascal and the inadequacy of reason when it comes to religious matters. However, as I have suggested, Kierkegaard seems to think that Pascal is not thorough enough in rooting out the rationalizing tendency, and this more extreme approach might not sit well with Nietzsche. One of the things Nietzsche admires about Pascal, even while criticizing him, is the way he struggled with sacrificing his reason (Birault, 1988, pp. 281–82). From Nietzsche's perspective (and Kierkegaard's as well), this was a hard-fought battle between Pascal's natural inclinations and a most unnatural Christian faith, and Nietzsche might argue that, when faced with the same conundrum, Kierkegaard is too quick to give up on his own extraordinary intellectual capabilities.

While it is not entirely clear whether Kierkegaard's struggle would make him more like Luther or more like Pascal, as Nietzsche understands them, it is quite apparent that Kierkegaard believes the outcome of this conflict must be more complete than Pascal does. Thus, depending on what he determined about the nature of Kierkegaard's personal struggle, Nietzsche may or may not see Kierkegaard as similarly victimized – but perhaps to an even greater degree – by Christianity. In any case, there is at least one element of

---

[28] Consider, for example, Luther's (1959, p. 374) famous diatribe: "But the devil's bride, reason, the lovely whore comes in and wants to be wise, and what she says, she thinks, is the Holy Spirit. Who can be of any help then? Neither jurist, physician, nor king, nor emperor; for she is the foremost whore the devil has."

Kierkegaard's approach to faith and reason that Nietzsche would likely appreciate. In one relatively early passage, he seems to reinforce Kierkegaard's claims about reason remaining silent when he mocks the insecurity in Pascal's apologetic need to speak "as loudly as he could" on doctrinal matters that make little sense (Nietzsche, 1997, p. 53). As dangerous (and yet, ridiculous) as he might find Christianity, Nietzsche tends to respect the kind of strength of will on display in refusing to justify one's beliefs to others.

4 Conclusion

It is difficult in a reception chapter like this – covering two centuries of great thinkers beyond the primary subject – to provide some grand unifying appraisal of the way an author, text, or idea has been seen by posterity. Kant's views on the practical necessity of belief in God have something in common with the strategy employed by Pascal's Wager, despite no indication that he was all that familiar with Pascal, let alone his famous argument. Kierkegaard, on the other hand, would seem to have very little patience for the Wager or other similar means of making religion appear more attractive, although he recognizes in Pascal a kindred spirit of sorts. The obvious divergence in these connections to Pascal is already noteworthy, and this is before taking into consideration Voltaire's rather harsh (and I would say, somewhat off base) treatment and Nietzsche's surprisingly gentle reception. In the latter case, given Nietzsche's extremely negative opinion of Christianity – not to mention Kant, "a *cunning* Christian" (Nietzsche, 1990, p. 49) – one might suspect that he would offer a scathing critique of Pascal, and yet, his brief comments on the Wager are fairly supportive, and his general assessment of the man is at least as sympathetic and respectful as Kierkegaard's. Setting aside the others, then, I suppose what I find most remarkable at the end of this exploration is that these two nineteenth-century luminaries harbor such affection for Pascal even while profoundly disagreeing with him, in profoundly different ways, about matters so central to his thought.

5   The Wager and William James

Jeffrey Jordan[*]

A few pages into his 1896 essay, "The Will to Believe," William James (1842–1910) mentions an objection to Pascal's Wager:

> You probably feel that when religious faith expresses itself thus, in the language of the gaming-table it is put to its last trumps. Surely Pascal's own personal belief in the masses and holy water had far other springs; and this celebrated page of his is but an argument for others, a last desperate snatch at a weapon against the hardness of the unbelieving heart. We feel that a faith in masses and holy water adopted willfully after such a mechanical calculation would lack the inner soul of faith's reality; and if we were ourselves in the place of the Deity, we should probably take peculiar pleasure in cutting off believers of this pattern from their infinite reward. [1956 [1896], pp. 5–6]

Some commentators suggest that James is here offering an objection to the Wager.[1] But this is doubtful as later in his essay James declares that "Pascal's argument, instead of being powerless, then, seems a regular clincher, and is the last stroke needed to make our faith in masses and holy water complete" (p. 11). Others, in part because of this later comment, hold that James's argument is best seen as a Pascalian wager.[2] Both interpretations, as we will see, are problematic.

In what follows, I argue for three propositions. The first is that James's Will to Believe (WTB) argument, while it is, in part, a pragmatic argument, is not a Pascalian wager. The second proposition is that conjoining three key elements of James's argument with a standard version of Pascal's Wager results

---

[*] I wish to thank Douglas Stalker, Abdulkadir Tanis, Scott Coley, and the editors of this volume for their generous comments and suggestions.
[1] See, e.g., Slater (2009, p. 57); and see Connor (2006, pp. 184–85).
[2] See, e.g., Wernham (1987, pp. 75–80). Michael Slater holds that James does develop a Pascalian wager but that is found in his 1911 essay, "Faith and the Right to Believe" and not his 1896 "The Will to Believe" essay. See Slater (2009, pp. 48–66).

in a hybrid Pascalian wager distinct from the versions of the Wager developed by Pascal. The third proposition is that this hybrid wager survives several strong objections lodged against Pascal's original wagers. Let us first seek an understanding of what a Pascalian wager is, and the nature of James's WTB argument.

1 The Wager and the Will to Believe Argument

A pragmatic argument is any argument intended to motivate an action, because of the benefits associated with the performance of that action. Pragmatic arguments are practical in orientation, justifying actions thought to facilitate the achievement of our goals, the satisfaction of our desires, or the demands of morality. While James's WTB argument is, in part, a pragmatic argument, it is not a Pascalian wager. A Pascalian wager has the structure of a gamble, a decision made in the midst of uncertainty about the future. Pascal assumed that a person, just by virtue of being in the world, faces an unavoidable betting situation whether God exists or not. A pragmatic argument is a Pascalian wager if it has both of two features, with the first being that it involves a decision situation in which the possible gain or benefit involved with one of the alternatives swamps all the others.[3] With Pascal's Wager, of course, the possible gain of theism is supposed to be not just greater than that of nonbelief, it is purportedly infinitely greater. One might contend that Pascalian wagers are found in domains that are not religious, so it is best to understand the swamping property as involving not an infinite value but as a gain vastly greater than any of its rivals – a gain so great as to render the probability assignments, unless they are known to approach zero, nearly irrelevant.[4]

The second feature has to do with what is at stake. The object of the gamble must be something of ultimate concern. For instance, one can imagine a Pascalian wager in which a person diagnosed with a terminal disease must decide whether to invest time and effort in unconventional therapies as a long-shot desperate last hope.[5] Pascalian wagers deal with subjects that are of momentous importance. As long as one's argument is pragmatic in nature, and an outcome associated with one of the alternatives is so stupendous as to swamp all the others, and that alternative has to do with something of ultimate concern, one is employing a Pascalian wager.

---

[3] The swamping property need not be a gain or benefit but could be a catastrophic loss.
[4] I will not try to remove the vagueness lurking in this claim.   [5] I owe this example to Doug Stalker.

Unlike the Wager, the range of James's argument extends far beyond the issue of the rationality of theism to include various philosophical issues, including whether to accept philosophical determinism or indeterminism; whether to embrace the idea that life has meaning; and whether the universe is rational. There are two major stages in James's argument. The first attempts to defeat the idea that one may believe only those propositions supported by sufficient evidence. The second stage seeks to situate theistic belief within the domain of the permissible, even if it lacks sufficient evidential support. The first stage of James's argument is not a pragmatic argument but an argument for the intellectual unacceptability of what we might call the "Agnostic Rule." Recognizing that the first stage of James's argument is not pragmatic makes clear why it is incorrect to see James's WTB argument simply as a Pascalian wager, and this is so despite the affinity of James's WTB argument with Pascal's Wager.

The Agnostic Rule asserts that rationality or morality requires withholding belief whenever the evidence is insufficient. James argues that the Agnostic Rule is unacceptable because:

> a rule of thinking which would absolutely prevent me from acknowledging certain kinds of truth if those kinds of truth were really there, would be an irrational rule. [1956 [1896], p. 28]

James's example of a certain kind of truth which the Agnostic Rule precludes one from acknowledging is that "there are ... cases where a fact cannot come at all unless a preliminary faith exists in its coming" (p. 25). Among other instances James provides of this kind of truth is that of social cooperation:

> a social organism of any sort whatever, large or small, is what it is because each member proceeds to his own duty with a trust that the other members will simultaneously do theirs. Wherever a desired result is achieved by the co-operation of many independent persons, its existence as a fact is a pure consequence of the precursive faith in one another of those immediately concerned. [1956 [1896], p. 25]

If there are propositions that are true, only if believed, even prior to an appreciation of the evidence, then a major step in James's WTB argument is well-supported.

The foil of James's essay, and a prominent proponent of what we are calling the Agnostic Rule, was the Cambridge mathematician, W. K. Clifford (1845–1879). Clifford argued that:

if I let myself believe anything on insufficient evidence, there may be no great harm done by the mere belief; it may be true after all, or I may never have occasion to exhibit it in outward acts. But I cannot help doing this great wrong towards Man, that I make myself credulous. The danger to society is not merely that it should believe wrong things, though that is great enough; but that it should become credulous, and lose the habit of testing things and inquiring into them; for then it must sink back into savagery. [1879, pp. 185–86]

Clifford famously presented the Agnostic Rule as a rule of morality: "it is wrong always, everywhere, and for any one, to believe anything upon insufficient evidence" (p. 185). According to Clifford's moral version of the Agnostic Rule, anyone believing a proposition that she takes as lacking at least a preponderance of evidential support is immoral.

James contends that the Agnostic Rule is but one intellectual strategy available to us. A proponent of the Rule advises, in effect, that one should avoid error at all costs, and thereby risk the loss of acquiring all the truths available to us. But another strategy open to us is to seek truth by any means available, even at the risk of error. James champions the latter via an argument that we might sketch as:

1. Two distinct intellectual strategies are available:
   - Strategy A: Risk a loss of truth and a loss of a vital good for the certainty of avoiding error.
   - Strategy B: Risk error for a chance at truth and a vital good.
2. The Agnostic Rule exemplifies Strategy A.

But,

3. Strategy B is preferable to Strategy A because Strategy A would deny us access to a certain class of truths.

And,

4. Any intellectual strategy which denies access to a certain class of truths is unacceptable.

Therefore,

5. The Agnostic Rule is unacceptable.

This argument consists of four premises, with two of them, (2) and (4), obvious enough. And if James is right that there is a kind of proposition which is true only if it is believed, then premises (1) and (3) look well-supported as well.

Of course, accepting proposition (5), and advancing a strategy of seeking truth via any available means, even at the risk of error, does not entail that anything goes. An important part of James's essay restricts what legitimately might be believed in the absence of adequate evidence. To understand these restrictions, let us paraphrase eight key concepts presented by James:

- *Hypothesis*: something that may be believed.
- *Option*: a decision between two hypotheses.
- *Living option*: a decision between two live hypotheses.
- *Live hypothesis*: something that is a real candidate for belief. Roughly, a hypothesis is live, we might say, for a person just in case the hypothesis has an intuitive appeal for that person. A hypothesis is maximally live for someone if there is a willingness to act irrevocably on that hypothesis.
- *Momentous option*: the option may never again present itself, or cannot be easily reversed, or something of importance hangs on the choice. It is not a trivial matter.
- *Forced option*: the decision cannot be avoided as the consequences of refusing to decide are the same as actually deciding for one of the alternative hypotheses.
- *Genuine option*: one that is living, momentous, and forced.
- *Intellectually open*: the evidence does not settle the issue.

James's contention is that any hypothesis, which is part of a genuine option, and which is intellectually open, may be believed, even in the absence of sufficient evidence. No rule of morality or rationality is violated if one accepts a hypothesis that is genuine and open. With the Cliffordian threat subdued, James moves in the second stage of his argument toward "a defence of our right to adopt a believing attitude in religious matters, in spite of the fact that our merely logical intellect may not have been coerced" (1956 [1896], pp. 1–2). This second stage of James's WTB argument is pragmatic.

Employing a very general and abstract understanding of religion, James claims that:

> Religion says essentially two things ... the best things are the more eternal things, the overlapping things, the things in the universe that throw the last stone, so to speak, and say the final word ... The second affirmation of religion is that we are better off even now if we believe [religion's] first affirmation to be true ... The more perfect and more eternal aspect of the universe is represented in our religions as having personal form. The universe is no longer a mere *It* to us, but a *Thou* ... We feel, too, as if the appeal of religion to us were made to our own active good-will, as if

evidence might be forever withheld from us unless we met the hypothesis half-way. [1956 [1896], pp. 25–27]

Notice what James calls the second affirmation of religion – we are better off even now if we believe. In his book, *The Varieties of Religious Experience*, James suggests that religious belief has what we might call "temporal benefits" – vital goods which are had in this world, and which are not gained only in a post-mortem existence. According to James, these temporal benefits include:

> A new zest which adds itself like a gift to life, and takes the form either of lyrical enchantment or of appeal to earnestness and heroism ... An assurance of safety and a temper of peace, and, in relation to others, a preponderance of loving affections. [1936 [1902], pp. 475–76]

We may understand James's point here as the claim that there are benefits even now to theistic belief quite apart from whatever benefits might accrue in an afterlife. If James is correct, there are vital goods, including temporal benefits, attached to theistic belief. With the appeal to temporal benefits, James has ventured into the realm of the pragmatic. Even though the second stage of the WTB argument is pragmatic, it is not a Pascalian argument as is clear when the structure of James's argument is presented in a step-by-step manner.

Given that theism is intellectually open and that the decision whether to believe theistically is part of a genuine option, and given that there are temporal benefits attached to theistic belief, there is, James argues, sufficient reason to believe. James's second main argument of the WTB essay might be sketched as:

6. The decision whether to accept theism is a genuine option.
And,
7. Theism is intellectually open.
And,
8. There are vital goods at stake in accepting theism.
And,
9. No one is irrational or immoral in risking error for a chance at truth and a vital good.
So,
10. One may accept theism.

With this argument, James seeks to support the second of the two primary theses of his essay: a religious commitment is permissible, given the fulfillment of certain conditions, as neither morality nor rationality rule it out. Why did James argue for the permissibility of accepting theism, rather than arguing that accepting theism is rationally mandated? There are at least two reasons. The first involves the person-relativity of theism being a genuine option. Not everyone will have theism as a live hypothesis, nor will everyone agree that there are vital goods at stake in the option involving theism. And, second, not everyone will agree that theism is intellectually open as some hold, for instance, that the probability of theism is zero. These two facts – there is debate whether theism is intellectually open and whether it is a genuine option – are enough to make the permissible the appropriate stopping point of the argument.

A common complaint about James's argument is that it presupposes doxastic voluntarism. Doxastic voluntarism is the thesis that persons can acquire beliefs at will; that persons have direct control over their beliefs. Perhaps the most prominent objection along these lines is due to Bernard Williams, who argued, in effect, that it is impossible to both believe that p, and to know that p is false. Doxastic voluntarism, however, could be true only if that were possible (Williams, 1973a, pp. 136–51). While Williams's argument may present a problem for doxastic voluntarism, it does not present one for James. For one thing, James's proposal is operative only under conditions of intellectual openness, and is not operative in the face of conclusive or strong adverse evidence. James does not countenance believing when the evidence is clear that the hypothesis is likely false. For another thing, James's talk of believing this or that hypothesis can be understood as accepting this or that hypothesis. Whether belief is under our control or not, acceptance surely is. Acceptance is a voluntary action consisting of a judgment that a particular proposition is true. One accepts a proposition when one assents to that proposition, and acts on it. James's WTB argument works perfectly well if we understand it as involving acceptance and not belief.

Another objection commonly leveled against James's argument is that "it constitutes an unrestricted license for wishful thinking ... if our aim is to believe what is true, and not necessarily what we like, James's universal permissiveness will not help us" (Hick, 1990, p. 60). That is, *hoping* that a proposition is true is no reason to think that it *is*. This objection is unfair. As we have noted, James does not hold that the falsity of the Agnostic Rule implies that anything goes. Restricting the relevant permissibility class to propositions that are intellectually open and part of a genuine option arguably provides ample protection against wishful thinking. Moreover, why think that believing what is true and believing what we like are necessarily mutually exclusive? Some philosophers have suggested that James thought that passional reasoning was, under certain circumstances, a reliable means of acquiring true beliefs.[6] If certain uses of the passions are a reliable means of acquiring true belief, then the wishful-thinking charge is not just unfair, but would wildly miss the mark.

2 The Many-Gods Objection

Pascal presents four versions of his wager in his *Pensées*, with the fourth found in the concluding remarks that Pascal makes to his interlocutor:

> Now what harm will come to you from choosing this course? You will be faithful, honest, humble, grateful, full of good works, a sincere, true friend…It is true you will not enjoy noxious pleasures, glory and good living, but will you not have others?
> I tell you that you will gain even in this life … [L418/S680]

The fourth version is best understood as a decision under uncertainty, as it employs no probability judgments but proceeds on the idea that the benefits of wagering for God vastly exceed those associated with wagering against, whether God exists or not. No matter what, wagering pro theism is one's best bet as it strongly dominates not wagering. This fourth version of the Wager is an argument from strong dominance:

11. When choosing among various alternatives, if every outcome of a particular alternative α is better than those of the other alternatives, one should choose α.

And,

12. Every outcome of wagering for God is better than those of not wagering for God,

Therefore,

13. One should wager that God exists.

One should understand (13) as asserting that rationality requires wagering. Premise (12) is true only if one gains simply by wagering that God exists. Only if, that is, wagering for God results in temporal benefits. Pascal apparently thought that this was obvious:

---

[6] See, e.g., Wainwright (1995), who argues that a right disposition is necessary for appreciating the evidence supporting theism; and that grasping the significance of the evidence in support of theism is influenced by one's passions.

> The Christian's hope of possessing an infinite good is mingled with actual enjoyment as well as with fear, for, unlike people hoping for a kingdom of which they will have no part because they are subjects, Christians hope for holiness, and to be free from unrighteousness, and some part of this is already theirs. [L917/S746]

A sincere Christian commitment results, he thought, in virtuous living, and virtuous living is more rewarding than vicious living.

The response of Pascal's interlocutor, we might easily imagine, would be that Pascal has made an illicit assumption: why think that virtuous living requires Christianity? And even if virtuous living requires Christianity, why think that being morally better is tantamount to being better off all things considered? Indeed, Pascal's interlocutor would charge Pascal with a second illicit assumption: Pascal wrongly assumes that the relevant alternatives are only wagering for the God of Christianity, and wagering that there is no god. But what of other possible deities? As early as 1762, Denis Diderot (1734–1784) objected that:

> Pascal has said if your religion is false, you have risked nothing by believing it true; if it is true, you have risked all by believing it false. An Imam could have said as much. [1875 [1746], LIX]

The complaint of Diderot is that the betting options are not limited to Christianity and atheism, since one could formulate a Pascalian Wager for Islam, or, say, the sects of Hinduism, or for any of the competing sects found within Christianity itself. The many-gods objection (MGO) asserts that Pascal's Wager is flawed because its alternatives are not jointly exhaustive of the possibilities. The MGO rests on two standard Pascalian assumptions. The first assumption is that there is an infinite gain for right belief; while the second is that there is only a finite cost attached to religious wagering. The MGO also depends upon a non-Pascalian assumption that *a religious option, if possible, has a positive probability*. The MGO exploits the idea of an infinite utility to create a kind of decision-theoretic impasse, since the expected utility of wagering, say, that a non-theistic deity exists would be the same as that of wagering that the God of theism exists, as long as there is a positive probability for every possible deity.

The observant reader will have noticed the shift from the Christian God to that of the theistic God. Diderot aside, Pascal arguably intended his wagers as support of theistic belief generally and not specifically in support of Christian theism, or any particular tradition found within Christianity.[7] It is likely that Pascal intended a two-part apologetic case. The first part involves the Wager motivating a commitment to theism; while the second part has arguments that Christianity is the "true religion" vis-à-vis its theistic rivals of Judaism and Islam.[8] Understood this way, Pascal's apology was in line with the standard seventeenth- and eighteenth-century apologetic strategy of arguing first that there is a god (or, in Pascal's case, that there is reason to wager that the theistic God exits), and then, second, identifying which god it is that exists. This is the strategy adopted by, among others, William Paley (1734–1805), who employed the design argument to argue for a divine designer, and then employed a different argument, an argument from miracles, to argue that the designer was that associated with the Christian religion. Arguably, then, there is no good reason to expect more from the Wager than is expected from, say, Paley's design argument, or Thomas Aquinas's five ways. These are arguments for theism, and not any specific brand of theism. While objections to these arguments are common, no one faults them for failing to specify which particular theistic tradition, denomination, or sect one should adopt. Demanding that a single theistic argument, if successful, must specify the particular sort of theism demands too much.

The MGO however need not limit the alternatives to the deities associated with various actual religious traditions. A proponent of the MGO could posit an imaginary deity as an alternative as long as one has reason to hold that the probability of the deity's existence is greater than zero. Michael Martin, for example, conjures up a "perverse master of the universe" deity who "punishes with infinite torment after death anyone who believes in God or any other supernatural being (including himself) and rewards with infinite bliss after death anyone who believes in no supernatural being" as an alternative with as much right to compete in a Pascalian wager as has Pascal's God of theism.[9] While there is no Church or creed of the "perverse master" as long as the non-Pascalian assumption that every possible religious option warrants a positive probability holds, Martin's exotic alternative gains a foothold.

Pascal's various versions are "2 × 2" wagers as each posits only two states of the world (this is a world in which God exists; and this is a world in which God

---

[7] A theist is anyone who accepts theism. Theism is the proposition that God exists. The theistic religions – Judaism, Christianity, and Islam – are each expansions of theism. That is, they are (in part) conjunctions of other propositions to the core proposition that God exists.
[8] See, e.g., L203–220/S235–253 and L298–322/S329–353.
[9] See Martin (1990, pp. 232–34).

does not exist); and only two alternatives (wager that God exists; and do not wager that God exists). In effect, the MGO contends that rather than a 2 × 2 wager, it is more appropriate to frame the Wager, at the least, as a 3 × 3 wager, with three states of the world (this is a world in which the theistic God exists; and this is world in which a non-theistic God exists; and this is a world in which no god exists); and three alternatives (wager that the theistic God exists; wager that the non-theistic God exists; and wager that no god exists). If the MGO is sound, then no particular religious option is recommended. It is here that James's WTB arguments provide a Pascalian with resources in eluding the MGO.

3 Three Jamesian Resources

The upshot of the MGO is that an infinite utility associated with more than one of the alternatives results in a tie, with no specific alternative being preferable to its rivals. The MGO seeks to inflate the Wager from a 2 × 2 wager to at least a 3 × 3 wager, with the associated outcomes arranged so that an agent would have no more reason to select one alternative over another.[10] The three Jamesian resources are intended to deflate the Wager back to a more manageable 2 × 2 size.

The first Jamesian resource is the idea of a forced choice. Now, arguably, the idea of a forced choice is found in Pascal's initial comments on the Wager as he says "you have to wager." Wagering is unavoidable and thereby forced, since refusing to wager is tantamount to wagering against. A decision is forced whenever deciding nothing is equivalent in practical effect to choosing one of the alternatives. Voltaire (1694–1778) objected that:

> 'Tis evidently false to assert, that, the not laying a wager that God exists, is laying that he does not exist: For certainly that man whose mind is in a state of doubt, and is desirous of information, does not lay on either side. [1994 [1734], p. 127]

Voltaire is no doubt correct that not laying a wager that God exists is not the same as wagering that God does not exist. But Pascal never asserted it was.

---

[10] Even though it is possible to imagine any number of exotic deities of the sort conjured by Martin, any extension beyond a 3 × 3 wager to a 4 × 4 wager, or a 5 × 5 wager, or, a $n × n$ wager, for any number n, is logically redundant. There either is a decision indeterminacy generated by including a third alternative, or not. The number of additional alternatives adds nothing. So, expanding beyond a 3 × 3 wager adds nothing of logical or dialectical consequence to the MGO.

When Pascal asserts that one must wager, he is not asserting that the refusal to do so is identical with wagering against, but rather that refusing to wager has the same practical consequence as wagering against. One remains in the same practical state by either wagering against or by laying no wager, as one is not wagering that God exists. What is it to wager that God exists? We might understand wagering that God exists as committing oneself to the existence of God. This would include reorienting one's goals, values, and behavior by including the proposition that God exists among one's most basic values and attitudes. Wagering may not require a belief that God exists, as, perhaps, accepting that God exists is enough for a pro-wager. Recall that acceptance is a voluntary action that consists of both assenting to a proposition, and acting on it. Pascal seems to employ this understanding of wagering when he says to those who seek to believe, "learn from those who were once bound like you and who now wager all they have" (L418/S680). Given the facts of human psychology, one who reorients her goals, values, and behavior by accepting the proposition that God exists will very likely come, over time, to believe that God exists. Wagering for God is, in effect, a regimen involving a sort of belief-inducing technology, a habitual and inclusive role-playing as if one already believed by engaging in the behaviors associated with believers. By doing so, one enhances the prospect that one will acquire theistic belief. Habitual role-playing, the idea goes, foreseeably eventuates in acquiring the belief.

The second Jamesian resource is arguably the most important anti-MGO weapon developed by James – the distinction between live hypotheses and dead ones:

> Let us give the name of hypothesis to anything that may be proposed to our belief; and just as the electricians speak of live and dead wires, let us speak of any hypothesis as either live or dead. A live hypothesis is one which appeals as a real possibility to him to whom it is proposed. If I ask you to believe in the Mahdi, the notion makes no electric connection with your nature, – it refuses to scintillate with any credibility at all. As an hypothesis it is completely dead. To an Arab, however (even if he be not one of the Madhi's followers), the hypothesis is among the mind's possibilities: it is alive. This shows that deadness and liveness in an hypothesis are not intrinsic properties, but relations to the individual thinker. They are measured by his willingness to act. The maximum of liveness in hypothesis means willingness to act irrevocably. Practically, that means belief; but

there is some believing tendency wherever there is willingness to act at all. [1956 [1896], pp. 2–3]

Instead of James's "hypothesis" let's talk of alternatives (what one may choose in a decision situation). Perhaps we might understand James's distinction between live alternatives and dead ones this way: there is a difference between assigning probability zero to a proposition, and assigning no probability to that proposition. For example, we are about to toss an ordinary coin which we believe is fair (or, perhaps, fair enough). We consider only two outcomes as relevant – heads or tails. But, of course, coins occasionally land on their sides.[11] Yet, even knowing this, we consider the partition of heads and tails to jointly exhaust the real possibilities. That is, while heads and tails in fact do not exhaust the possibilities, we may treat them, without fault, as if they do. It is not that we assign a zero probability to the coin landing on its side; it is rather that we assign it no probability. It is not a live alternative, while heads and tails are live alternatives. Live alternatives are those we assign a probability, while dead ones are assigned no probability, or are assigned a zero probability. Facing limitations of resources, knowledge, and time, practical parsimony – assigning no probability to otherwise possible alternatives – is unavoidable, even if we are not always aware of it. A live alternative, then, is one we consider a real possibility as we assign it a probability; while a dead one has either a zero probability, or no probability assigned to it.[12] Importantly, the liveliness of alternatives is person-relative. An alternative live for you may not be for me, so while we cannot draw a bright line separating the domain of the living from the dead, presumably the domain of live alternatives will be shared by those situated in similar intellectual and social contexts.

One might object that if one knows that something is logically possible, then one should not ignore it by assigning no probability value, or by assigning a zero probability. But, consider the distinction between logical possibility and metaphysical possibility. This distinction, very roughly, is between that which is consistent or which entails no contradiction (logical possibility), and that which is in fact possible given the metaphysical structure of reality (metaphysical possibility). If something is metaphysically possible, then it is logically possible, but the converse does not hold, as the domain of the

---

[11] An ordinary coin has about 1/6000 chance of landing on its edge. See Murray and Teare (1993, pp. 2547–52).

[12] Are infinitesimals viable probability values? If so, then a dead alternative could be any that receives no probability assignment, a zero probability, or an infinitesimal probability assignment.

logically possible is more populous than that of the metaphysically possible. To illustrate the distinction with a common example: perhaps we can imagine water being made of carbon and chlorine (logical possibility), but it may be metaphysically impossible that water be anything other than H_2O, since that is what water essentially is. So, what is the probability that water is anything other than H_2O? Zero, it seems, given what we know about the universe. In like manner, a hypothesis like Martin's "perverse master" may be logically possible as we can consistently describe a deity who punishes all and only theists. That is no reason to hold that such imaginary constructions are also metaphysically possible given what else we know about the universe. While it may be that probability flows from metaphysical possibility, clearly probability cannot flow from logical possibility alone, since there could hardly be things which are probable and yet metaphysically impossible.

With the Jamesian distinction between live and dead alternatives in hand, the Pascalian can credibly reject many and perhaps all of the non-theistic deities found in historical religions, or found in the imaginations of philosophers, as dead alternatives – alternatives that properly receive either a zero probability assignment, or no probability assignment at all.[13] Perhaps the opportunity costs of believing in such a deity are too high, or perhaps the novelty of a certain deity, having made its first appearance as a recent product of someone's imagination, puts it beyond all credibility.[14] And it is not implausible to think that the same would hold for however one describes the third alternative, whether involving a deity freshly baked in the fertile imagination of a philosopher, or a deity from an actual non-theistic religion, since it would not be surprising that the vast majority of those encountering

---

[13] Let's distinguish between exclusivist religions and universalist religions. A universalist religion holds that those outside of the true religion nonetheless gain salvation; while an exclusivist religion holds that only devotees of the true religion gain salvation (rather than religions, we could speak of traditions within a religion). As regards the Wager, universalist religions can be ignored, as they have no relevant place within a Pascalian calculation.

[14] Why consider Martin's "perverse master" a dead alternative? Wagering on the existence of the perverse master, a supernatural being, requires accepting that there are no supernatural beings. With the perverse master wager, one seems ready to both accept that there are supernatural beings, while accepting that there are no beings like the perverse master. This may not just be odd but incoherent. Perhaps Martin's "perverse master" is not logically possible. In addition, deities imagined by critics such as Martin face what may be an insurmountable practical hurdle as there are no communities of the like-minded in which to immerse oneself, no sympathetic literature to read, and no traditions or social practices connecting one to a larger community. This point holds for philosophers' fictions like a "perverse master" but not for versions of the MGO employing actual and exclusivist religious traditions as alternatives.

the Wager today occupy a position similar to that described by James: "if I say: 'Be an agnostic or be Christian,' ... trained as you are, each hypothesis makes some appeal, however small, to your belief" (1956 [1896], pp. 2–3).[15] If this point is correct, then the alternative of wagering that a non-theistic god exists could be properly set aside by many. Put another way, if the class of relevant alternatives may be pared via the distinction between live and dead alternatives, the Wager may reduce from a 3 × 3 back to a 2 × 2 matrix.[16]

In addition to metaphysical possibility, part of what distinguishes live and dead alternatives resides with James's observation that religious commitment brings a "new zest which adds itself like a gift to life." This observation serves as the third Jamesian resource, although, of course, not all agree with James that temporal benefits accompany a religious commitment, as some find a religious commitment burdensome. But, there are many who have a natural affinity for a life of religious commitment and who agree with James that a religious commitment adds a depth to life which would otherwise be lacking. Among metaphysically possible alternatives, some will be seen as associated with temporal benefits, and some will be seen as lacking that association. Those lacking the association, even if metaphysically possible, are yet dead. The judgment as to which metaphysically possible alternatives are associated with temporal benefits is, clearly, person-relative. People with different interests, histories, and beliefs, may well arrive at different judgments. While person-relativity narrows the number of persons who will find the Wager rationally credible, it is hard to see that this is a telling objection since every argument carries presuppositions that limit the class of those who find it credible to those sharing its presuppositions. No argument regarding a controversial topic will be credible to all persons. Although a wager predicated upon the three Jamesian resources may lack universal credibility, this in no way implies that it lacks a legitimate inferential role, or has no apologetic use.

---

[15] Live alternatives today, for many, are some kind of atheism (whether it is a hard-core variety of Metaphysical Naturalism (roughly: every fact is a physical fact; and science alone provides knowledge), or a softer Naturalism which permits non-physical facts but no personal god), agnosticism, or some version of theism.

[16] Earlier, we saw that, according to Pascal, habitual role-playing foreseeably eventuates in acquiring the desired belief. One might wonder whether persons can adjust their person-relative judgments of live alternatives or dead ones in much the same manner. Could habitual role-playing, that is, eventuate in dead alternatives now seeming alive? Perhaps. But one should recall the context of Pascal's advice: it is to those who find the Wager argument rationally persuasive, and nonetheless find themselves lacking belief. That context is not incidental, and it is hard to see any comparable context motivating one to seek to adjust their judgments of liveness and deadness.

In any case, let all the metaphysically possible alternatives with the association constitute what we might call class A; and let all the alternatives lacking that association constitute class B. Assuming that all else is equal as regards the alternatives found within classes A and B, selecting from class A is rationally preferable to selecting from class B. The appeal to temporal benefits, then, helps to narrow even further the range of live alternatives from which to pick. If we keep in mind that the Agnostic Rule is false, and that a persistent uncertainty shrouds our current state, it is hard to see what would be objectionable for those who find theism a live alternative that exceeds in temporal benefits its alternatives, wagering on Pascalian grounds.[17]

We have, then, a hybrid Pascalian wager, incorporating the Jamesian ideas of forced decisions, live and dead alternatives, and temporal benefits, that focuses on theism and not on any particular theistic tradition. We will distinguish between alternatives that are live by calling them "available alternatives" as opposed to those that are not live:

14. For any person S making a forced decision under uncertainty, if one of the available alternatives, α, has an outcome as good as the best outcomes of the other available alternatives, and never an outcome worse than the worst outcomes of the other available ones, and otherwise has only outcomes better than those of the other available alternatives, then S should choose α. And,
15. The best outcomes of wagering that God exists are as good as the best outcomes of the other available alternatives, and the worst outcome of wagering that God exists is no worse than the worst outcomes of the other available ones.

Therefore,

16. Any who hold that the temporal benefits associated with wagering that God exists exceeds those associated with other available alternatives may wager that God exists.

Notice there is no assertion that the temporal benefits associated with wagering for theism in fact exceed the temporal benefits associated with nontheistic wagering, or with wagering on either agnosticism or philosophical atheism. Sustaining such a claim would take us away from the logic of the

---

[17] Might the MGO be revived by arguing that one should assign a positive probability to any alternative relevantly similar to those alternatives that one finds as live alternatives? This is not so clear, since, if it is true that the live alternatives are some version of atheism, agnosticism, or theism, it is hard to see what relevantly similar alternatives are overlooked.

hybrid Pascalian wager and into the murky realm of the social sciences – a region we lack the space and time to explore.[18]

We can however say this much: if (14) and (15) are true, and if one holds that the temporal benefits associated with a theistic commitment exceed those associated with the other live alternatives, then neither the MGO, nor the challenge posed by the Agnostic Rule, is a threat to a Pascalian wager equipped with the Jamesian armaments of a forced decision and the distinction between live and dead alternatives. Are there other objections lurking? Let us briefly examine one other objection leveled against the extended family of Pascalian wagers and ascertain whether this objection is fatal to the hybrid wager.

4 The Charge of Irrationality

Daniel Garber (2009) has objected that no one should accept a Pascalian wager as a sound argument for theistic belief, since, having accepted the Wager argument and undertaking to induce belief as Pascal recommends, one ends up in an irrational state, as self-inducing belief is rationally problematic. Garber's objection is based on a distinction between first-order rationality for believing *p* (the belief that *one has good reason to believe p*), and second-order rationality for believing *p* (the belief that *one has good reason to deny that one is deluded at the first order*). Garber argues that a Pascalian can enjoy first-order rationality for theistic wagering, but would thereby lack second-order rationality because the regime one undertook to inculcate theistic belief may have induced a delusion. In short, while theistic wagering may be rationally permissible, the theistic belief that eventuates from that wagering is not rationally permissible:

> If I follow Pascal's program, I will indeed land in a state in which I believe, and in which I am genuinely convinced that I can give a good reason for what I believe, if challenged. But am I entitled to trust my confidence when I am in that state? After all, I deliberately preformed a series of steps that I knew would, if I followed them, put me into exactly that state. Now it is one thing if, in the course of events, I find myself in that epistemic state. But it would seem to be quite another if I am deliberately going about deceiving myself, believing because I want to believe. The process by which I attain

---

[18] Explorations into the relevant social science in the context of a Pascalian wager, include Rota (2016a, pp. 33–48) and McBrayer (2014, pp. 130–40).

the rational belief would seem to undermine the rationality of the final outcome. [2007, p. 39]

According to this objection, wagering that God exists likely results in an irrational belief. How so? Wagering that God exists is, in practical terms, as we have seen, a commitment to reorienting one's goals, values, and behavior by including the proposition that God exists among one's most basic values and attitudes. And, as we noted, given the facts of human psychology, anyone who reorients her goals, values, and behavior, by accepting the proposition that God exists will very likely, in time, come to believe that God exists. Garber's objection is that while wagering for God may be rational (in the first-order sense), any belief that God exists which flows out of that wagering would be irrational (in the second-order sense). In short, the problem is that a deliberate but deceptive process was employed to bring about a belief, independent of the evidence. This original objection seems to respect the dialectical force of Pascalian wagers, but undercuts the point of those wagers.

Garber's objection however is indictable on two counts. First, it is far from clear that wagering that God exists requires self-deception, and even if it does, doing so is problematic. Self-deception may be a serious problem with regard to inculcating a belief that one takes to be false, but it does not seem so with the inculcation of a belief that is intellectually open. What is belief? We can say that believing a certain proposition, *p*, just is *being disposed to feel that p is probably the case*. Clearly enough, however, it does not follow that believing that *p* involves *being disposed to feel that p is probably the case based on the evidence at hand*. The latter does not follow from the former since the latter is more complex than the former. If this is correct then self-deception does not seem particularly problematic in cases in which dispositive evidence is lacking.

In addition, Garber inverts the Pascalian advice. Pascal did not advise self-inducing belief in order to arrive at what one takes to be good reason to believe. Pascal advised seeking to self-induce belief when one finds oneself with what one takes as good reason to believe, but nonetheless lacks belief. That is, if upon reflection one holds that a wager argument works, but one still lacks belief, then Pascal advises that one "learn from those who were once bound like you and who now wager all they have. These are people who know the road you wish to follow, who have been cured of the affliction of which you wish to be cured ..." (L418/S680). Taking oneself to have good reason to believe comes before seeking to self-induce belief in the Pascalian scheme, not after. Considered in the light of the order envisioned by Pascal, the Pascalian advice lacks the sinister color described by Garber.

5 The Hybrid Wager

The hybrid wager differs from Pascal's original wagers in at least two significant ways. The latter argued that persons are rationally mandated to wager that God exists, as it would be positively irrational to be irreligious. The hybrid wager however does not generate a rational mandate, but a rational and moral permission. Given that certain conditions hold, one may wager that God exists without moral or rational fault. Rationality may not require wagering, even if it permits it. One might think of the difference this way: the strategy of Pascal's original wagers is that of going on the offensive to convert non-theists to theism. The strategy of the hybrid wager however may be seen as a defensive bulwark intended to justify or rationalize a theistic commitment. In any case, the hybrid wager has a different force than Pascal intended for his wagers.

However, it is not just the force that differs, as the audience of the hybrid wager will be more limited than that envisioned by Pascal. Pascal intended his wagers to ensnare any who encounter and understand them. The hybrid wager however is equipped not just with assumptions standard to Pascal's wagers, but also with certain assumptions (think of the liveness condition for instance), which are not as widely distributed. The scope of the hybrid, therefore, is narrower than that of the originals. Importantly, however, one should recall that a majority of the world's population is estimated to have some sort of theistic affiliation or other, so an argument, which may be useful to a majority of the world's population, is not trivial.

Finally, we should recall that less than maximal scope and force do not imply unsoundness. Indeed, if the hybrid wager eludes the Agnostic Rule, the MGO, and the charge of irrationality, as argued here, then this is a significant result. This significant result goes a long way in support of a variation of one of the claims of William James with which we began: a hybrid of Pascal's argument, instead of being powerless, then, seems a regular clincher, and is the last stroke needed to make our faith complete.

Part II
Assessment

6   The (In)validity of Pascal's Wager

Alan Hájek[*]

Pascal (in)famously argues that you should wager that God exists – bet your life that he exists. In fact, he gives three main arguments for this conclusion (and perhaps even a fourth). Legions of critics have objected to one premise or another; scores of advocates have defended the premises, or have offered close variants of them. However, a different line of objection questions the *validity* of the arguments. Grant Pascal all his premises; does it follow that you should wager that God exists? Hacking (1972) emphatically says yes, for all three arguments; equally emphatically I say no, for all three arguments.

I have been saying this for a long time (see my 2003 and 2012a). However, I think there is still more to be said, on both sides. I will work my way through Pascal's text, reconstructing his arguments. I will investigate the kind of dominance reasoning that he employs. Next, I will piece together his formulation of expected utility theory, which is often credited to him but not carefully analyzed. I will then assess the validity of his arguments – unfavorably. Hacking (and also McClennen) will be both my inspiration and foil. I will then offer ten variants of Pascal's most important argument, the one from generalized expectations, all of which are at least prima facie valid, and most of which are clearly so. These arguments all modify Pascal's decision matrix. I will then turn to valid arguments that modify his decision theory. I will conclude with a brief discussion of what I take to be Pascal's fourth argument.

1 The Argument from (Super)dominance

Pascal begins formulating the decision problem that you face: "Either God is or he is not." This determines the states. "How will you wager?" This

---

[*] Many thanks to Paul Bartha, Eddy Chen, Nick DiBella, Yoaav Isaacs, Graham Oppy, and Hayden Wilkinson for very helpful comments.

Table 6.1 (Super)Dominance Wager

	God exists	God does not exist
Wager for God	Win everything (salvation)	Earthly happiness, lose nothing
	Best	Second best
Wager against God	Wretchedness (damnation)	Earthly happiness
	Worst	Second best

determines your options: either you wager for or against God. So, we have a 2 × 2 decision matrix. Next, Pascal indicates how much "happiness" is associated with the possible outcomes. Regarding wagering for God, Pascal writes: "Let us weigh up the gain and the loss involved in calling heads that God exists. Let us assess the two cases: if you win, you win everything, if you lose, you lose nothing." He certainly has salvation in mind for winning everything if God exists, and presumably just earthly happiness with no loss if he does not. This allows us to complete one row of the matrix. As for wagering against God, he speaks of "wretchedness," which surely corresponds to damnation if God exists. So, with a little filling in on Pascal's behalf, we get the matrix laid out in Table 6.1, with the outcomes ranked.

Pascal concludes: "Do not hesitate then; wager that he does exist." That is, you should wager for God. In his important article, Hacking (1972) calls this "the argument from dominance," and maintains that "[t]he argument is valid" (p. 188). Let us grant Pascal this matrix; our question is whether his conclusion follows from it.

It is not clear that considerations of *happiness* settle the decision problem. As we would now put it, your *utilities* may not be so simple, and they are needed to determine what you should do. For example, Pascal speaks of "the true" as something "to lose," and of "your reason" and "your knowledge" as things "to stake" as well as "your happiness." This suggests that there are also *epistemic* inputs into your utilities. In that case, Pascal's conclusion about what you should do – *all things considered* – does not follow; at best, we get a conclusion only about what best serves your happiness.

Now, perhaps we could spot him a hedonic account of utility for the sake of the argument, and from now on I will. But does the conclusion follow even then? In explaining "dominance," Hacking writes: "The simplest special case occurs when one course of action is *better no matter what the world is like*" (p. 187, my emphasis), and he claims that dominance yields one of "[t]he valid argument forms investigated by decision theory" (p. 187). This both *overstates* and *understates* Pascal's case for wagering for God over wagering against God.

It *overstates* Pascal's case: in fact, as Hacking says himself, "If God is not, both courses of action are pretty much on a par" (p. 188). If the world is like *that*, wagering for God is *not* better than wagering against; we have so-called "weak dominance" rather than the kind of "strong dominance" promised in Hacking's explanation. But it also *understates* Pascal's case: the argument is not merely an instance of dominance reasoning. Such reasoning (even strong) is *invalid* without a further assumption of probabilistic independence of actions and states, or dependence of the right kind. If they are suitably correlated, dominance reasoning can yield poor advice.[1] Yet Pascal's argument does not need that assumption. Whatever the correlation between God's (non-)existence and your choice may be, you cannot go wrong by wagering for God.

Closer to the mark is McClennen's (1994) analysis of Pascal's argument as one from "superdominance":

> [T]his is really an argument from a principle of superdominance: each of the outcomes associated with betting on God is at least as good as, or better than, each of the outcomes associated with betting against God. In such a case, we are spared the burdensome business of sorting out notions of probabilistic and causal dependence or independence between our choice of an action and the relevant states ... [p. 118]

This suggests that the argument is valid as it stands.

But again, this both *understates* and *overstates* Pascal's case for wagering for God over wagering against God. It *understates* Pascal's case: by McClennen's definition, we would have superdominance even if all four outcomes in the decision matrix were exactly the same in value. Indeed, we would then have superdominance in both directions: wagering for God over wagering against God, and vice versa (a *very* weak superdominance, to be sure). Hence superdominance alone, so defined, is too weak a relation to yield a valid argument for favoring one option over another. Pascal's case is better than that: winning everything is *better than* (not merely at least as good as) each of the outcomes associated with wagering against God, and wretchedness is *worse than* (not even as good as) each of the outcomes associated with wagering for God. But it

---

[1] I used to give my Caltech students the following specious dominance argument for why they should not submit their essays in their pass/fail course: "Submitting an essay is a pain. Either you will pass the course or you won't. If *you will pass*, then not submitting is better than submitting; if *you will not pass*, then not submitting is better than submitting. Not submitting is better no matter what the world is like." Of course, they got the joke: the probability of passing given submitting is much greater than the probability of passing given not submitting. The states are highly dependent on the actions.

also *overstates* Pascal's case: even his reasoning is *invalid* without the further assumption that it is possible that God exists. A strict atheist might insist that God's existence is impossible (for example, by regarding his putative omniscience, omnipotence, and omnibenevolence to be jointly inconsistent). Then there is no reason to prefer wagering for God over wagering against God. You cannot go wrong by wagering for God; but nor can you go wrong by wagering against God. After all, the two remaining possible outcomes are assumed to be "pretty much on a par" according to this first argument.

2 The Argument from Expectation

In fact, Pascal then immediately allows that they may not be. An imaginary interlocutor challenges this assumption: "Yes, I must wager, but perhaps I am wagering too much." Perhaps in wagering for God, one does give up something after all – perhaps too much compared to wagering against God. Then the decision matrix above misrepresents the situation. If God does not exist and you wager for God, the outcome is really: earthly life, lose *something*. We no longer have a case of (weak) dominance, nor superdominance. However, the outcome still has *finite* value, as does earthly life simpliciter.

Moreover, Pascal is not concerned that God's existence might be impossible – on the contrary, he claims that it has probability ½: "there is an equal chance of gain and loss." Probabilities are now on the table for the first time, and his gambling language goes into high gear. His unit of utility is *a life* (a finite, earthly life, I assume) and he claims that there is "an infinity of infinitely happy life to be won." We will understand this as salvation having infinite utility.

Let's further assume that the payoff if God exists and one wagers against God is infinitely worse than that of salvation – a safe assumption for the "wretchedness" of damnation. Some commentators take the utility in that case to be finite, as I do. Pascal writes: "[God's] justice towards the damned is less vast ... than his mercy towards the elect." (Later we will let that utility be negative infinite.) We have all the materials we need to construct a new decision matrix (Table 6.2), where the f_i are finite numbers):

Such is our decision problem; in the following paragraph, Pascal gives us the tool to solve it. Indeed, the tool was newly fashioned, and his Wager may have been its first ever application (Hacking, 1975). He presents what we now recognize as a formulation of *expected utility theory*. "Every gambler takes a certain risk for an uncertain gain, and yet he is taking a certain finite risk for an uncertain finite gain without sinning against reason." How much, then,

Table 6.2 Wager with Probability ½

	Probability ½	Probability ½
	God exists	*God does not exist*
Wager for God	∞	f_1
Wager against God	f_2	f_3

should he be prepared to risk? Or as we would now say it, what is the fair price for a gamble? The Krailsheimer translation of the *Pensées* puts Pascal's answer this way:

> [T]he proportion between the uncertainty of winning and the certainty of what is being risked is in proportion to the chances of winning or losing. (L418/S680)

But this makes no sense. It has the form: "the proportion ... is in proportion to the chances of winning or losing." What did Pascal really say? Here's the original text:

> [L]'incertitude de gagner est proportionnée à la certitude de ce qu'on hasarde selon la proportion des hasards de gain et de perte.

I offer this translation:

> The uncertainty of winning is proportioned to the certainty of what is being risked according to the proportion of chances of winning and of losing.

We can make sense of this, but it will take some work.

Consider a gamble that pays you W units (for 'win') if X, and nothing otherwise. We are seeking the fair price f, at which you should be indifferent between playing the gamble and not. According to expected utility theory, $f = P(X)W$. Do we get that answer by following the Pascal quote? You will pay f with certainty: this is "the certainty of what is being risked" (the certain amount that is being risked). If you win, your gain is $W - f$: this is "the uncertainty of winning" (the uncertain amount that you stand to win). The latter is proportioned to the former – that is,

$$\frac{f}{W-f}$$

according to

$$\frac{P(X)}{1-P(X)}.$$

That is,

$$\frac{f}{W-f} = \frac{P(X)}{1-P(X)}.$$

Rearranging, we have

$$f(1 - P(X)) = P(X)(W - f)$$

and hence

$$f - fP(X) = P(X)W - fP(X).$$

Adding $fP(X)$ to both sides we get

$$f = P(X)W.$$

Voilà!

Given Pascal's pioneering use of decision theory in the Wager, I have wanted to make that use explicit: you should value a gamble at its expected monetary value (in the terminology of this volume's introductory chapter. If we identify utility with monetary value, then this is to value a gamble at its expected utility. We need the generalization to expected utility to evaluate "gambles" in which there are non-monetary payoffs, such as lives (as that chapter explains). Given a choice among various options, each of which can be regarded as a "gamble," you should choose so as to maximize expected utility. This is modern decision theory's central maxim of rational choice, which Pascal anticipated.

In summary, I take the premises of his second argument to be:

1. The probability that God exists is ½.
2. The decision matrix is as given.
3. You should maximize expected utility.

Hence,

Conclusion. You should wager for God.

Hacking calls the first premise "monstrous" (1975, p. 189). It is another sign of just how seriously Pascal takes his gambling paradigm: "a coin is being spun which will land heads or tails." Monstrous or not, let's do the putative expectation calculations on Pascal's behalf:

The expectation of wagering for God is

$$\infty(\tfrac{1}{2}) + f_1(\tfrac{1}{2}) = \infty.$$

This exceeds the expectation of wagering against God, namely,

$$f_2(\tfrac{1}{2}) + f_3(\tfrac{1}{2}) = \text{some finite value.}$$

You maximize expectation by wagering for God, so it appears that by premise 3, that's what you should do. Despite his misgivings about the probability assignment, one can see why Hacking maintains that "[t]he argument...is valid" (p. 189).

However, I demur. The main point is that there are options besides wagering for God that Pascal has neglected, which also have infinite expectation. I develop this point after the next section, since his third formulation of the Wager is equally susceptible to it.

3 The Argument from Generalized Expectation

Monstrous the premise may be that the probability of God's existence is ½, but Pascal realizes that it plays an inessential role in his argument. *Any* positive (and finite[2]) probability will do the job just as well. After all, for any such probability p,

$$\infty p + f_1(1-p) = \infty$$

$$> f_2 p + f_3(1-p) = \text{some finite value.}$$

It all happens in this short passage, as important as any in the history of philosophy of religion:

> [T]here is an infinity of infinitely happy life to be won, one chance of winning against a finite number of chances of losing, and what you are staking is finite. That leaves no choice; wherever there is infinity, and where there are not infinite chances of losing against that of winning, there is no room for hesitation, you must give everything. (L418/S680)

I interpret Pascal as replacing premise 1 with:

1. (generalized). The probability that God exists is positive (and finite).

So the argument for wagering for God becomes one from *generalized* expectation. When we speak simply of "Pascal's Wager," this is usually the

---

[2] As opposed to infinitesimal.

argument that we mean. Hacking calls it an "argument from dominating expectation":

> If in some admissible probability assignment, the expectation of A_1 exceeds that of any other act, while in no admissible assignment is the expectation of A_1 less than that of any other act, then A_1 has dominating expectation. The argument from dominating expectation concludes, "*Perform an act of dominating expectation.*" [1975, p. 187]

Again, he explicitly judges the argument to be valid.

And again, Hacking both *understates* and *overstates* Pascal's case for wagering for God over wagering against God. He *understates* it. It is not merely an argument from dominating expectation, as he defines that. That allows some of the admissible assignments to yield *the same* expectation for A_1 and any other act. What he has defined should be called "*weak* dominating expectation." Curiously, his definition of "dominating expectation" is at odds with his earlier definition of "dominance," which said that one course of action is *better no matter what the world is like* – a *strict* sense. To be clear, Pascal's argument here is one from *strict* dominating expectation: in *all* admissible probability assignments, the expectation of wagering for God *exceeds* that of wagering against God. So, Hacking's classification of the argument from dominance and this one is back to front: the "dominance" in the former argument is in fact *weak*, and in the latter *strict*, while his definitions go the other way around.

Hacking also *overstates* Pascal's case for wagering for God, since he maintains that it is valid. I submit that it is not.

4 The Invalidity of the Argument from Generalized Expectation

Grant Pascal his premises: the probability of God's existence is positive (and finite), the decision matrix is as he claims, and you should maximize expected utility. It does not follow that you should wager for God.

Even when you maximize expected utility by Φ-ing, it does not follow that *you should Φ*. When two or more options are tied for maximal expectation, it does not follow that you should perform *any particular one* of them – another option also maximizes expectation, after all. Buridan's ass, equally hungry and thirsty, is poised midway between a bale of hay and a pale of water, and it maximizes expectation by going to either. It does not follow that it *should go to the hay*.

This primes us for the fundamental problem for the validity of Pascal's Wager. Granting him his premises, it's true that wagering for God maximizes expected utility. The trouble is that *other* options besides wagering for God do so too. Here's one: toss a coin; if it lands heads, wager for God; if it lands tails, wager against God. With probability ½, the coin lands heads, and you perform an action that has expectation ∞; with probability ½, the coin lands tails, and you perform an action that has expectation, $f_2 p + f_3(1-p)$. So, the expectation of this option is:

$$\infty(½) + [f_2 p + f_3(1-p)](½) = \infty.$$

Pascal should be careful what he wishes for: ∞'s swamping effect, which was crucial to his argument for wagering for God, is just as evident here! This *mixed strategy* between wagering for and wagering against God also has infinite expectation. So, you could just as well maximize expected utility by following *it*.[3]

This suffices to show that the argument from generalized expectation is invalid. (And so too the argument from expectation, of course.) But while we're at it, let's open the floodgates. Suppose you buy a ticket in a million-ticket lottery, and you wager for God if and only if your ticket wins. With probability 1/1,000,000 your ticket wins, and you perform an action with infinite expectation; with probability 999,999/1,000,000 your ticket loses, and you perform an action with finite expectation. Again, this mixed strategy has infinite expectation, a mixture of ∞ and something finite. And so it goes. Wager for God if and only if a monkey randomly hitting the keys of a typewriter perfectly types the entire text of *Hamlet*. With some positive probability it will do so. Again, this mixed strategy has infinite expectation.

But this still understates the problem. Suppose you don't want to wager for God; you'd rather live a life of sin instead. With some positive probability (plausibly many orders of magnitude greater than the *Hamlet* probability), you will nonetheless wind up wagering for God. So even this turns out really to be a mixed strategy between wagering for God and wagering against God,

---

[3] You might say that Pascal's decision problem is limited to just the two options, wagering for and wagering against God – after all, they are the only options in the decision matrix. You might add that I am questioning that decision matrix, which is to question the truth of that premise, rather than the validity of the argument. I reply that the mixed strategies do not need to be explicitly presented in the decision matrix in order to be counted as options. It is standard in decision theory to regard mixed strategies as coming "for free," automatically assumed to be options once a set of pure strategies has been fixed.

and thus one with infinite expectation. Pascal's imaginary interlocutor laments: "I am so made that I cannot believe." Pascal counsels him to behave like a believer, "taking holy water, having masses said, and so on." But even those who do the exact opposite, doing all that they can to *avoid* belief, nonetheless have some positive probability of wagering for God despite that. So even such an *avoidance* strategy's expectation is some weighted average of infinity and a finite value, for suitable weights – which is infinite. Indeed, it looks like *anything* that you do has infinite expectation by Pascal's lights. Not only is his argument invalid; it is invalid in the worst possible way. Pascal should again be careful what he wishes for: in generalizing to cover all positive probabilities for the existence of God, and hence of salvation, he unwittingly generalizes to all positive probabilities for getting it – that is, to everything.

5 Valid Reformulations

Yet surely there is still something compelling about Pascal's reasoning. One feels that it should not be dismissed so quickly. It seems like something of an artefact of the mathematical representation of infinity that all distinctions among expected utilities collapse in the way they did. Perhaps we can do better on his behalf. Some modification of the premises is required. A natural place to start is with the decision matrix, and in the next section I offer no fewer than ten reformulations that revise it. (I considered several of them in my (2003) and (2012a), but I have some new perspectives on them, and I add some further reformulations here.) Expected utility theory might also be adjusted, and in the subsequent section I consider reformulations that do that.

In both sections I will present the reformulations in decreasing order of faithfulness to the spirit of Pascal's argument. Indeed, the first reformulation to which I now turn is not a reformulation at all according to some commentators: they regard it is the proper formulation of his argument in the first place.

6 Modifying the Decision Matrix

6.1 Salvation Has Infinite Utility, Damnation Has Negative Infinite Utility

What is the utility associated with damnation according to Pascal? I have supported my reading of it being a finite value. But a popular understanding is

Table 6.3 Damnation Has Negative Infinite Utility

	God exists	God does not exist
Wager for God	∞	f_1
Wager against God	$-\infty$	f_3

that it is negative infinity – perhaps an infinity of infinitely *unhappy* life, or perhaps an infinity of finitely unhappy life. Then the decision matrix looks like that laid out in Table 6.3.

How could this help with the previous trouble with infinity utility – won't this be double trouble? Perhaps not. Wagering for God has infinite expectation, and wagering against God has negative infinite expectation. What about all the mixed strategies? Each has an expectation of $\infty - \infty$. And given a choice between an option with expectation ∞, and any with expectation $\infty - \infty$, the former seems preferable. Imagine yourself facing such a choice; I bet I know how *you* would choose! So, this reformulation appears at first to be valid.

6.1.1 $\infty - \infty$ Is Undefined?

However, there are subtle problems here. We might regard $\infty - \infty$ as being simply *undefined*, much as 1/0 is. Then I think that this reformulation is invalid, for an interesting reason. It is not the case that wagering for God maximizes expected utility, for ∞ cannot be compared to *undefined*. In particular, it is not the case that

$$\infty \geq \text{undefined},$$

much as it is not the case that

$$x \geq 1/0,$$

for any number x. An inequality does not hold when one side of it is undefined. If any of one's options has undefined expectation, one cannot maximize expectation – period. All the more, one cannot do so when infinitely many of them have undefined expectation! In particular, one cannot maximize it by wagering for God. So as before, Pascal's conclusion does not follow.

6.1.2 $\infty - \infty$ Is Indeterminate?

However, the status of $\infty - \infty$ does not seem as bad as that of 1/0. Suppose you tried to assign a value to 1/0 – say, 17. That can't be right, for then we would have

$$1 = 17 \times 0,$$

which is false. And so it goes for any putative value; so there is determinately no such value. By contrast, it seems we *can* legitimately assign values to $\infty - \infty$ (think of them as precisifications, permissible candidates for the difference). 17 is one such value:

$$\infty - \infty = 17$$

seems to be legitimate, since it is true that

$$\infty = 17 + \infty.$$

π is another such value, since it is also true that

$$\infty = \pi + \infty.$$

And so it goes for each such value assignment. Indeed, even ∞ itself is such a value, since

$$\infty = \infty + \infty.$$

There seems to be exactly one value from the extended real line that is not (determinately) legitimate: $-\infty$. Then we would have

$$\infty - \infty = -\infty,$$

and hence

$$\infty = \infty - \infty,$$

which is indeterminate, by the very fact that there are many precisifications of the right-hand side, only one of which is equal to the left-hand side! So, it seems that for any $x \in (-\infty, \infty]$, x is a legitimate value, and it is indeterminate that $-\infty$ is.

So, we may regard $\infty - \infty$ as *indeterminate*, as Jeffrey (1983a) does, and we may regard all of the (determinately) legitimate values as admissible precisifications of it. (We have higher-order indeterminacy thanks to $-\infty$: it is indeterminate whether it is an admissible precisification. But that will not affect our reasoning from now on.) We can then supervaluate over all of them to determine what is true, or indeterminate, or false.[4] In particular, it is indeterminate that $\infty - \infty$ has any particular value from the extended real line, since that is true on one precisification, and false on the rest.

---

[4] This is reminiscent of the supervaluational treatment of a vague predicate, in which we consider the set of all of its admissible precisifications. We then ask what is true on all of them ("supertrue," hence true simpliciter), what is false on all of them ("superfalse," hence false simpliciter), and what is true on some and false on others ("indeterminate," in the usual interpretation).

6.1.3 $\infty - \infty$ Is Indeterminate, and a Valid Reformulation of the Wager with a Further Decision Rule?

Again, on this reading, this reformulation of the Wager appears at first to be valid. It seems that an option that has infinite expectation should be preferred to any that has indeterminate expectation. (I bet I know how *you* would choose!) So, it seems that outright wagering for God wins.

That sounds like the right thing to say, and I have previously (2012a) said it. I also argued for it another way, as follows:

> Consider all the ways that the indeterminacy could be resolved. On one resolution [precisification] it is infinite, the same as the expected utility of wagering for God. On all other resolutions it is finite, or negative infinite, strictly less than the expected utility of wagering for God. On no resolution is wagering for God worse than wagering against God, and on all but one resolution it is better. [p. 181]

We might call this an argument from *weak dominance of expectation* (not to be confused with what Hacking called "dominating expectation"). And we might call the underlying decision rule a "weak precisificational dominance rule":[5]

> For all acts *A* and *B*: If there are (i) some precisifications on which *A* has greater expected utility than *B* and (ii) no precisifications on which *B* has greater expected utility than *A*, then *A* should be preferred.

To be sure, our earlier discussion of the argument from (super)dominance cautioned us that dominance reasoning must sometimes be handled with care: when it can be undermined by correlations between acts and states. But notice that *this* kind of dominance reasoning cannot be so undermined. So, it appears that this reformulation of the Wager is valid if we add the weak precisificational dominance rule as a premise.

6.1.4 $\infty - \infty$ Is Indeterminate, and This Reformulation Is Invalid without the Further Decision Rule?

Intriguingly, however, it appears to be invalid without this further premise. For premise 3 says that you should *maximize* expected utility: perform an action whose expected utility is at least as great as that of any other. But is infinite utility at least as great as indeterminate utility? What is the status of the inequality:

$$\infty \geq \text{Indeterminate}?$$

---

[5] Thanks to Nicholas DiBella for the name and statement of the rule.

Again, it is tempting to supervaluate. The admissible "precisifications" are all of the values in $(-\infty, \infty]$. However, then it is not supertrue that you should wager for God – this recalls my criticism of Hacking's discussion of "the argument from dominating expectations." Wagering for God is not determinately better than any particular mixed strategy, with its indeterminate expectation. After all, there is one precisification of the indeterminate expectation, namely ∞, which also accords it maximal expectation. It is indeterminate, rather than true, that wagering for God is better than something with indeterminate expectation! So, it is not true that *you should wager for God*. (Compare: Suppose that it is indeterminate whether it is better for Buridan's ass to go to the hay than to the water. Then it is not true that it *should go to the hay*.) So, it is not true that the argument is valid; at best, it is indeterminate that it is.

In sum, if we take $\infty - \infty$ to be undefined, this reformulation of the Wager is invalid as it stands. If we take $\infty - \infty$ to be indeterminate, then it is at best indeterminate that it is valid. Either way, it is not *true* that it is valid. However, it becomes valid if we add as a further premise the weak precisificational dominance rule.

It comes as something of a relief, then, to turn to another reformulation that is more straightforwardly valid.

6.2 Salvation Has Surreal Infinite Utility

We know that Pascal thought that salvation has infinity utility; it is less clear that we should represent this as "∞." He never used that symbol. However, we do get some insights into his understanding of infinity. He writes: "Unity added to infinity does not increase it at all, any more than a foot added to an infinite measurement: the finite is annihilated in the presence of the infinite and becomes pure nothingness." This fits well with what we may call *the reflexivity of ∞ under addition*:

$$\infty + 1 = \infty,$$

and more generally,

$$\infty + r = \infty, \text{ for all } r \in (-\infty,\infty].$$

However, a closely related property of ∞ led to the third argument's downfall – its *reflexivity under multiplication* (by positive numbers):

$$k.(\infty) = \infty \text{ for any } k > 0.$$

Table 6.4 Infinite Surreal Utility

	God exists	God does not exist
Wager for God	ω	f_1
Wager against God	f_2	f_3

So it seems we would do better to represent the utility of salvation with a better-behaved infinity: halving it, or millionth-ing it, etc. should make a difference. In particular, we want an infinity such that

$$p.(\text{infinity}) < \text{infinity, for any } 0 < p < 1.$$

There are various number systems that provide such infinities, among them those of the hyperreal numbers as we find in non-standard analysis (Robinson, 1966), and of numerosity theory (Benci et al., 2018). In my (2003), I suggested that hyperreal numbers could be used to tell apart the infinite expectations of wagering for God and the various mixed strategies, which previously came out the same.[6] However, I focused on the *surreal numbers* (introduced by Conway, 1976), and I will do so again here.

I do not have the space here to repeat my exposition of how Conway constructs them. Suffice to say that they include many infinite ordinals. Some of them are familiar from Cantor: e.g., ω, ω + 1, ω², and ω^ω. But crucially, they also include *smaller* infinite ordinals, such ω − 1, and √ω. Better yet, they include numbers that we will need, like ω/2, ω/1,000,000, and so on, which will figure in the expectations of the various mixed strategies that we have countenanced.

Now replace ∞ in the decision matrix with some infinite surreal number. But which one? It doesn't matter: whichever one you pick, the resulting argument will be valid.[7] For definiteness, let's pick ω (on a given utility scale). The decision matrix is presented in Table 6.4.

The expectation of wagering for God is

$$\omega p + f_1.(1 - p) \quad ☺$$

which is infinite. The expected utility of wagering against God is

$$f_2.p + f_3.(1 - p),$$

---

[6] See also Sobel (1996). Herzberg (2011) presents a representation theorem for hyperreal expected utilities, and offers a valid reformulation of Pascal's Wager with such utilities.

[7] We are keeping Pascal's premise that the probability of God's existence is positive, *and finite*. Now I want to highlight the assumption that the probability is not infinitesimal. That way we are sure that when ω is multiplied by it, the result is still infinite. This guarantees that the expectation of wagering for God is uniquely maximal.

which is finite. Happily, the expected utilities of the various mixed strategies are less than that of wagering for God. For example, for the coin-toss strategy it is:

$$\tfrac{1}{2}[\omega.p + f_1.(1-p)] + \tfrac{1}{2}[f_2.p + f_3.(1-p)],$$

which is easily shown to be less than ☺. In general, a mixed strategy, with probability $q > 0$ for wagering-for and probability $(1 - q) > 0$ for wagering-against, has expectation

$$q[\omega.p + f_1.(1-p)] + (1-q)[f_2.p + f_3.(1-p)],$$

which is less than the expectation of wagering-for, and indeed is a strictly increasing function of q. Our reformulation of Pascal's Wager is valid.

6.3 Vector-Valued Value: Salvation Has Finite "Heavenly" Value

It may be a drawback of the surreal infinite reformulation that according to it, one can reach and even exceed the utility of salvation by adding enough finite units of utility. For example, suppose that $1 provides 1 unit of utility, and that utility is linear in dollar amount. Then if you accrue ω (no mean feat, I admit!), you will achieve the utility of salvation, and by adding another dollar you will exceed it. Indeed, you don't even need to achieve such wealth with certainty. A St. Petersburg game in which a coin can be tossed up to ω times will do just as well. But salvation doesn't seem to be the sort of thing whose utility can be reached or exceeded just by piling on enough money, or other earthly goods. It seems to be of another order altogether.

My (2003) tries to capture this thought with a vector-valued reformulation. Suppose there are two sorts of value: call them "earthly value" and "heavenly value." Expected utility is a two-dimensional (vector) quantity, of the form (x, y). Salvation has 1 unit of heavenly value, the maximal amount. A probability p of salvation corresponds to p units of "heavenly expectation." Suppose that any increase in heavenly expectation trumps any increase in earthly expectation. We have a lexicographic ordering: when choosing between two actions, we compare first their heavenly expectations, preferring the action with greater heavenly expectation; if these are tied, we then prefer the action with the greater earthly expectation. Heavenly expectation exceeds even infinite earthly expectation: a tiny chance at salvation – even infinitesimal – is better than a guarantee of playing the St. Petersburg game. Picture an expected utility as a point in the Cartesian plane, in the infinite strip bounded by the

Table 6.5 Vector-Valued Utilities

	God exists	God does not exist
Wager for God	$(e_1, 1)$	$(e_2, 0)$
Wager against God	$(e_3, 0)$	$(e_4, 0)$

x axis ($y = 0$) and the horizontal line $y = 1$. Heavenly value, appropriately enough, sits at the top. The higher up your utility is, the better off you are.

The decision matrix is now as laid out in Table 6.5, with e_1, e_2, e_3 and e_4 amounts of earthly value.[8]

The expectation of any action can now be calculated by finding the earthly and heavenly expectations. Wagering for God has heavenly expectation

$$1.p + 0.(1 - p) = p > 0.$$

This beats wagering against God, which has heavenly expectation 0. It also beats any mixed strategy. The coin-toss strategy has heavenly expectation ½, the lottery strategy has heavenly expectation 1/1,000,000, and so on. The higher the chance of wagering for God, the better the expectation. Wagering for God uniquely maximizes your expectation: the reformulation is valid.

6.4 Salvation Has Finite Utility for an Infinite Period of Time

Pascal thought of salvation as "an infinity of infinitely happy life." This is infinite twice over. But one infinity will suffice to generate infinite utility. Infinitely happy life for finite time – even a nanosecond – has a total utility of infinity. And so does a finitely happy life for infinite time, assuming either that total utility is linear in amount of happiness, or just that marginal utility does not diminish too rapidly with increasing total happiness. So even earthly happiness, enjoyed forever, may suffice. This recalls the concern at the beginning of the last section: salvation doesn't seem to be the sort of thing whose utility can be reached just by piling on sufficient *duration* of such happiness. However, at least it offers the prospect of a valid argument.

---

[8] Admittedly, $(e_1, 1)$ can be improved by adding something to the first component – that is, by improving one's earthly life. But that is no concern for Pascal's idea of salvation: the second component cannot be improved. Indeed, it is an advantage of this representation that it respects that idea, while allowing differences in the quality of one's earthly life to make a difference to one's overall utility. Salvation and a happy earthly life is better than salvation and an unhappy earthly life.

Table 6.6 Long-Run Average Utilities

	God exists	God does not exist
Wager for God	1	0
Wager against God	0	0

How do we compare finite goods enjoyed for infinite time, when both yield infinite total utility? We may take the *long-run average utility* of each of them. Find the total utility of each up to time t, for various t; divide this in each case by t; then take the limit as t tends to infinity.

Suppose that salvation consists of some finite happiness over infinite time – something with 1 unit of utility for each unit of time (on a suitable choice of utility scale and time unit). Then the long-run average utility of salvation is 1. And suppose that otherwise, you have your earthly life for finite time and then you die. Forever after you get zero units of utility for each unit of time (on a suitable choice of utility scale). Then whatever happens up till your death makes no contribution to the long-run average: the subsequent infinite period of zero utility overwhelms it.

Table 6.6 shows the decision matrix, now with long-run average utilities rather than total utilities.

This makes vivid Pascal's idea that "the finite is annihilated in the presence of the infinite and becomes pure nothingness." Note that the long-run averages in this matrix have the same values as the "heavenly" components in the previous matrix. And the calculations look similar here: the expected utility of wagering for God is p, which exceeds the expected utility of wagering against God, 0. The mixed strategies all have intermediate expected utility: the coin-toss strategy has expected utility $p/2$, and so on. Even if Pascal would not like this conception of salvation, he would at least like the ordering that it induces.

6.5 Salvation Has Finite Utility

Our wagers are getting progressively less Pascalian. But they are also getting progressively more tractable, and more susceptible to the standard decision-theoretic machinery. The next one can be formulated entirely within von Neumann and Morgenstern's (1944), Savage's (1954), or Jeffrey's (1983a) theory. It avoids a major objection that decision theorists such as Jeffrey and McClennen have to the original wager: that the very notion of infinite utility is suspect.

Table 6.7 Finite Utility

	God exists	God does not exist
Wager for God	f	f_1
Wager against God	f_2	f_3

Pascal's Wager promised to work for all positive probabilities for God's existence, however small. In this sense, it promised to catch in its net all possible rational agents. (Pascal would presumably regard strict atheists as irrational; in any case, they were always going to be beyond its reach.) However, it is really addressed to *people* like us, and it will have served its purpose if it catches all the rational members of its potential audience. That potential audience is finite. As such, there is the *smallest* positive probability that any of its members assigns to God's existence. Call it $p_{\min}$. The Wager will have done its job if it works for $p_{\min}$; all the more, it will work for all larger probabilities. But then the utility of salvation need not be infinite; a sufficiently large finite utility f will do. It needs only to be large enough to satisfy this inequality:

$$f \cdot p_{\min} + f_1 \cdot (1 - p_{\min}) > f_2 \cdot p_{\min} + f_3 \cdot (1 - p_{\min}),$$

and hence

$$f > [f_2 \cdot p_{\min} + f_3 \cdot (1 - p_{\min}) - f_1 \cdot (1 - p_{\min})]/p_{\min}.$$

So f could be, say, $1 + [f_2 \cdot p_{\min} + f_3 \cdot (1 - p_{\min}) - f_1 \cdot (1 - p_{\min})]/p_{\min}$.

Table 6.7 gives the decision matrix.

We have tailored f so as to guarantee that wagering for God surpasses wagering against God. It follows that it also surpasses all the mixed strategies.

Salvation can involve a finite, finitely happy life, and yet figure in a valid argument for wagering for God. It is remarkable that it does *better* in this regard than Pascal's infinity of infinitely happy life. Offhand, one would have expected the exact opposite: intuitively, the greater the utility of salvation, the stronger the case for wagering for God. Not so – but then we are used to our intuitions involving infinity being overturned.

6.6 Salvation Has Finite Utility, Damnation Has Negative Infinite Utility

We can let salvation be even worse, and have it still figure in a valid argument for wagering for God. In fact, any finite utility, however low, will do the job.

Table 6.8 Negative Infinite and Positive Finite Utilities

	God exists	God does not exist
Wager for God	f_1	f_2
Wager against God	$-\infty$	f_3

We can then produce a "mirror image" of Pascal's Wager, in which the infinitude moves from salvation to damnation, and turns negative. Table 6.8 displays the decision matrix.

Now, a concern to gain an infinite reward is replaced with a concern to avoid infinite loss of utility. The expectation of wagering for God is finite, and that of wagering against God is negative infinite. Moreover, wagering for God, with its finite expectation, is the unique optimum, so once again we can validly conclude that rationality requires you to wager for God. All of the mixed strategies get dragged down to the level of wagering against God, where previously they all got raised up to the level of wagering for God. Infinity swamps just as effectively in either direction.

6.7 Four Hybrids

Now that we see the trick, we can rerun it several times. We keep the utility of salvation finite, and we replace $-\infty$ with various surrogates, paralleling our previous replacements for ∞. So, we start with the decision matrix of the previous section, and replace $-\infty$ successively with:

- surreal negative infinite utility: $-\omega$
- vector-valued utility with *hellish* value in the second component: $(e_1, -1)$
- finite misery for an infinite period of time, with a long-run average of -1
- a sufficiently low finite utility.

We have come a long way from both Pascal's own conception of the Wager and its underlying theology, but at least we have valid arguments for his conclusion.

7 Modifying the Decision Rule

Now let's change tack. Premise 3 says that you should maximize expected utility. But as we have seen, with Pascal's decision matrix you could do that without wagering for God. Let's modify the decision rule and produce valid arguments that way.

7.1 Increase the Probability of Salvation

The main problem for Pascal's Wager is that the expectation of wagering for God is tied with that of other options. The natural remedy is to add a tie-breaking rule. Schlesinger (1994) offers this one: "try and increase the probability of obtaining the prospective prize" (p. 97). Of course, "the prospective prize" here is salvation. I suggest restating the proposal: you should perform the action that maximizes your probability of salvation. This eliminates the various mixed strategies that I considered, since these have lower probabilities of your achieving salvation than outright wagering for God does. We might think of the resulting decision rule as inducing a lexical ordering: when one action uniquely maximizes expected utility, perform it; when two or more actions are tied for maximal expectation, apply the tie-breaking rule. Supplementing Pascal's Wager with this rule, the resulting argument is valid.

Schlesinger's rule is undoubtedly intuitively appealing. Indeed, he gets the desired result by fiat. There's nothing wrong with that, but it would be more satisfying if the rule could be given foundational support. (Think of how the rule of maximizing expected utility is supported by a representation theorem, and long-run justifications.) Relatedly, one wonders how the rule generalizes to other cases, where it is less clear what "the prospective prize" is – for example, where there are many "prizes," and maximizing the probability of achieving an especially good one may increase the probability of some bad outcome.

7.2 Relative Utilities

Fortunately, Bartha (2007) gives Schlesinger's rule a secure foundation. The rule falls out as a special case, but Bartha's theory is more general. He observes that so far we have been assuming *one-place* utility functions. He proposes *relative utilities*, which involve *utility ratios*, as a way of representing non-Archimedean preferences, and thus extending orthodox decision theory to take infinite utility in its stride.

The basic quantity of relative utility theory is a three-place function of the familiar one-place utility function $U(\_)$. Specifically,

$$U(A, B, Z) = (U(A) - U(Z))/(U(B) - U(Z)).$$

Here, A and B are the options whose utilities we want to compare, and Z is a "base point," ranked no higher than either of them, from whose perspective they are compared. The idea is to measure how large a proportion of the

Table 6.9 Relative Utilities

	God exists	God does not exist
Wager for God	1	0
Wager against God	0	0

interval from Z to B is the interval from Z to A. We can define infinite relative utility in terms of the Pascalian preference structure.

Let "$X \succeq Y$" denote that X is preferred to or viewed indifferently to Y, let "$X \succ Y$" denote that X is strictly preferred to Y, and let "$[pA, (1-p)C]$" denote a gamble that yields A with probability p and C with probability $1-p$. Suppose $A \succeq B$ and $B \succeq Z$. Then

$$U(A, B; Z) = \infty \longleftrightarrow [pA, (1-p)Z] \succ B \text{ for all } 0 < p < 1.$$

Bartha goes on to prove the important result that the relative utility of a gamble is its expected relative utility.

We may interpret premise 3 of the original wager as:

3. You should maximize *one-place* expected utility.

Bartha effectively replaces it with:

3'. You should maximize *three-place* expected relative utility.

The upshot is that we get a relative utility matrix (see Table 6.9) that looks just like the one involving long-run averages, above.[9]

Relative to the prospective prize of infinity, earthly prizes vanish (much as in the calculation of long-run averages, any contribution of one's earthly life vanishes). The utility of salvation relative to any other pure (unmixed) strategy is infinite. And when we calculate expected relative utilities, we have wagering for God as the unique maximum, and all of the mixed strategies ordered below it. The resulting argument is thus valid.

It is worth emphasizing how Bartha's approach reinterprets the superiority of the pure strategy of wagering for God over any mixed strategy.[10] Pascal started with utilities and derived rational preferences from them. Relative utility theory reverses this conceptual order: it starts with preferences and offers a way of representing them. Rather than determining what Pascalian preferences should be, our goal is to rationalize the preferences that Pascal

---

[9] I must elide over some details of what determines a suitable base point for evaluating the optimal option.

[10] Thanks here and until the end of this section to Paul Bartha.

already has. In particular, he prefers any gamble with a positive (and finite) chance of salvation to any finite prize. Rather than treating numerical utilities as the starting point for Pascal's Wager, we view them as a useful representation of a Pascalian agent's preferences between different outcomes.

Going back to von Neumann and Morgenstern, there is a set of axioms on preferences that ensures the existence of a utility function that exactly mirrors the agent's preferences: the best option in any decision is always one that maximizes expected utility. There is an important special case: given two outcomes X and Y with X preferred to Y, and a choice between two actions that will each produce either X or Y (with given chances for each), you should always choose the action that has a higher chance of producing X than Y. Resnik (1987) calls this the *Better-Chances Condition*.

In order to represent the infinite utility of salvation in Pascal's Wager, we need to give up one of the von Neumann–Morgenstern axioms: Continuity. However, as the relative utilities approach shows, we can keep all of the others, and the Better-Prizes Condition still follows from these other axioms because it does not depend upon Continuity. The superiority of the pure wager over any mixed-strategy can then be understood as a direct application of the Better-Prizes Condition: the pure wager gives you a higher chance than any mixed strategy of producing infinite utility rather than a finite outcome. (Thus Schlesinger's rule is put on firm footing.) The relative utility matrix makes this clear, without making the role of the axioms explicit.

For further details about the relative utilities approach, see Bartha (2007; and Chapter 12 of this volume).

8 A Fourth "Wager"?

In his "End of this address," Pascal has one last twist in store for us:

> Now what harm will come to you from choosing this course? You will be faithful, honest, humble, grateful, full of good works, a sincere, true friend ... It is true you will not enjoy noxious pleasures, glory and good living, but will you not have others?
>
> I tell you that you will gain even in this life, and that at every step you take along this road you will see that your gain is so certain and your risk so negligible that in the end you will realize that you have wagered on something certain and infinite for which you have paid nothing. (L418/S680)

Table 6.10 Superduperdominance

	God exists	God does not exist
Wager for God	Win everything (salvation) Best	Earthly happiness, gain something Second best
Wager against God	Wretchedness (damnation) Worst	Earthly happiness Third best

This passage appears to summarize Pascal's personal viewpoint of the decision problem. It makes two further striking claims regarding wagering for God: "you will gain even in this life," and "you have wagered on something certain." In that case, "the Wager" dissolves, twice over: utilities alone definitively settle that you should wager for God, and in any case it is not really a gamble at all, since your gain is certain! So the decision matrix becomes that of Table 6.10.

Now we need not even worry that God's existence might be impossible, as we did for the argument from superdominance. We have here an argument from *superduperdominance*: the worst outcome associated with wagering for God is *strictly better* than the best outcome associated with wagering against God. This yields a valid argument for wagering for God, even if we allow that God's existence might be impossible.

Pascal has come full circle back to the first wager, and he now goes even beyond it. There is no need for the tool that he developed – decision theory – to solve *this* decision problem. However, all the more one may question its premises.

9 End of This Address

This brings out the tightrope that Pascal must walk in order to provide a sound argument for his conclusion – one that is both valid *and* that has true premises. It's a balancing act familiar to philosophers: strengthening a premise in an argument may conduce to the validity of the resulting argument, but it may come at the cost of the plausibility of that premise. We interpreters of Pascal's arguments face a balancing act of our own: seeking to remain as faithful as we can to his original text, while charitably interpreting him so as to give them their best shot at soundness.

In this chapter, I have concentrated on the validity – or otherwise – of his arguments. I believe that the closest I came to presenting a valid "wager" argument with high fidelity to the text was in the first reformulation: salvation has infinite utility, damnation has negative infinite utility, supplemented by an additional decision rule. The subsequent reformulations, while valid, involved progressively greater departures from what I took to be his reasoning; I hope that they are still of independent interest. However, I have mostly set aside questions of the truth – or otherwise – of the premises of Pascal's own arguments and of the reformulations. As we tell our students: the validity of an argument is one thing, its soundness is quite another.

7 The Many-Gods Objection to Pascal's Wager: A Defeat, then a Resurrection

Craig Duncan[*]

Famously, Pascal's Wager purports to show that a prudentially rational person should aim to believe in God's existence, even when sufficient epistemic reason to believe in God is lacking. Perhaps the most common view of Pascal's Wager, though, holds it to be subject to a decisive objection, the so-called many-gods objection, according to which Pascal's Wager is incomplete since it only considers the possibility of a Christian God. I will argue, however, that the ambitious version of this objection most frequently encountered in the literature on Pascal's Wager fails. In the wake of this failure I will describe a more modest version of the many-gods objection and argue that this version still has strength enough to defeat the Canonical Wager.

The essence of my argument will be this: the Wager aims to justify belief in a context of uncertainty about God's *existence*, but this same uncertainty extends to the question of God's *requirements for salvation*. Just as we lack sufficient epistemic reason to believe in God, so too do we lack sufficient epistemic reason to judge that believing in God increases our chance of salvation. Instead, it is possible to imagine diverse gods with diverse requirements for salvation, not all of which require theistic belief. The context of uncertainty in which the Wager takes place renders us unable to single out one sort of salvation requirement as more probable than all others, thereby infecting the Wager with a fatal indeterminacy.

1 The Wager Defined

I will assume at the outset that prudential rationality requires one to perform the act, from among those acts open to one, with the highest expected utility. The techniques of decision theory allow us to encapsulate Pascal's Wager in the decision matrix laid out in Table 7.1.

[*] I would like to thank Paul Bartha and Lawrence Pasternack for extensive feedback on drafts of this chapter. Additionally, Rick Kaufman, Stephen P. Schwartz, Jonathan Peeters, and Paul Saka provided valuable feedback. Any mistakes that remain are my own.

Table 7.1 Pascal's Wager

	God exists (p)	God does not exist (1 − p)	Expected Utility (EU)
Choose God (G)	∞	f_1	$p \cdot \infty + (1-p) \cdot f_1$
Do Not Choose God (N)	f_2	f_3	$p \cdot f_2 + (1-p) \cdot f_3$

In this matrix, the variable p represents the probability that God exists. The quantities in each cell of the matrix represent the "utility payoff" of that cell. The variables f_1, f_2, and f_3 represent finite numbers, whereas ∞ stands for infinity. According to this matrix, God grants the reward of salvation to those who believe in God, and *only* to them. The matrix thus models the case of a god who practices *salvific exclusivism*, a term I will use to refer to a policy of excluding some people from salvation on account of their beliefs (or lack of beliefs). Note, too, that this matrix models a god who does *not* punish nonbelief with eternal torment, but rather simply annihilates (or fails to resurrect) non-theists. Supposing that p > 0, the expected utilities of the two options are

$$EU(G) = p \cdot \infty + (1-p) \cdot f_1 = \infty$$

$$EU(N) = p \cdot f_2 + (1-p) \cdot f_3 = \text{some finite value.}$$

Thus, according to the decision matrix above, choosing God is an infinitely superior option to not choosing God. And what is more, the option of believing in God is infinitely superior *regardless* of the value of p, so long as p > 0 – that is to say, so long as one cannot be *absolutely certain* that God does not exist.

That is a surprisingly strong conclusion, and indeed, one source of appeal of Pascal's Wager is that, if successful, it can rationally license belief in God even in the absence of sufficient epistemic reason to believe in God. I will call "canonical" any version of the Wager that purports to license a type of theistic belief despite there being insufficient evidence for that belief. I turn now to the question of whether the many-gods objection can defeat the Canonical Wager.

2 The Ambitious Many-Gods Objection

One of the earliest formulations of the many-gods objection to Pascal's Wager comes from Denis Diderot, who famously dismissed Pascal's Wager with the

Table 7.2 Two-God MGO

	Christian-favoring god exists(p_c)	Muslim-favoring god exists(p_i)	No god exists(p_n)	Expected Utility
Choose Christianity(C)	∞	f_1	f_2	$p_c \cdot \infty + p_i \cdot f_1 + p_n \cdot f_2 = \infty$
Choose Islam (I)	f_3	∞	f_4	$p_c \cdot f_3 + p_i \cdot \infty + p_n \cdot f_4 = \infty$
Choose Non-Theism (N)	f_5	f_6	f_7	$p_c \cdot f_5 + p_i \cdot f_6 + p_n \cdot f_7 =$ some finite #

brusque remark that "[a]n Imam could reason just as well this way."[1] Suppose, then, we were to transform the original Pascalian matrix into a two-gods matrix by distinguishing between a "Christian-favoring god" who saves only Christians and a "Muslim-favoring god" who saves only Muslims. Table 7.2 above is the result.[2]

The two-gods matrix appears to justify Diderot's point. Anyone who accepts this matrix (and supposes non-zero probabilities) will conclude that Muslim belief, like Christian belief, has infinite expected utility. Thus, this wager fails to single out Christian belief as uniquely rational. However, although this wager cannot discriminate between Christian and Muslim belief, it *can* convict the *non-theist* of prudential irrationality, since non-theism has only finite expected utility.

In reply, the non-theist could propose expanding the two-gods matrix so that it contains an extra column for an inclusive god who saves non-theists as well as Christians and Muslims. This makes the expected value of non-theism infinite, like that of Christian and Muslim belief. However, the defender of the Wager has a plausible response. For the defender may insist that non-theism

---

[1] Cited in Jordan (1994b, p. 101).
[2] A point of clarification: the "Choose Christianity" option in this matrix and others should be understood to be choosing any sort of recognizably Christian belief. Therefore, the Christian believer may believe in an exclusivist Christian god who favors only Christians, or (say) the Christian believer may believe in an inclusivist Christian god who saves all virtuous individuals regardless of their beliefs. Likewise, the Christian-favoring god is a god who favors Christians of any kind; thus, this god favors both Christians who believe in a Christian-favoring god and Christians who believe in an inclusive Christian god. We could in principle imagine more discriminating kinds of gods, such as a god favoring only those Christians who believe specifically in an exclusivist Christian-favoring god, or a god favoring only Catholic Christians, or Baptist Christians, etc. Indeed, one problem that supporters of Pascal's Wager face is the problem of keeping the relevant matrix down to a tractable size. However, rather than press this point against the Wager, I will keep things simple by assuming the Christian-favoring god in this matrix to be favorably disposed to all types of Christians.

Table 7.3 Three-God MGO

	Christian-favoring god exists(p_c)	Muslim-favoring god exists(p_i)	Skeptic-favoring god exists (p_s)	No god exists(p_n)
Choose Christianity (C)	∞	f_1	f_2	f_3
Choose Islam (I)	f_4	∞	f_5	f_6
Choose Non-Theism (N)	f_7	f_8	∞	f_9

remains a riskier bet than either theistic alternative. After all, only *one* state of the world generates an infinite "win" for the non-theist – namely, the state in which an inclusive god exists – whereas a theist "wins" in the inclusive god case PLUS the case of a god who favors his or her sort of theism.

Thus, a better reply from the non-theist is to expand the two-gods matrix by adding a "skeptic-favoring god," who at death saves non-theists and annihilates theists.[3] This leads to the perfectly symmetrical matrix in Table 7.3.

Note that of course the non-theist does not *believe* in the existence of a skeptic-favoring god; after all, such a belief would turn him or her into a type of theist rather than a non-theist. Instead, all the non-theist has to believe, in order to license the inclusion of this new column in the matrix, is that there is some non-zero *chance* that a skeptic-favoring god exists. And that is easy to believe, since it is commonly held that only logical contradictions should be assigned zero probability, with all other propositions having non-zero probability. The non-theist can then simply observe that, so long as the probability of a skeptic-loving god is greater than zero – even a mere speck greater than zero – the expected utility of non-theism is infinite. Thus, the non-theist can conclude that rejecting religious belief is just as prudentially rational as embracing it.

I will call "ambitious" any version of the many-gods objection that claims to be able to defeat Pascal's Wager so long as the probability of a skeptic-favoring god is non-zero. The ambitious many-gods objection thus tries to beat the Wager at its own game: just as the strongest version of the Canonical Wager purports to

---

[3] For examples of this reply, see Mackie (1982, p. 203), Martin (1983, p. 59), Oppy (1991, p. 165), and Blackburn (1999, p. 188). Jordan (2006, pp. 74–75) traces the first appearance of a skeptic-favoring god as a reply to Pascal's Wager to Leslie Stephen (Stephen, 1898, pp. 274–75). Although not explicitly addressing the Wager, David Hume (1998 [1779], p. 129) makes a similar point about whom God would favor. In his *Dialogues on Natural Religion* he writes: "And were that divine Being disposed to be offended at the vices and follies of silly mortals ... the only persons entitled to his compassion and indulgence would be the philosophical sceptics, a sect almost equally rare, who, from a natural diffidence of their own capacity, suspend, or endeavor to suspend all judgment with regard to such sublime and such extraordinary subjects."

show that belief in God is rationally required so long as God has a non-zero probability of existing, so too the ambitious many-gods objection claims to defeat the Wager so long as the probability of a skeptic-loving god is likewise non-zero.

3 Making Probabilities Relevant

I believe that the ambitious many-gods objection fails, for I believe that the theist can plausibly reject the claim that the precise probability of a skeptic-loving god is irrelevant so long as it is non-zero. I begin my argument for this claim with an analogy. Suppose as a gift you are given a choice between two lottery tickets, SAFE BET and LONG SHOT. In both lotteries, a ball is drawn from an urn containing a million marbles, only one of which is white, with the rest black. In the SAFE BET lottery, drawing a black ball earns you a million dollars; a white ball earns you nothing. LONG SHOT is the opposite: a white ball earns you a million dollars; a black ball earns you nothing. The clear rational choice is SAFE BET. Now suppose the odds remain the same but the prize is a trillion units of personal happiness. SAFE BET is now even more clearly the uniquely rational choice. Finally, suppose that the prize increases to ∞ units of happiness. Intuitively speaking, SAFE BET remains clearly rationally superior to LONG SHOT. And yet, owing to the infinite stakes, SAFE BET and LONG SHOT both have infinite expected utility. Thus, in terms of expected utility, decision theory tells you nothing about which is better.

Surely, though, it is absurd to regard the relative probabilities of an infinite win to be irrelevant. To avoid this absurdity, rational choice theorists need to find *some* way to show that the the relative probabilities of SAFE BET and LONG SHOT matter for the rationality of a choice between them. This need, moreover, has important implications for the Wager, since whatever method is found to show that probabilities matter in the choice between SAFE BET and LONG SHOT will surely also show that probabilities matter in the choice between the skeptic-favoring god, the Christian-favoring god, and the Muslim-favoring god. And then it will follow that non-theism is rational only if a skeptic-favoring god is more probable than both other gods, thereby defeating the ambitious many-gods objection.

But can a method be found of preserving the relevance of probabilities when infinite payoffs are involved? Unfortunately for supporters of the ambitious many-gods objection, I believe so. At least two ways suggest themselves: one might either modify standard rational choice theory or reject the claim that the happiness of salvation is equal to the extended real number ∞. Some defenders of the Wager have opted for the former option. For instance, Schlesinger (1994, pp. 89–90) and Jordan (2006, p. 104) propose adding a rule to rational choice theory stipulating that, when choosing among options that all have infinite expected utility, rationality requires one to choose the option most likely to yield infinite utility. This rule is quite plausible insofar as it instructs one to choose SAFE BET over LONG SHOT.

However, I believe theists would do just as well to consider the second option mentioned above, namely, rejecting the claim that the happiness of salvation is equal to ∞. As a lead-in to my description of this option, I suggest that the root intuition at the heart of the Wager is the idea that salvation is *incomparably good*.[4] By this I mean that the goods of this world allegedly pale into insignificance when compared with the good of salvation, or any chance thereof. Setting the utility of salvation equal to the mathematical notion of ∞ is *one way* to attempt to model this root notion of incomparable goodness. We must ask, though, whether there is another way of modeling this idea.

Indeed there is. My suggestion is that the idea of incomparable goodness can be adequately modeled by using an "arbitrarily large" finite number. Consider again the original Pascalian matrix in Table 7.1, and suppose we were to set the value of salvation at, say, $100^{100^{100}}$. Then even if the odds of God existing were quite small (say, 1.0×10^{-10}), the expected utility of theism would remain extremely large – large enough to decisively outweigh the relatively paltry this-worldly expected utility of non-theism. Thus, this-worldly utilities could still be said to be swamped into insignificance, thereby preserving the root intuition that salvation is an incomparable good. Moreover, setting the good of salvation equal to an arbitrarily large finite number does not make probabilities irrelevant when competing options each offer a chance of salvation. For imagine God were to appear to you and allow you to choose either the SAFE BET or LONG SHOT lottery ticket, with the prize being salvation. Then the expected utility of the SAFE BET ticket would be $0.999999 \cdot 100^{100^{100}}$ compared to $0.000001 \cdot 100^{100^{100}}$ for the LONG SHOT lottery ticket. Both have very high expected utility, of course – it is good to be offered either lottery! – but the SAFE BET lottery has nearly a million times more expected utility, and thus is clearly the rationally superior choice. I conclude that the idea of an "arbitrarily large" finite number works better than ∞ as a method of modeling the root intuition that salvation is incomparably good.

At least two objections to this way of modeling the incomparable good of salvation deserve consideration. According to the first objection, when setting

---

[4] Here I follow the method employed in Duncan (2007).

the good of salvation equal to (say) $100^{100^{100}}$, it doesn't necessarily follow that the finite utilities of this world are swamped into insignificance, as I earlier claimed. For what if, say, the probability of God existing is a miniscule $\frac{1}{100^{100^{100}}}$? Then the expected utility associated with salvation will be a paltry 1 unit, and thus salvation is hardly an incomparable good. However, this objection attaches too much weight to a specific assignment of value for the good of salvation. Instead, the good of salvation is meant to be "arbitrarily" large. This means that one is simply to choose a finite number large enough that any utilities associated with this-worldly happiness become insignificant. If $100^{100^{100}}$ does not achieve this, then simply increase it (e.g., use $100!^{100!^{100!}}$).[5] This reply, though, leads in turn to a second objection.

In an influential paper on Pascal's Wager, Alan Hájek briefly considers and rejects the method of assigning the good of salvation an arbitrarily large finite value.[6] His ground for rejecting this method is that it appears to transform God into a mere satisficer, which is inconsistent with the perfection that necessarily belongs to God's nature (Hájek, 2003, p. 45). After all, no matter what number one chooses for the value of salvation, there is always a higher number, and thus God could have constructed an even better salvation than the one on offer. Moreover, there also is the possibility of competing religious sects falling into a never-ending bidding war as each sect claims that its god offers a better salvific payout than its rival sects' gods.

This objection suffers from the same flaw as the first objection, namely, attaching an overly literal significance to the particular number assigned to represent the value of salvation. Whatever particular number is used, it is important to note that this number's only role is to *model* the incomparable goodness of salvation; the number is not meant to denote the precise quantity of this good. As such it is more accurate (that is, truer to the root intuition of incomparability) to think of this number as simply "arbitrarily large" rather than fix its value to a precise quantity. This blocks any charge that God is guilty of satisficing, and thwarts any attempt to initiate a salvific payout bidding war.

---

[5] What, though, if the probability of God's existence is to be judged "arbitarily small" (i.e., arbitrarily close to zero)? My preference is to regard the expected utility as indeterminate in such a case; another possibility is to judge the expected utility as equal to 1 (by using the same large finite number in the numerator and denominator). Neither result strikes me as an unintuitive verdict in the case of an arbitarily small chance of an arbitrarily large payout. Note too that models of the Wager that use the extended real number ∞ face a similar quandry with infinitesimals (Oppy, 1991, p. 163), so this is not an issue unique to my model.

[6] In addition to arbitrarily large finite numbers, Hájek also considers alternative methods such as using "surreal numbers" or assigning lexical priority to the good of salvation. My preference for arbitrarily large finite numbers over these alternatives is largely on grounds of simplicity.

4 The Modest Many-Gods Objection

I believe that the method of using an arbitrarily large finite number to represent the incomparable good of salvation is plausible enough that we have reason to explore its implications for the Wager. To begin with, we have seen that this method preserves the relevance of salvific probabilities, which in turn entails that the ambitious many-gods objection fails. In the remainder of this chapter, I will argue that in the wake of this defeat we can "resurrect" a successful version of the many-gods objection, which I will call the "modest many-gods objection." This is a more modest version since, unlike the ambitious many-gods objection, it does not aspire to establish the very strong claim that non-theism is prudentially rational so long as a skeptic-loving god has *any* non-zero probability.

Table 7.4 Three-God MGO with Finite Utilities

	Christian-favoring god exists(p_c)	Muslim-favoring god exists(p_i)	Skeptic-favoring god exists(p_s)	No god exists(p_n)
Choose Christianity (C)	H	f_1	f_2	f_3
Choose Islam (I)	f_4	H	f_5	f_6
Choose Non-Theism (N)	f_7	f_8	H	f_9

The matrix in Table 7.4, which I will use to explore the modest many-gods objection, uses H (as in "Heaven") instead of ∞ to represent the incomparable good of salvation; I stipulate that H represents an arbitrarily large finite number.

The resulting expected value equations are as follows:

$$EU(C) = p_c(H) + p_i(f_1) + p_s(f_2) + p_n(f_3)$$

$$EU(I) = p_c(f_4) + p_i(H) + p_s(f_5) + p_n(f_6)$$

$$EU(N) = p_c(f_7) + p_i(f_8) + p_s(H) + p_n(f_9).$$

Let us consider under what conditions Christian belief has greater expected utility than non-theism.[7] Given the above equations, it follows that:

$$EU(C) - EU(N) > 0 \text{ iff } [p_c(H) + p_i(f_1) + p_s(f_2) + p_n(f_3)] - [p_c(f_7) + p_i(f_8) + p_s(H) + p_n(f_9)] > 0.$$

---

[7] A full investigation would of course also have to compare EU(N) with EU(I), the expected utility of choosing Islam. I believe the arguments I use in my investigation of EU(C) versus EU(N) carry over to this case.

Thus,

$$EU(C) - EU(N) > 0 \text{ iff } H(p_c - p_s) + p_i(f_1) + p_s(f_2) + p_n(f_3) - p_c(f_7) - p_i(f_8) - p_n(f_9) > 0.$$

Let $f_a = p_c(f_7) + p_i(f_8) + p_n(f_9) - p_i(f_1) - p_s(f_2) - p_n(f_3)$. Then it follows:

$$EU(C) - EU(N) > 0 \text{ iff } (p_c - p_s) > \frac{f_a}{H}.$$

Note that f_a is a function purely of the this-worldly utilities associated with losing bets, discounted by the probabilities associated with the relevant gods. Thus, f_a will simply be some non-arbitrarily-large finite number. As a result, f_a will be dwarfed in size by the arbitrarily large value H, so that $\frac{f_a}{H}$ is in fact arbitrarily close to zero. Therefore, for all practical purposes, EU(C) is greater than EU(N) if and only if the quantity $(p_c - p_s)$ is greater than zero – that is, if and only if $p_c > p_s$.[8]

It will be my contention below that the defender of the Wager is unable to show that p_c is greater than p_s. That may seem surprising. After all, from a common-sense standpoint, there is surely reason to believe that p_s is extremely low. I will not dispute this, but in fact I will claim that an epistemically responsible agent should not judge p_c to be any higher than p_s.[9]

Two sorts of considerations, scriptural and moral, are especially relevant to an investigation of the relative values of p_c and p_s. Christian-favoring exclusivist gods of course have a scriptural advantage over a skeptic-favoring god, since the New Testament purports to provide evidence of the former, while no scriptures at all exist in favor of the latter. A full investigation of the question, then, would need to assess the quality of the New Testament's testimonial evidence. That is a large task, and I will not attempt it here. Instead I will content myself with noting that a number of reasons tell against trusting the New Testament's testimonial claims: the existence of various internal inconsistencies in the Gospel narratives;[10] the fact that the Gospel narratives were

---

[8] Note that this means the essence of the Canonical Wager is preserved, since theistic belief can be prudentially required even when the odds of God existing are less than 50%. For instance, if (say) $p_c = 0.00002$ and $p_s = 0.00001$, then Christian belief generates incomparably more expected utility than non-theism.

[9] Recall that the Schlesinger/Jordan rule mentioned in the previous section is an independent way of establishing the relevance of probabilities when payoffs are infinite. Adopting it instead of the method of arbitrarily large finite numbers would thus be an independent way of motivating the inquiry that I now begin, namely the inquiry into whether $p_c > p_s$. As such, the argument of the remainder of this chapter is independent of my specific claims in this section regarding arbitrarily large finite numbers.

[10] Cf. Ehrman (2009).

written down by non-eyewitnesses, and only decades after the occurrence of the events they purport to describe;[11] the all-too-human process by which some texts were canonized for inclusion in the New Testament while other texts (including alternative "gospels") were rejected;[12] the general Humean argument against believing testimonial reports of miracles;[13] and the existence of competing scriptures and miracle claims in other religions. Rather than explore the intricacies of these considerations, I will simply register my belief that one can, with full epistemic propriety, judge that the Scriptures constitute, at best, *extremely* weak evidence for the existence of an exclusivist Christian god. Readers may disagree with this assessment of Scripture, but for the sake of argument I invite such readers to see what follows if my assessment is correct. Prima facie, even weak evidence for Christianity appears to be a fatal blow to the non-theist, since even a slight advantage of p_c over p_s is enough to generate incomparably greater expected utility for the option of believing in a Christian-favoring god. Thus, the non-theists' only hope for acquitting themselves of the charge of prudential irrationality is to find a compensating advantage that p_s enjoys over p_c, enough to cancel out Christianity's scriptural advantage.

I believe this hope can be fulfilled via a moral case against salvific exclusivism. In particular, I will argue that the favoritism shown by the salvific exclusivist gods in the Many-Gods matrix is an immoral sort of favoritism. Of course, even if I am right, this moral objection cuts against *both* the Christian-favoring and skeptic-favoring gods, since both types of gods are exclusivist. However, I will argue in the next section that a skeptic-favoring god's exclusivism is in fact *less* immoral than a Christian-favoring god's exclusivism. The greater moral plausibility of the skeptic-favoring god, I will argue, raises its odds of existing relative to the Christian-favoring god, thereby potentially counteracting the scriptural advantages of the Christian-favoring god. As a result, the most sensible response is to refuse to judge either one of p_c or p_s to be greater than the other, and judge instead that the task of determining their relative value is beyond our epistemic ability.

5 The Morality of Salvific Exclusivism

Let us examine, then, the moral case against salvific exclusivism. This is easiest to see in the case of a punitive god who condemns to eternal torment those

---

[11] Cf. Ehrman (2016, 2014). [12] Cf. Ehrman (2003).
[13] Hume (2000 [1748], pp. 109–31, Section X, "Of Miracles").

whom he does not save.[14] It is extremely doubtful whether even the most grievous wrong by any *finite* human being could warrant a strictly *infinite* punishment. But if even the most execrable wrong-doer fails to deserve eternal punishment, how could any person deserve such a fate merely on account of his or her morally innocent beliefs? That seems impossible, and thus a punitive exclusivist god, in virtue of meting out undeserved punishment, is an unjust god. (I assume for the time being that both non-theism and theism are morally innocent forms of belief. Later in this chapter I will scrutinize this assumption.) The immorality of punitive salvific exclusivism in turn contradicts the traditional idea of God as a perfect being: the idea of a morally perfect being who eternally torments people on account of their morally innocent beliefs is no more coherent than the idea of an omnipotent being who cannot make a pizza, or an omniscient being who does not know the thousandth digit of Π. Therefore, a punitive, salvifically exclusivist god cannot be *God*.

What, though, about a *non*-punitive salvifically exclusivist god, who merely annihilates the unsaved? Such are the gods in the matrices above. Might *that* sort of god be moral? I do not believe so. For starters, it is quite easy to conceptualize annihilation – and the consequent loss of an incomparable good – as itself a type of punishment. (Loss of privileges is a quite familiar form of punishment, after all.) Moreover, although it is true that such a god does not eternally *torment* disfavored people, let us note that such a god does refuse to grant the incomparably good reward of eternal salvation to some people on account of their morally innocent beliefs. That is morally arbitrary, and thus morally imperfect. Therefore, while a policy of differential salvific reward based on morally innocent beliefs is certainly *less* morally bad than eternally punishing such people, it is by no means clear that a morally *perfect* being could enact such a policy.

However, let us consider one interesting argument to the contrary due to Philip Quinn (1994, pp. 74–78).[15] Quinn begins by agreeing with critics of the doctrine of hell that no finite being can deserve infinite punishment. Quinn then cleverly inverts this claim: just as no finite being can deserve infinite punishment, it is likewise true that no finite being can deserve *infinite reward*.

---

[14] We could expand the matrices above to include *both* gods who annihilate the unsaved *and* gods who condemn the unsaved to hell. However, the moral argument that I present below, which blocks the conclusion that $p_c > p_s$, applies with even more force to the appalling case of a god who sends the unsaved to hell. Thus, including such a god does not help the defender of the Wager.

[15] Quinn is responding to Terence Penelhum, who argues that it is immoral to wager in Pascalian fashion, on the grounds that Pascal's exclusionary god is immoral and thus a wagerer is complicit in that divine immorality (Penelhum, 1971, pp. 216–18). I agree that a salvific exclusivist god is

Salvation is thus in every case a gracious, undeserved gift. As such, says Quinn, considerations of desert do not constrain the distribution of salvation. No one who is denied salvation can rightly complain that he or she failed to get a deserved reward, and hence, no one who is denied salvation can rightly complain that he or she was treated unjustly. Quinn concludes from this that God does not act unjustly in denying the gift of salvation to non-theists.

Quinn's argument, though, presumes that the act of giving gifts in excess of desert is immune to moral evaluation. But this is false. For example, a teacher may graciously distribute Halloween candy to her students. If the students cannot be said to *deserve* this treat, then it is true that the teacher could not have been faulted had she chosen not to distribute any candy to anyone. But suppose she gives the candy to all the kids except for the kid in the corner with red hair (she dislikes red hair, say). Then the teacher has acted arbitrarily and thus can be morally faulted. For imagine there is some threshold on a scale of deservingness above which a kid may be said to deserve some reward such as Halloween candy in the classroom. Though we have supposed that all students fall *below* this threshold, I suggest that to deny candy only to the red-haired kid is in effect to treat her, merely on account of her red hair, as falling *further* below the threshold of desert than the others. But that would be a moral mistake, for red hair makes one no less deserving of treats. Similarly, I suggest that an individual with a morally innocent belief is no less deserving of a gracious gift of salvation than any other finite being, and thus that any god who denies such a person salvation while granting it to others has made a moral mistake.[16] I conclude that a morally perfect god cannot be a salvifically exclusivist god.

If I am right that the idea of a morally perfect, salvifically exclusivist god is a contradiction in terms, then it may seem that such gods must have strictly *zero* probability of existing, since logical contradictions have zero probability of being true. And if that is the case, then it would seem that the moral objection voiced above to salvific exclusivism is enough to defeat Pascal's

---

immoral. But this does not necessarily make the wagerer likewise immoral, since (as I noted earlier in fn. 2) one can wager for God without being an exclusivist oneself; e.g., one can be an inclusivist Christian who also believes there is merely a small (but non-zero) chance that God is an exclusivist who favors Christians of all stripes (including inclusivist Christians).

[16] In the literature on desert it is now common (since Feinberg, 1974) to distinguish between *non-comparative desert* (which judges people against some absolute standard of deservingness) and *comparative desert* (which judges people relative to others' success or failure in heeding the relevant standard). I am suggesting that although moral perfection may permit merciful departures from non-comparative justice, it requires that such departures comply with comparative justice. (If God saved all but redheads, say, that would be unjust!)

Wager by itself, without the need for any reasoning along the lines of the many-gods objection. However, things are not so simple. For at this point we should distinguish *objective probability* from *epistemic probability*. If a state of affairs is logically impossible then it has objective probability 0, to be sure. But it may be that an agent does not have enough evidence to *know* that a given state of affairs is logically impossible. If so, then although in fact that state of affairs has objective probability 0, the agent should not assign it epistemic probability 0, but rather should assign it some non-zero probability. By way of illustration, consider a mathematical analogy. Goldbach's Conjecture states that every even integer greater than 2 can be written as the sum of two primes. If it is true, then it is necessarily true, and thus has objective probability 1. However, although we have strong inductive evidence for the truth of the conjecture (computers have so far failed to find a disconfirming instance), as of yet mathematical proof is lacking. As a result, no human agent is in a position to assign *epistemic* probability 1 to Goldbach's Conjecture. At most, an epistemic probability very close to 1 is warranted.

To clarify: by "objective probability" I mean a *wholly* mind-independent type of probability that does not vary at all from agent to agent. By "epistemic probability" I mean a probability assignment that reflects the degree to which the proposition in question is supported by the evidence that the judging agent possesses.[17] Since different agents often possess different evidence, epistemic probabilities will often vary from agent to agent; to that extent they are "subjective." However, epistemic probabilities are not *wholly* subjective, since they are constrained by plausibility arguments and logic. For instance, if agent A possesses evidence E that strongly supports proposition p, but A assigns p low epistemic probability, then A's probability assignment is flawed.[18]

Since the arguments for the immorality of salvific exclusivism are very strong, the arguments for the impossibility of a morally perfect, salvifically exclusivist god are likewise very strong. Such gods may indeed have zero *objective* probability of existing. However, I am not willing to say that the anti-exclusivist arguments leave absolutely no room for even the tiniest speck of

---

[17] For further discussion of kinds of probability, including epistemic probability, see Skyrms (2000, pp. 23–26, 137–50) and Mellor (2005, pp. 11–12, 80–90).

[18] This potentially blocks Jeff Jordan's (Ch. 5 of this volume) appeal to the idea of a "live option" as a means of rebutting the many-gods objection. If agent A has *any* evidence at all in support of p, then A rationally ought to assign p epistemic probability >0. Thus, in such a case A is *not* rationally permitted simply to declare that p "is not a live option for me" and assign it probability 0.

doubt. For starters, many religious believers make the (implausible, but not *certainly* false) claim that "God's morality" is wholly distinct from human morality. Additionally, Quinn's argument defending the propriety of salvific exclusivism is not *wholly* unreasonable. While I have provided a rebuttal to his argument, I cannot claim to have shown beyond all doubt that a morally perfect, salvifically exclusivist god is a logical impossibility. Thus, I conclude that although the moral argument against salvific exclusivism drastically lowers the values of p_c and p_s, understood as *epistemic* probabilities, it fails to shrink these quantities all the way to zero.

In one way that is a disappointment to the non-theist (it would have been nice if the moral argument against salvific exclusivism had been able to defeat the Wager all by itself). However, the non-theist should still be heartened by the fact that the moral argument lowers the probabilities p_c and p_s very close to zero – a small fraction of 1 percent at most, I would say. We might express this point by saying that there is an extremely low "probability ceiling" on all types of salvifically exclusivist gods.[19] Therefore, the odds of a Christian-favoring god, if they are indeed greater than the odds of a skeptic-favoring god, could only possibly be greater by just a tiny speck – some fraction of the distance between zero and the extremely low probability ceiling just mentioned. Still, the scriptural advantages of the Christian-favoring god arguably provide just that speck of increased probability, and the incomparable good of salvation then converts this slight probability advantage into an incomparable advantage for Christian belief in terms of expected utility. Non-theists thus need to find a compensating epistemic advantage enjoyed by skeptic-favoring gods, enough to create a comparable speck of increased epistemic probability for p_s.

A further stretch of moral argument provides this needed speck. Key to this argument is the dialectical context of the Canonical Wager, which assumes a lack of sufficient evidence for belief in God (so that the pragmatic considerations of the Wager are needed to lead rational people to such belief). In this context, a noteworthy asymmetry between the Christian believer and the non-theist exists. For in such a context the non-theist is complying with epistemic rationality by refusing to form theistic beliefs in the absence of sufficient evidence, whereas the Christian is forging ahead and embracing such belief despite this absence. This gives the non-theist a credible claim to be more

---

[19] Note that I am only officially committing myself here to *salvifically exclusivist forms of Christianity* being much less than 1% likely to be true, not all forms of Christianity.

epistemically virtuous than the Christian believer.[20] Thus, a skeptic-favoring god's salvific exclusivism is in fact a policy that rewards a type of virtue. As a result, I suggest that this god's salvific exclusivism, while still immoral, is *less* morally appalling than the exclusivism of the Christian-favoring god. By pursuing a less morally appalling salvific policy, the skeptic-favoring god is in that respect a more plausible god than the Christian-favoring type.

By way of illustrating this point, consider the following contrasting pair of questions:

A. Would a morally perfect god (i) provide insufficient evidence for the truth of Christianity and then (ii) make humans' chance of salvation depend on their possessing Christian beliefs?
B. Would a morally perfect god (i) provide insufficient evidence for his own existence and then (ii) make humans' chance of salvation depend on their respecting this lack of evidence as they form their beliefs?

While I believe both questions warrant an answer of No, surely it is *less* preposterous to answer Yes to B than to A. This relative boost in plausibility for the skeptic-favoring god, I suggest, is the needed offset to the scriptural advantage enjoyed by the Christian-favoring god.

Thus, the overall situation is this: the extremely low probability ceiling that stems from the moral implausibility of all types of salvific exclusivism means that, if a skeptic-favoring god and a Christian-favoring god differ at all in their probability of existing, then this difference is at most a tiny fraction of a percentage point. The question then becomes whether in this very narrow space one type of god is more likely than the other. On the one side of the balance, scriptural considerations favor a Christian-favoring god, whereas on the other side of the balance, moral considerations favor a skeptic-favoring god. I do not believe that the human ability to estimate probabilities is capable of the exquisite precision it would take to judge with any confidence that, say, the scriptural considerations weighed against the moral considerations leave a 0.001 percent greater chance of the

---

[20] For an overview of "virtue epistemology," see Greco and Turri (2016). Perhaps futher investigation within "the ethics of belief" would reveal that epistemic virtues and vices are simultaneously *moral* virtues and vices. If so, then my argument grows stronger, since in the dialectical context of the Wager, Christian belief would then be a moral vice, and the skeptic-favoring god would thus not be a god who excludes people from salvation on account of morally innocent beliefs (theism not being morally innocent). However, my argument does not *require* this claim. Even if epistemic virtue is exclusively a non-moral excellence, simply by being an excellence it remains relevant to moral desert. E.g., a teacher whose judgments of deservingness track student academic excellence is morally preferable to a teacher who grants and withholds rewards independent of *any* excellence-tracking criteria.

Christian-favoring god existing than the skeptic-favoring god, or vice versa. Instead, the only epistemically responsible verdict is to judge that, given how extremely unlikely each god is, we cannot reliably know which type of god is slightly more likely than the other. Thus, the modest form of the many-gods objection achieves the same result that the ambitious form sought, namely, it shows Pascal's Wager to be indeterminate. The Wager does not single out either wagering for God, or wagering against God, as prudentially superior.

6 Objections

There is bound to be resistance to my argument that we are not epistemically equipped to settle the probabilities between a skeptic-favoring god and a Christian-favoring god. So much history exists in support of the latter god that many readers will naturally think it superior in probability to the former. In this section I consider possible objections to my argument that might occur to such readers. In replying to these objections, my goal is not to show the superiority of p_s over p_c, but rather to raise enough doubts about p_c to lead readers who would otherwise favor p_c to accept that it is no easy matter to determine its relative size compared to p_s.

6.1 Objection 1: Forgiveness, Not Belief

According to this objection, the test for salvation is not fundamentally *belief-based*, but rather is *forgiveness-based*. That is to say, what matters for salvation is whether you have asked for and received God's forgiveness for your sins. Belief in God is a necessary *element* of this process (one doesn't ask for forgiveness from an entity one does not believe exists), but it is the forgiveness itself that is saving, not the belief. And since the presence of forgiveness *is* a change in creaturely moral status, it is thus *not* morally arbitrary for God to favor those who have sought and received forgiveness from God, and to reserve salvation exclusively for them.

This argument deserves a fuller inquiry than I have space to provide here. I remain skeptical of its promise, however. What matters most from the moral point of view, I suggest, is that a wrong-doer recognizes the wrong done and shows contrition/remorse. A divine being who views all such contrition as nugatory unless supplemented with an explicit request to itself (the divine being) for forgiveness is a divine being in the grip of either a morally dubious legalism or a morally unbecoming narcissism. As such, it is highly doubtful that a morally perfect being would place such a requirement at the heart of its salvific policy. Moreover, the non-theist can plausibly argue that since an immoral act wrongs the party who is mistreated by the act, morality demands that forgiveness for one's wrongs should be sought from the fellow beings whom one wronged, to the extent that this is possible. No divine intermediary is needed for a morally worthy form of forgiveness to be possible.[21]

6.2 Objection 2: A Theist-Favoring God

Might a supporter of the Wager do well to consider a distinct type of salvifically exclusivist god, namely, one who favors *all theists* (on some suitably inclusive definition), not just Christians? Such a god is less arbitrarily exclusionary than a Christian-favoring god, it might be said, and therefore the moral objections against such a god are correspondingly lessened.

Perhaps so, but not all moral objections disappear, and I believe that the moral objections which remain against a theist-favoring god are still stronger than the moral objections against a skeptic-favoring god. After all, the dialectical context of the Canonical Wager assumes that arguments of natural theology (e.g., the cosmological, design, and ontological arguments) are insufficient to establish God's existence. And surely it is highly unlikely that a morally perfect god would (i) provide insufficient evidence for the truth of theism and then (ii) make humans' chance of salvation depend on their belief in theism. By contrast, in the sphere of religious belief, non-theists possess more epistemic virtue than theists, and thus the skeptic-favoring god's favoritism is not *wholly* morally arbitrary.

Additionally, note that the scriptural advantages enjoyed by a Christian-favoring god are significantly sacrificed in the case of a switch to a theist-favoring god. After all, the arguments of natural theology do *nothing* to support the claim that the god who exists saves all and only theists in the next life. And the scriptures of revealed religions typically describe a deity who shows favoritism toward people who believe

---

[21] Theologians have formulated detailed doctrines of God's saving grace that differ significantly from popular "ask for forgiveness and be saved" versions of Christianity. For instance, on
a strict Augustinian version of the the Fall, humans in their corrupted nature can do nothing themselves to be saved, and instead are wholly dependent on a prior and entirely unmerited act of God's grace. "Asking forgiveness" on this view is not the *cause* of one's salvation; rather, it is the *effect* of God's saving and unmerited grace already at work within oneself, for it is only God's prior act of grace that leads one to seek forgiveness at all. The logic of this doctrine, though, leads to
a predestinarianision vision of salvation that seems at odds with a presupposition of the Wager, which supposes that there are indeed actions that one can take to increase one's chances of salvation.

in those particular scriptures, as opposed to a favoritism directed toward all theists but no non-theists. Therefore, all things considered, it is by no means clear that a theist-favoring god is any more probable than a skeptic-favoring god.

6.3 Objection 3: Morally Superior Christians?

According to this objection, even if non-theists are superior to theists in terms of epistemic virtue, it is possible that along other dimensions of virtue Christians possess more virtue on average than skeptics possess. Christianity, the objection claims, offers time-tested moral instruction and motivation, and so we should expect Christians on average to be more morally virtuous than non-theists. From this claim, the objection concludes that (taking all types of virtue into account) a Christian-favoring god is no less rewarding of virtue than a skeptic-favoring god is, and thus, a Christian-favoring god is no more morally objectionable than a skeptic-favoring god. As a result, and contrary to what I have argued above, moral considerations confer no probability advantage on the skeptic-favoring god.

Several problems beset this line of reasoning, however: (i) It does not accord well with traditional Christianity, which does not typically promise to transform sinners into moral exemplars, but instead is all too conscious that sinners remain sinners. (ii) It is empirically doubtful that non-theists on average have lower levels of moral virtue than theists.[22] (iii) Even if this objection works in its own terms, it still fails to show that non-theism is *inherently* prudentially irrational. At most, this objection shows that serious vice is inherently prudentially irrational, but leaves it up to individuals to take effective practical steps to avoid such vice – and these practical steps needn't involve inculcating religious belief in oneself, if one succeeds in finding other means of avoiding vice.

Thus, an objection alleging moral superiority on the part of theists will successfully rescue the Wager only if it can be shown that non-theism *itself* is inherently immoral. Some theists do claim this. For instance, William Lane Craig (1994, pp. 35–36) writes:

> When a person refuses to come to Christ it is never because of a lack of evidence or because of intellectual difficulties: at root, he refuses to come because he willingly ignores and rejects the drawing of God's spirit on his heart. No one in the final analysis fails to become a Christian because of lack

---

[22] I review some of this evidence in Duncan (2013), pp. 392–93. Cf. Decety et al. (2015).

of arguments; he fails to become a Christian because he loves darkness rather than light and wants nothing to do with God.

This strikes me as a grossly implausible and unfair generalization. Far from revealing the immorality of non-Christians, it instead casts Craig's own brand of Christianity in a morally dubious light. What is more, even if correct, it would be irrelevant in the dialectical context of the Canonical Wager. For Craig's charge is that non-Christians reject the ample evidence that God furnishes, but we have noted that the Wager's context presupposes a lack of rationally sufficient evidence for God's existence. If by stipulation we agree that the evidence for God's existence is inconclusive, then I fail to see how non-theism *per se* could possibly be immoral.

A general lesson to draw from Craig's quotation is that Christian salvific exclusivism typically supposes that God has provided sufficient proof of Christianity. This observation in turn helps us to diagnose a chief failing of the Canonical Wager: it aims to combine (i) an assignment of significant probability to Christian salvific exclusivism – which only really makes sense if sufficient evidence for Christianity *does* exist – with (ii) an assumption that sufficient evidence for Christianity does *not* exist (thereby necessitating a pragmatic argument for becoming Christian). Unsurprisingly, this combination proves to be untenable. This same general observation also helps offset a possible incredulous reaction to my argument: "You ask us to agree that Christian salvific exclusivism is obviously unjust," a critic might say. "And yet that implausibly implies that for thousands of years Christian theologians – some of the brightest thinkers of their times – overlooked this obvious injustice, and indeed, organized their lives around this injustice instead of calling it out." My response is that these thinkers did not suppose themselves to be in the dialectical context of the Wager, but instead supposed themselves to have sufficient evidence for the truth of Christianity. Had they themselves judged the proofs of Christianity to be inadequate, many of them would likely have taken note of the injustice of Christian salvific exclusivism.

6.4 Objection 4: Imperfect Gods

This objection questions whether moral objections to exclusivist gods can really acquit the non-theist of a charge of prudential irrationality. According to this objection, we must consider the hypothesis of *imperfect* deities, or even demons, who favor Christians in the afterlife, or who favor skeptics in the afterlife. A morally flawed afterlife policy doesn't count against the existence of an imperfect deity, after all. In particular, moral considerations do not make an imperfect skeptic-favoring god any more likely than an imperfect Christian-favoring god; the skeptic-favoring god thus loses its probability advantage in this regard.

However, while true, this fact is of little help to the Wagerer, for an imperfect Christian-favoring god also loses its scriptural advantage over the skeptic-favoring god. After all, an *imperfect* deity who favors Christians in the afterlife is distinct from the god of Christian tradition (what Christian would say that the god whom he or she worships is morally flawed?). Christian Scriptures thus don't count in favor of such a god. Accordingly, there is no reason to think that either type of imperfect god is more likely than the other type.

7 Conclusion

Let us take stock. My exploration of the many-gods objection led ultimately to the matrix in Table 7.4, which uses arbitarily large finite values to represent the incomparable good of salvation. Rejecting the ambitious form of the objection (which claims that Pascal's Wager is defeated so long as a skeptic-favoring god has *any* non-zero probability), I defended a modest form of the objection. This form entails that the prudential rationality of non-theism turns on the question whether a Christian-favoring god is more likely to exist than a skeptic-favoring god, that is, on whether $p_c > p_s$. However, the probabilities p_c and p_s describe something quite preposterous (and hence highly doubtful), namely, a morally perfectly being who excludes some people from salvation on morally innocent grounds. As a result, if one of p_c or p_s is greater than the other, then the absolute difference between the two quantities will be extremely tiny, since both quantities are themselves already very tiny. Compounding the difficulty of judging which quantity is greater is the fact that competing evidential considerations pull in opposite directions. I have suggested that the scriptural advantage enjoyed by the Christian-favoring god is weak evidence at best, and that this advantage is in any case cancelled by the moral advantage enjoyed by the skeptic-favoring god who, in the dialectical context of the Canonical Wager, is a rewarder of epistemic virtue. All things considered, then, epistemically responsible agents are in no position to judge one of the probabilities p_c and p_s to be greater than the other. Therefore, skeptics can rationally reject the Wager's conclusion that the prospect of salvation makes non-theism imprudent, and believers are unable to rationally rely on the Wager to justify their theistic views.

8 The Wager as Decision under Ignorance: Decision-Theoretic Responses to the Many-Gods Objection

Lawrence Pasternack

1 Introduction

The many-gods objection (MGO) is among the earliest, best-known, and most frequently discussed objections to Pascal's Wager. In the mid-eighteenth century, both Diderot and Voltaire raised the concern that there is more than one theological hypothesis upon which to wager (Diderot, 1875 [1762], ¶59; Voltaire, 1971 [1764], p. 280) and in more recent years, MGO has been expanded to include "cooked-up" hypotheses and "philosophers' fictions":[1] gods who require us to step on every third sidewalk crack (Gale, 1991), gods who punish us in proportion to the number of insects we have killed (Saka, 2001),[2] gods who only reward those who prefer Chardonnay to all other wines (Anderson, 1995), and gods who require that we reject theism (Kauffman, 1978; Mackie, 1982; Martin, 1983).

There have been various attempts to defend the Wager against MGO, charging that "cooked-up" hypotheses (a) do not uphold one or another metaphysical feature associated with authentic religion such as simplicity, perfection, worship-worthiness, or some other "exalted notion" (Lycan and Schlesinger, 1996; Schlesinger, 1994; Golding, 1994; Anderson, 1995; Nemoianu, 2010); (b) do not have the backing of tradition (Jordan, 2006); or (c) would not be taken as a "live hypothesis" by a sincere wagerer (Jordan, Chapter 5 of this volume).[3]

Rather than turning to (a)–(c), or other considerations outside the Wager itself, this chapter will explore a different kind of strategy: viz., whether characteristics already built into the Wager's decision-theoretic approach to religious belief can be used to fend off various categories of "cooked-up"

---

[1] These are terms used by Jeff Jordan. He migrates from the former to the later over time. See fn. 8.
[2] Saka's example is taken from DeGeneres (1995).
[3] For related discussions on limiting the Wager to a specific audience, see Rescher (1985); Franklin (1998). See also Holyer (1989).

hypotheses and "philosophers' fictions." We will begin with a brief presentation of the Wager and the many-gods objection. Next, we will briefly review the strategies (a)–(c) mentioned above. The remainder of the chapter will then examine some features of the Wager's decision-theoretic structure and how they might be used against MGO.

2 Pascal's Wager and the Many-Gods Objection

The canonical formulation of Pascal's Wager follows what Ian Hacking calls "dominating expectation" (Hacking, 1972). An act is said to have a dominating expectation when:

(α) The probabilities of each outcome in a decision matrix are > 0, but no specific probability value is assigned to them.
(β) On one admissible probability assignment, one act has an expected utility greater than any other act.
(γ) There is no admissible probability assignment on which that act has an expected utility less than any other act.

Item (α) further reflects a feature of the Wager important for this chapter: it is classically represented as a "decision under ignorance," as we do not know or have to know the relevant probabilities aside from the claim that the one at issue is greater than zero. As we will see, this feature is very much what is at stake in the difference between the strategies listed as (a)–(c) in the Introduction versus the decision-theoretic strategies developed later in this chapter. We will return to this issue in sections 3 and 4. But for now, let us formulate the basic decision matrix for the Wager (see Table 8.1).

The theistic choice[4] results in either $+\infty$ (a heavenly afterlife of infinite value) if one is correct or, if one is incorrect, some finite value which our mortal lives

Table 8.1 Pascal's Wager

	God exists	God does not exist
One chooses God	$+\infty$	f
One does not choose God	f	f

---

[4] What is being chosen may be a belief, or the intention to cultivate a belief, to enact various rituals, to adhere to moral rules, etc. Note, however, that not all alternatives to belief preserve the Wager's validity. See Hájek's discussion of "mixed strategies" in Chapter 6. Further, Pascal recognizes that we cannot merely choose to believe. We can, however, choose to take actions which may engender a belief and take actions which could make one a recipient of grace (see L418/S680, L7/S41, L944/S767). See Moriarty (2003, pp. 152–57). The latter issue is discussed in Chapters 2 and 3 of this volume.

are assumed to ordinarily have. The atheistic choice results, if incorrect, in either eternal suffering if hell is assumed, or some finite value if a cessation of existence at one's death (for simplicity, let us assume the latter). The atheistic choice results, if correct, in some finite value that our worldly existence is assumed to ordinarily have. Lastly, let the finite value in the bottom right cell be greater than the one above so we do not have a simple dominance argument for choosing God.[5]

Assuming that the probability of God's existence is > 0, then the expected utility of the choice to be a theist is $+\infty$ whereas the expected utility of the choice to not be a theist is either $-\infty$ if hell is assumed, or some positive finite value if non-existence is assumed. Thus, the Wager offers a prudential argument for theism: however improbable God's existence is, so long as it is not impossible, then the slimmest chance of an infinite reward makes its pursuit preferable to a choice to not accept God.

The best-known objection to the Wager is that it should not be presented as a binary choice: there are many religions and many claims about what is required to receive a heavenly reward. Let us use the term "theological hypothesis" (TH) to refer to the set of claims pertaining to supernatural beings who putatively play a role in our afterlife fate, the nature of that fate, and what is within our power with respect to that fate (rituals, beliefs, moral conduct, etc.). Further, at least some theological hypotheses carry exclusivity clauses such as the requirement that all our acts of piety must be directed toward a particular deity and a prohibition against performing any actions for the purpose of winning favor with any other deity.

With such clauses in effect, let TH_1 be the traditional theological hypothesis broadly associated with Abrahamic religions[6] and let TH_2 be the theological hypothesis of a divinity who requires that we always wear purple slippers while indoors and prohibits us from wearing any other footwear (or none at all) while indoors.[7] Of course, MGO's matrices can stretch to infinity, given all the theological hypotheses that can be "cooked up." But for the sake of simplicity, let us limit our portrayal of the MGO decision matrix to just three options (see Table 8.2).

---

[5] One assumption which will not be discussed is the relative finite value of life for the theist and the atheist (the so-called "mundane wager"). We grant the relative finite values above in order to follow the standard dominating expectation formulation of the Wager.

[6] There are, of course, important differences between Christianity, Judaism, and Islam. Pascal intended (Augustinian/Jansenist) Christianity. Contemporary versions of the Wager often default to an ecumenical theism.

[7] Through this chapter, I will use this theological hypothesis as our default example, capable of standing in for most other "cooked-up" hypotheses and "philosophers' fictions."

Table 8.2 MGO

	TH$_1$ is true	TH$_2$ is true	TH$_1$ and TH$_2$ are false
One chooses TH$_1$	+ ∞	f	f
One chooses TH$_2$	f	+ ∞	f
One does not choose any TH	f	f	f

According to MGO, if the probabilities of TH$_1$, and TH$_2$ are each > 0, and f again reflects some finite value, then the expected utility of the top two rows are equivalent. To simplify the math, if we assume non-existence rather than hell if one is wrong, the two top rows would each sum to + ∞ and the third row would sum to some finite value. Thus, theism is supported over atheism. However, there appears to be no decision-theoretic solution as to which deity, religion, or set of rituals to choose, leaving the wagerer without the means to determine which theological hypothesis is in his/her best interest. If limited to just what is contained in the Wager, it appears as if the wagerer is stuck, unable to choose between a traditional theological hypothesis and an outlandish "cooked-up" hypothesis about wearing purple slippers.

3 Wager Buttressing

In section 4, we will explore how the decision-theoretic structure of the Wager can itself exclude various MGO hypotheses. But first, let us consider the more common approach, where these hypotheses are barred from the Wager's matrix due to their own shortcomings, independent of the Wager itself. As noted in the Introduction to this chapter, these strategies include arguments to the effect that: (a) "cooked-up" deities do not uphold some metaphysical feature associated with authentic religion; (b) they do not have the backing of tradition; and (c) they would not be taken as "live hypotheses" by a sincere wagerer. We will briefly consider two general difficulties with (a)–(c) before discussing a third in more detail.

The first of these difficulties is simply that they are inelegant: rather than having the Wager itself manage MGO hypotheses, they instead seek to bar them through some independent consideration. Behind this, however, is a deeper concern: if the Wager needs such independent considerations, there seems to be an implicit concession to MGO, granting that the Wager on its own is not able to make the case for traditional theism. Lastly, and through the remainder of this section, we will consider whether the strategies used in (a)–(c) to bar MGO's "cooked-up" hypotheses may likewise bar atheism as well from the Wager's matrix.

With regards to (a), perhaps the most serious objection is that appeals to metaphysical considerations in order to support traditional theism dampen the Wager's form as a decision under ignorance. But in addition to this, the use of such considerations may very well militate against atheism as well, for just as the deities of "cooked-up" hypotheses lack some "exalted notion," atheism likewise presents a cosmology where the traditional theist's "exalted notions" are still absent. In other words: strategy (a) may have the unintended consequence of not merely barring "cooked-up" hypotheses from the matrix, but on the same grounds that they are barred, atheism may be as well.

What then of (b), the appeal to tradition? Presumably, such appeal is not merely to whether or not there is a tradition backing some hypothesis, but it is rather about whether or not that hypothesis has the merit of having been vetted by our "epistemic peers." For otherwise, the Wager could include the numerous traditional deities who have demanded human sacrifice or other such terrors – returning us, in effect, to just the sort of objectionable gods found in MGO's "cooked-up" hypotheses.

By contrast, if we understand the appeal to tradition as one which concerns epistemic peerage, an important question emerges for our inquiry: do the intellectuals of centuries past count as our epistemic peers? If so, then it seems atheism is again on the chopping block, for the great thinkers through history have nearly one and all rejected atheism. On the other hand, if we regard epistemic peerage as involving an understanding of modern science (Evolutionary Biology, Cosmology, Neuroscience, etc.), then epistemic peerage and the authority of tradition actually come apart: appeals to tradition would end up promoting beliefs that would be regarded as absurd by our epistemic peers, while the beliefs of our epistemic peers would have led to their censure (or worse) in centuries past.

Lastly, what of (c), the attempt to defend the Wager by denying that MGO hypotheses are "live" options for sincere wagerers? This strategy has the merit of trying to employ a criterion not so removed from the Wager itself. However, as we will see, the appeal to "live" hypotheses, once duly considered, is quite out of place when trying to capture the perspective of sincere wagerers.

In Chapter 5 of this volume, Jeff Jordan draws upon the Jamesian notion of a "live hypothesis," and describes this notion through primarily psychological and prudential terms: it is a hypothesis which has "intuitive appeal" for a person (p. 105), makes an "electric connection" with one's nature (p. 112), and is one where the perceived opportunity costs are not too high (p. 114).[8] Consequently, where the line falls between live and dead hypotheses will be "person-relative" (p. 115), as it is based upon one's existing belief-set, interests, risk tolerance, etc.

Jordan thus recognizes that there will not be uniform agreement over the hypotheses that can enter the Wager's matrix, for there will be some who regard MGO's "cooked-up" hypotheses as live options. This, however, is not for Jordan "a telling objection since every argument carries presuppositions which limit the class of those who find it credible to those sharing its presuppositions" (p. 115). We may, however, wonder whether he is too quick to dismiss this concern.

While it is true that all (or at least nearly all) arguments have presuppositions that are not universally accepted, an important measure of an argument's worth is its ability to persuade those who are not already committed to its conclusion. This is of particular relevance in the case of the Wager, for it is supposed to move not just those already disposed toward traditional theism, but even the atheist who finds theism highly unlikely. Accordingly, we must wonder whether Jordan's appeal to the live/dead hypothesis distinction suits the Wager's aspirations. For we may suspect that the more staunch the atheist, the less likely they would be to draw the line between live and dead hypotheses where Jordan needs it to be.[9]

Consider further that a wagerer who is considering not just a vague ecumenical theism, but a particular religious tradition, will likely be aware of some of the latter's core doctrines. In the case of Christianity, this would include: the Trinity (three-in-one), the Incarnation (fully human–fully divine), and Vicarious Atonement (we bear an infinite debt of sin which is paid for on our behalf by the "son" of God). My point here is not that these doctrines are unworthy of belief, but rather that the Jamesian category of "live hypotheses" does not suit them. They are, rather, traditionally represented as "Holy Mysteries," truths whose "very nature lies above the finite intellect"

---

[8] Jordan has commented that the concept of a live hypothesis also involves the non-psychological, non-prudential concept of metaphysical possibility. However, while an agent's belief that a hypothesis is metaphysically possible is a necessary condition for its being "live," metaphysical possibility or a belief therein is not sufficient. As such, the range of "live hypotheses" for an agent will thus tend to be more narrow than the range of what the agent takes to be metaphysically possible.

[9] The same concern holds for the staunch theist, who finds atheism no more tenable than MGO's "cooked-up" hypotheses. Jordan's strategy thus limits the Wager's audience much more than he suggests, as it would exclude not only staunch atheists and theists, but also anyone who is open minded enough to accept MGO's "cooked-up" hypotheses as "live."

(Herbermann et al., 1913, v.10, p. 662);[10] and thus are not likely to carry "intuitive appeal," particularly for the non-believer.

Let us now tie this point to the Wager. Consider that the live/dead distinction is inherently conservative since intuitively appealing hypotheses will tend to be those which cohere with one's existing presuppositions, interests and habits. Yet if the Wager is supposed to persuade those who see an "infinite distance between the certainty of what you are risking and the uncertainty of what you may gain," it does not seem an argument where the most intuitively appealing hypotheses are to be given the greatest weight. Quite the opposite in fact!

Jordan's use of the Jamesian category of "live hypotheses" imparts the wrong picture of the Wager. It is, rather, an argument meant to change one, to challenge the status quo, to upturn the comfortable equilibria of secular life, and risk losing everything one has for some *far-fetched* alternative, steeped in impenetrable mysteries, whose possibility one cannot even begin to measure. The Wager is thus better seen as an argument where traditional theism is cast as a *dead* hypothesis, one we are asked, nonetheless, to embrace and, with due time and effort, will spawn within us life anew.[11]

4 Decision-Theoretic Responses to MGO

The many-gods objection is typically presented as a decisive blow against the Wager, and the common rebuttals to MGO are supposed to offer decisive responses to it. The alternative strategy we shall here consider is not that of knock-out punches, but rather of attrition. To achieve this, we will not turn to principles that stand outside the decision-theoretic structure of the Wager. Instead, we shall see that there are numerous untapped resources built into the Wager that, one by one, can be paired against various categories of "cooked-up" hypotheses and "philosophers' fictions." Through the following subsections, we shall draw out some of the formal features and presuppositions of decision-theoretic approaches to religious belief and see that, through them, many theological hypotheses can be rebutted.

---

[10] While there is, of course, a long tradition of examining these Mysteries, in part to demonstrate that they, at least, are not contradictory (e.g., Aquinas *de Trinitate* 1.1 and 1.3), the point here is that belief in them is not in virtue of their intuitive appeal.

[11] We may also add that the more intuitively appealing theism is made, the less doxic significance is left to the Wager itself. The doxic role of the Wager is discussed elsewhere in this volume, including the chapters by Wood, Moser, and Golding.

4.1 The Stability Constraint

Let us begin with a formal feature of decision-theoretic matrices in general. One condition that a hypothesis must meet for inclusion within a matrix is that there is a stable choice–outcome relation. That is, under the condition that the hypothesis is true, it must be the case that anyone who chooses it will receive what everyone else who chooses it will receive. We may call this the *Stability Constraint* and if it does not obtain, then the expected utilities falling under that hypothesis cannot be calculated.

Consider, for instance, theological hypotheses employing "trickster" deities like the Norse god Loki or Akba-atatdia, the Native American Coyote god. Such divinities are erratic in behavior and if they were placed in charge of what we will receive in the afterlife, they would not hold to any stable principle when allocating those fates. They may offer a heavenly reward to one of their devotees, but then on a whim, choose to deny that reward to another who is just as devoted.

Advocates of MGO would grant the above category of hypotheses (i.e., trickster deities) logical possibility, and so we may then ask how to render them within a decision matrix. The problem here is that without a stable choice–outcome relation, we cannot assign values to some of the matrix cells. At best, a matrix that tried to include a hypothesis from this category would appear as laid out in Table 8.3 (let TH_1 = a traditional theological hypothesis where afterlife outcomes will vary based upon one's choice; and TH_2 = a trickster deity hypothesis).

Table 8.3 Trickster Deity

	TH_1 is true	TH_2 is true	TH_1 and TH_2 are false
One chooses TH_1	$+\infty$	?	f
One chooses TH_2	f	?	f
One does not choose any TH	f	?	f

The rational merits of choosing TH_2 cannot be evaluated. Even if TH_2 and its kin are possible, this category of theological hypotheses lacks a feature that is necessary for decision-theoretic analysis. Thus, someone evaluating what theological hypothesis to adopt on strictly decision-theoretic grounds cannot choose TH_2. Without a stable decision-outcome relation, TH_2 lacks, we may say, one of the structural elements necessary for a hypothesis to be "well-formed."[12]

---

[12] Paul Bartha has suggested that we could restore the outcome stability by way of a probability distribution over possible outcomes. The intent behind TH_2, however, is that (barring an ability to

The Stability Constraint illustrates one way in which a category of theological hypotheses can be rejected without either claiming that the probability of the hypothesis is zero or by using a principle that is external to the Wager. Instead, it serves as a formal principle for the inclusion of hypotheses within a matrix, analogous to the syntactic principles of formal logic which govern well-formed formulas. Accordingly, theists and atheists alike can accept the Stability Constraint as it is merely a formal requirement for the inclusion of hypotheses within a decision matrix and does not presuppose any further epistemological or metaphysical doctrines. It does not pre-filter hypotheses on grounds that already favor theism and neither begs the question of the Wager nor threatens its standing as a decision under ignorance.

Of course, this single constraint does not offer a definitive rebuttal to MGO. It is not meant to. It is, however, our first strike against the objection and is illustrative of how an incremental assault can be waged. By drawing upon specific features of decision-theoretic approaches to religious belief and showing how each feature can vanquish specific categories of theological hypotheses, we can, bit by bit, subdue the many-gods objection.

4.2 The Outcome Plurality Constraint

A second feature of the Wager we can bring to bear against MGO is the *Outcome Plurality Constraint*. A theological hypothesis can still be well-formed if it violates this constraint, but if it is the case that regardless of what choice is made, there is no change in outcomes, a rationally self-interested wagerer will opt for a choice whose outcomes do vary based upon what choice is made. To illustrate this constraint, let us consider the theological hypothesis of Universalism (i.e., the doctrine that all receive salvation). This hypothesis does not have the quality of being a mere "philosopher's fiction" like other hypotheses used in MGO. In fact, it is a hypothesis that has the backing of the tradition. Versions of Universalism were endorsed by figures of the Early Church (Origen, Clement of Alexandria, Gregory of Nyssa) and the hypothesis re-emerged after the Reformation among, for example, Quakers and Unitarians. It has, thus, been vetted by the tradition and so at least according to this standard, it merits inclusion with the Wager's matrix. But, as we shall see, there is a way to eliminate it by way of the Wager itself.

---

sample the actual afterlife fates that have been assigned) there are no metrics to be used to set this distribution.

Table 8.4 Universalism

	TH$_1$ is true	TH$_3$ is true	TH$_1$ and TH$_3$ are false
One chooses TH$_1$	+ ∞	+ ∞	f
One chooses TH$_3$	f	+ ∞	f
One does not choose any TH	f	+ ∞	f

For Table 8.4, let TH$_1$ = a traditional theological hypothesis where afterlife outcomes will vary based upon one's choice; and let TH$_3$ = a Universalist hypothesis.[13]

The prudential logic of the above should be obvious. The best outcomes for TH$_1$ and TH$_3$ are the same, but a choice of TH$_3$ carries a risk that is not present if one chooses TH$_1$. TH$_3$ can thus be rejected solely by a decision-theoretic analysis: there is nothing to be gained by choosing it, but something to lose if one does not choose TH$_1$. In short, TH$_3$ is a wasted bet. If the hypothesis is true, whether or not one chooses it, the expected utility will always be the same. Accordingly, the believer and non-believer alike could dismiss TH$_3$, not because of any commitments antecedent to or independent of the Wager, but merely from the logic of the matrix alone.

We thus have a second general criterion for theological hypotheses: they must meet the Outcome Plurality Constraint. This is something that traditional Universalism fails to do. Likewise, more "cooked-up" hypotheses where there is no variation in outcomes would also be rejected merely by the logic of their matrices. Consider for example, diabolical Universalism, that is, a hypothesis where regardless of what is chosen, if the hypothesis is true, infinite suffering awaits everyone. Once again, nothing is to be gained by choosing this hypothesis, but there is something to lose by not choosing the more traditional hypothesis where outcomes do vary based upon what choice is made.

4.3 The Anti-Skepticism Constraint

A third constraint upon theological hypotheses can be generated from a very general presupposition. As with other methodologies we may use when making a decision, when one chooses to explore theological hypotheses through a decision matrix, the wagerer assumes his calculations of utility and his practical deliberations are accurate and reliable. That is, the wagerer

---

[13] Following our previous discussion of exclusivity, let us assume that the deity in TH$_1$ will not reward those who choose TH$_3$.

puts trust in his arithmetic ability as well as his grasp of such concepts as choice, infinity, death, and so forth. However, many "cooked-up" hypotheses and "philosophers' fictions" are not compatible with such trust and choosing them yields what others have called a "performative contradiction."

A performative contradiction occurs when you commit to a claim that contradicts one or more presupposed claims underlying the means you have used to commit to it. Jaakko Hintikka has famously used this strategy to interpret Descartes's Cogito, thereby showing that when we doubt our own existence, we are doubting something that doubting presupposes: the existence of the consciousness engaged in doubting (Hintikka, 1962). Jürgen Habermas has used it in order to challenge some of the Postmodern critiques of reason and the meaningfulness of language (Habermas, 1990, pp. 116–19). And Hilary Putnam has used it to refute the brain-in-a-vat conjecture (Putnam, 1981, pp. 1–21). In our case, the wagerer would perpetrate a performative contradiction if he were to choose a hypothesis that commits him to the rejection of the reliability of the cognitive apparatus he employs to bring him to the determination that the chosen hypothesis rationally merits its selection over others in the matrix. In other words, the sincere wagerer would place himself in a performative contradiction if he were to commit to a hypothesis that precludes his taking the Wager seriously.

We may cook-up our first example of a theological hypothesis that would lead to a performative contradiction by placing Descartes's Evil Deceiver in charge of our afterlife fates. Let us assume that if we choose to affirm the existence of this deity, he will offer us a reward of infinite worth when we die. But in choosing to affirm this hypothesis, we are choosing to affirm that there is a being who actively manipulates our cognitive functioning so that we make arithmetic errors and/or distorts our thoughts about various concepts key to wagering. The evil deceiver may, for instance, lead us to misunderstand choice–outcome relations or the difference between infinite and finite utility. We might think we are making the choice that maximizes self-interest, but that's only because the evil deceiver has twisted our thought processes.

Such a theological hypothesis is well-formed and could be represented in a standard matrix, for we may assume that regardless of his deceptive activity, if we satisfy what he demands of us, we will be rewarded. Accordingly, let TH_1 continue to be a traditional theological hypothesis and let TH_4 = a theological hypothesis that undermines our confidence in decision theory (see Table 8.5).

This matrix offers the appearance of a standard MGO scenario, one that typically is thought to undermine the Wager. But if TH_4 involves the

Table 8.5 Evil Deceiver

	TH$_1$ is true	TH$_4$ is true	TH$_1$ and TH$_4$ are false
One chooses TH$_1$	$+\infty$	f	f
One chooses TH$_4$	f	$+\infty$	f
One does not choose any TH	f	f	f

acceptance of an Evil Deceiver, by choosing it, one would be accepting a hypothesis which holds that we cannot trust that the above matrix accurately represents the intended wager. Thus, a wagerer would commit a performative contradiction if he were to adopt a hypothesis that undermines his trust in the faculties he uses to evaluate the merits of the hypothesis. He cannot both take his wagering seriously and also accept that he cannot rely upon his wagering in pursuit of what is in his rational self-interest.

4.4 Philosophers' Fictions and the Anti-Skepticism Constraint

Let us now turn to such "philosophers' fictions" as the sidewalk-crack god, the cockroach god, the Chardonnay god, and the purple-slipper god. These hypotheses allocate our afterlives according to requirements that, prima facie, seem extremely petty. The sidewalk-crack god requires that we step on every third sidewalk crack, the cockroach god demands that we avoid killing insects, the Chardonnay god is concerned with our white wine selection, and the purple-slipper god restricts access to heaven based upon the color and style of our indoor footwear.

Yet, however petty these requirements seem to be, let us for the purpose of this subsection entertain the possibility that they are, rather, of eminent intrinsic value and so are quite appropriate for determining what we deserve in the afterlife. It may be more intuitively appealing to render them as petty requirements imposed by an unjust deity, but we will save that interpretation for the next subsection. For the moment, let us instead understand the "philosophers' fictions" as stipulating requirements that, contrary to our intuitions, are of profound intrinsic worth and completely appropriate for the allocation of postmortem rewards and punishments. So, following this rendering of the purple-slipper hypothesis, there is nothing more intrinsically important than the selection of our indoor footwear and it is perfectly just for the deity to reward and punish us accordingly.

Of course, many people think that what happens to us in the afterlife should be determined by how loving and generous we are; many Christians hold that it depends whether or not we have accepted Christ's death as atonement for our sins; and many Muslims maintain that our afterlife fate is based upon our observance of Sharia. But according to the purple-slipper hypothesis, they are all wrong. They are wrong about what is required for our entry into heaven. They are also wrong about what is truly important in this life. The traditional precepts are of either no or of just minor importance, and despite their popularity, it is inappropriate for our afterlives to depend upon them. By contrast, only those who wear purple slippers are worthy of eternal joy while the sock-wearers, the bare-footed, the flip-floppers, or the most despicable of all, the apostate pink-slipper wearers deserve nothing less than an eternity of the most grievous torment.

Our theological hypothesis is thus that the deity is just and that our afterlives depend upon our fulfilling the most intrinsically important of all duties, the wearing of purple slippers. We may place this hypothesis (TH_5) in the decision matrix along with a traditional theological hypothesis (TH_1) and thereby craft a standard MGO matrix. See Table 8.6.

Although this matrix is well formed, for the wagerer to choose TH_5, he must commit to a wildly implausible hypothesis. But we will not here challenge it for its implausibility or its non-traditionality. Instead, let us consider whether in choosing TH_5, the wagerer perpetrates a performative contradiction.

Since TH_5 takes purple-slipper wearing to be of profound intrinsic value and the appropriate basis for the allocation of our afterlife fates, a wagerer who accepts this hypothesis is foregoing the ordinary conception of what is and is not of value. Insofar as he came to the Wager with such ordinary views, he would have to accept that his (and the mainstream) conception of value is deeply askew. It is not just that we are wrong about some detail or nuance of moral concern. Rather, when we look at what counts as right and wrong, worthless and valuable, just and unjust, from the standpoint of TH_5, we've got it all terribly off.

Table 8.6 Our Judgment Flawed

	TH_1 is true	TH_5 is true	TH_1 and TH_5 are false
One chooses TH_1	$+\infty$	f	f
One chooses TH_5	f	$+\infty$	f
One does not choose any TH	f	f	f

The wagerer, thus, if he were to choose TH$_5$, would have to accept that his capacity for recognizing what is and is not of value is severely flawed. But this is not something that the wagerer would have any positive reasons for choosing, and in fact he has a powerful reason to not choose it: if his understanding of value is so deeply flawed, insofar as the Wager is an argument involving expected utility, he cannot reasonably trust his ability to assess the merits of each hypothesis within the Wager or even be sure that pleasure rather than pain, happiness rather than sorrow ought to be pursued.

Thus, so long as the classic "philosophers' fictions" are understood as stipulating something that appears to us to be of trivial value is rather of profound intrinsic value properly suited to being the requirement for what happens to us in our afterlives, we have grounds internal to the Wager for rejecting this category of hypotheses. Such hypotheses imply that we cannot trust an important aspect of our cognitive activity; and without such trust, we cannot consider ourselves fit to engage in wagering.

4.5 Philosophers' Fictions and the Practical Reason Constraint

Unlike our preceding treatment of the "philosophers' fictions," where their afterlife requirements were assigned great intrinsic value, let us now shift to a second interpretation, one that takes their afterlife requirements as they appear – i.e., of little to no intrinsic worth and unjust bases for the determination of what we should reap in our afterlives.

This category of hypotheses will satisfy all our preceding constraints: the *Stability Constraint*, the *Outcome Plurality Constraint*, and the *Anti-Skepticism Constraint*. For example, the purple-slipper deity will stably hold to the stipulated requirement, will grant a heavenly reward to those who choose the right type of indoor footwear, and deny this reward to those who do not. Further, there is nothing in this hypothesis that jeopardizes our trust in our cognitive faculties since, unlike the previous rendering of the "philosophers' fictions," values are here as they appear.

There are countless hypotheses of this sort that we may cook-up and if no further constraint is available to eliminate them, they would devastate the Wager. We may, for instance, modify the basic purple-slipper hypothesis so that the deity does not require just any shade of purple, but one of an exact wavelength. From this, we may project an infinite number of different purple-slipper hypotheses, each deity demanding a minimally different hue. A purple whose wavelength is just one nanometer off would violate the afterlife requirement and thus the wagerer who chooses slippers ever so slightly too blue or too red would be damned to hell for all eternity.

There is nothing internally inconsistent in the above, and as noted, the infinite series of purple-hue hypotheses can satisfy all the constraints we have so far explored. But there is something within this category that may yet generate another constraint. They all present something of infinite value to be dependent upon our observance of a requirement that in itself is of little to no value. They forgo such traditional requirements as moral obedience, virtuous character, faith in Christ, observance of Sharia, etc. – that is, requirements whose intrinsic worth is either conventionally accepted or may be explained through further theological principles. Instead, "philosophers' fictions" present the universe as ultimately unjust since those who do live a life of genuine intrinsic worth will still be damned simply for their failure to wear the right hue of purple slippers, while those who are evil and the cause of great suffering in the world, if they happen to satisfy this petty requirement, will be granted eternal joy.

Such a universe would strike most as morally objectionable. But on its own, this is not enough to challenge this category of hypotheses. What we need is to look at our interest in a just universe and consider whether it may somehow be a necessary presupposition of practical reason. This has, in fact, been done by Kant, captured most recognizably in his doctrine of the Highest Good.[14] But before we move to his account of why we must presuppose a just universe, it would be helpful to explore another strand of performative contradictions.

More often than not, performative contradictions are employed either epistemically or semantically. In the previous sections, we examined hypotheses of the former type, hypotheses whose acceptance is not compatible with one or more cognitive activities used when choosing the hypothesis. A semantic performative contradiction would be one where a claim uttered is not compatible with one or more conditions upon which the utterance can have meaning. A helpful example is Hilary Putnam's attempt to refute the possibility that one is just a brain in a vat. His argument is that the utterance "I am a brain in a vat" is nonsensical (given his extensional theory of meaning) for if one is a brain in a vat, then one lacks the appropriate relation with the world such that "vat" can be used meaningfully (Putnam, 1981). This argument is (ironically) similar to George Berkeley's critique of physicalism, for if all meaning stems from our mental content and there is no mental content about mind-independent

---

[14] Interpretations of Kant's Highest Good are notoriously divided. I explore the causes for such disputes and seek to overcome them in Pasternack (2017).

entities, the utterance "there exists an extended world of matter" either reduces to a peculiar way of expressing *esse est percipi* or is gibberish.

A third type of performative contradiction may be titled "practical." One example is choosing to affirm strong determinism. One who makes this "choice" is committing to a worldview where choosing is not possible. Though idiomatically we may still say that someone has chosen to believe in determinism, and, of course, allow within a determinism that unchosen processes may simulate what we call "choice," in contexts where free will is presupposed, it becomes a performative contradiction to choose strong determinism. A theological hypothesis that includes strong determinism is thus not one that a sincere wagerer would choose for, in doing so, they are abandoning that genuine wagering is going on.[15]

Similarly, consider a theological hypothesis where the afterlife requirement holds that we must not wager. A sincere wagerer cannot choose this hypothesis, for in choosing it, he is going against the point of the Wager. By making this choice, he is violating the requirements for receiving a heavenly reward and so it is a hypothesis incompatible with his self-interest. The same holds for hypotheses that prohibit setting ends, making choices, contemplating the afterlife, etc. In each of these cases, a performative contradiction takes place that is practical in nature. A wagerer cannot affirm a hypothesis that prohibits wagering without going against his own self-interest. Of course, we can in most contexts choose contrary to self-interest without committing a performative contradiction, but for a wagerer to make such a choice is to abandon the particular activity of wagering. The Wager is an argument to one's self-interest and so to choose a hypothesis incompatible with self-interest is to no longer play along with the Wager. That is, one cannot be a sincere wagerer and choose a hypothesis that either prohibits wagering as such, or prohibits a particular feature of what one must do when wagering.

Collectively, these examples present what we may call a practical variant of performative contradictions. Some choices are not compatible with a genuine engagement in an activity. These activities can be quite incidental, such as playing a game, or can be more integral to our practical lives. In the case of the former, a practical performative contradiction would arise if one tries to win a game by cheating. For examples of the latter, the natural figure to turn to is Kant.

---

[15] See Chapters 2 and 3 for a discussion as to whether or not Pascal's Jansenism is compatible with the sort of choosing that is usually assumed for the Wager.

Performative contradictions arise whenever we attempt to justify a maxim that cannot be made into a universal law. The reason for this, according to Kant, is that justification is itself a rational procedure which requires that the grounds one uses to justify the action also hold for all other rational agents. When we make an exception for ourselves, we are pretending to justify our maxim, but the making of an exception is formally incompatible with the universal character of practical justification.

For our purposes, let us consider another aspect of Kant's practical philosophy. In addition to various other presuppositions built into the nature of practical reason, he maintains that we must also presuppose that the universe is just – and how he presents this presupposition is particularly germane to the category of hypotheses that include "cooked-up" hypotheses and "philosophers' fictions."[16]

Kant argues that in addition to reason's ability to determine the proper means for a desired end, it can also determine our ends, for practical reason is itself normative, in that it provides for our practical deliberation an *a priori* norm by which we can evaluate ends. Kant further argues that practical reason prescribes as our end not simply that we act from duty (i.e., observe the moral law) but, more comprehensively, that we take up an end that also includes our interest in happiness. More precisely, Kant distinguishes between the "supreme good" of morality, and the "complete good" that includes both morality and happiness. The Highest Good is thus defined by Kant as an ideal state of affairs in which happiness is distributed in "exact proportion" to moral worth. He regards this as the "totality of the object of pure practical reason" (5:108) and the proper "point of reference for the unification of all ends" (6:5).[17] Lastly, Kant advances very much the classic theological hypothesis of a just God and a "future life" where rewards and punishments are meted out.

I will not here review Kant's arguments for the Highest Good or his arguments for how our duty to promote this end commits us to a "rational faith" in the objective reality of God and a "future life."[18] What instead I here

---

[16] Note that this turn to Kant is not an appeal to tradition in the manner used by Jordan 2006 since the analysis here is not based upon there being some traditional view that runs counter to MGO hypotheses. This is not an appeal to a "great mind" of the tradition, or a positive *ad hominem*. The issues here do not have to do with who claimed what. Rather, they have to do with *what practical reason may actually be like* and its philosophical (vs. historical) implications for dealing with MGO hypotheses.

[17] Kant citations refer to the Akademie-Ausgabe by volume and page number.

[18] Note that Kant's argument for God is not one based upon metaphysical or epistemic considerations (such are rejected in the *Critique of Pure Reason*). His argument is rather one intertwined with our choice to observe the moral law. It is thus aptly called a "moral argument" in that we are given a moral reason (one however that is already intrinsic to practical reason) for religious belief. Note

propose is that *if* Kant is correct about the nature of practical reason – if it (a) involves more than means–ends reasoning, (b) governs our end-setting, and (c) prescribes that we adopt the Highest Good as our end, then the wagerer would commit a performative contradiction if he were to select a theological hypothesis incompatible with the Highest Good.

Thus, at least as understood by Kant, our faculty of practical reason proscribes against the adoption of a worldview at odds with the Highest Good. The point here is not that practical reason provides a proof that the Highest Good will obtain, but rather that one would be engaging in a performative contradiction analogous to the adoption of an immoral maxim if one were to commit to the sort of worldview which would make our pursuit of the Highest Good a pursuit of the impossible.

For Kant, traditional theism provides the postulates through which we can grasp how the Highest Good can obtain.[19] But we may also imagine how a distribution of happiness in accordance with moral worth can be made possible through the societal and technological advances that are the stuff of utopian literature. That is, while there are reasons we need not here discuss why Kant turns to theism, neither the Highest Good nor one's commitment to it as an end is incompatible with atheism. What, however, is incompatible with the Highest Good is the category of "cooked-up" hypothesis exemplified by the purple-slipper deity, for such hypotheses would have us forsake even the quantum of solace which remains still for the atheist, accepting instead a universe both unjust and absurd.

Hence, following Kant's conception of practical reason, there are internal reasons for rejecting the category of hypotheses that stipulate that our afterlives are determined by way of some petty requirement, such as those found in the typical "philosophers' fictions." By making such things as wearing purple slippers the condition for our receiving a positive afterlife, they push morality and happiness apart, and forgo the justice demanded by a Kantian interpretation of practical reason.[20] If this interpretation of practical reason is correct, we have found our way to block the category of hypotheses that most perniciously

---

also that in line with section 3's criticism of Jordan, we can see Kant's moral argument as not one where we are to simply adopt a comfortable hypothesis, but rather is portrayed as involving an inward "revolution" (6:47) of eminent difficulty (6:66–67). There are, in fact, a number of interesting philosophical parallels between Pascal's Wager and Kant's Moral Argument. These are briefly touched upon in Chapter 4.

[19] Note that Kant's account of the Highest Good results as well in a rejection of many core Christian doctrines. I discuss these in Pasternack (2015).

[20] It may be further argued, independent from Kant, that the utility of a *deserved* post-mortem reward exceeds the utility of one that is undeserved. Bartha's employment of relative infinite utilities in Chapter 12 aligns with this point.

affects the Wager. Of course, there are other conceptions of agency that compete with Kant's. But his happens to offer one that can bar hypotheses incompatible with ultimate justice, and if it or something like it is correct, we can legitimately reject the classic "cooked-up" hypotheses and "philosophers' fictions" on grounds built into the faculties used while wagering.

5 Conclusion

As MGO has been understood as a definitive knock-out blow to the Wager, so many of the Wager's defenders have tendered what they regard as knock-out rebuttals to MGO. By contrast, this chapter has offered a different sort of strategy, one that challenges individual MGO categories. Further, unlike other rebuttals to MGO which draw upon criteria independent of the Wager itself, this chapter has looked to formal features and presuppositions already within the Wager in particular and within practical reason more generally. By drawing out what is already built into wagering, we have been able to block numerous categories of hypotheses, including those employing capricious deities, deceptive deities, Universalist soteriologies, and two different renderings of the more notorious "philosophers' fictions," all without threatening either the status of the Wager as a decision under ignorance or its function as an apologetic directed toward atheists and agnostics.

Of course, this chapter has not addressed every possible threat that MGO can offer, nor has it demonstrated that ultimately there is only one viable hypothesis with an infinite expected utility. As such, it has not offered a complete rebuttal to MGO. But that was not its aim. The purpose of this chapter has rather been to harness the formal decision-theoretic features already internal to the Wager in order to show that the wagerer is not simply left in an aporetic state, unable to adjudicate between a plurality of theological hypotheses with infinite expected utility, but rather has resources at his disposal to rebut scenario after scenario employed by MGO. This is, therefore, a shift in the balance of power between the Wager and MGO. Because these resources can be extended as needed to further hypotheses that may be "cooked-up," a new burden is placed upon MGO. It is no longer free to merely "cook-up" new hypotheses, but must further test them against the formal features and presuppositions that this chapter has identified, and perhaps, further features that, in time, may be identified by others.

9 Rationality and the Wager

Paul Saka[*]

1 Introduction

Is it, or is it not, rational to believe in God? More pertinently, is it rational to adopt a religious faith on the basis of Pascal's Wager? By Pascal's Wager I mean any reasoning that uses the uncertain prospect of rewards to justify believing in God, attempting to induce belief in God, accepting God (q.v. Golding, Chapter 10 of this volume), or practicing religion (praying, attending church, etc.).[1] Very approximately, the Wager goes something like this:

(A) Action A is rational iff A promotes a certain end, happiness, well-being, or "the good."
(B) Believing in God promotes some given good.
(C) Therefore believing in God is rational.

"Promotion of the good" here is an intensional concept that factors into account comparative costs, benefits, and likelihoods. Risk or uncertainty is essential, making the argument a *wager*.

The answer to my opening question will depend on one's theory of rationality. For hardcore evidentialists, adopting any belief on the basis of a wager is irrational. Mostly setting that aside, however, I shall focus primarily on where instrumental rationality takes us. I shall begin by briefly characterizing its structure and then parse Pascal's Wager, distinguishing between metaphysical and mundane varieties. In subsequent sections my aim is not so much to rule categorically on the rationality of the Wager – though I sometimes share my own views on the matter – as it is to inventory theories of rationality and say how the Wager fares under them.

---

[*] I am grateful to Lawrence Pasternack, Paul Bartha, Emil Badici, and Paul Jorgensen for valuable comments. This work is dedicated to the memory of Jonathan Adler.
[1] Due to space limitations I shall refer to theism and religious practice almost interchangeably.

1.1 The Relativity of Rationality

The idea behind (A) can be formulated more precisely in a variety of competing ways. According to standard decision theory:

(A′) Action A is rational iff A maximizes so-called expected utility, *EU*.

This formulation occludes the fact that rationality is relative: what is rational for you to think and do is not necessarily what is rational for me. In particular, rationality depends on both a given subject S's beliefs about outcome probabilities and S's valuation of outcomes. These two parameters are reflected in credence function p and utility function u:

(A″) It is rational *for S* to perform A iff S's performing A maximizes
$p(S, outcome_1) \cdot u(S, outcome_1) + \ldots + p(S, outcome_n) \cdot u(S, outcome_n)$.

The relativity of p will be addressed in section 2, the relativity of u in section 3, the additional relativity of *perspective* in section 4, and relativity induced by one's choice of decision rule in section 5. Different relativizations, as we shall see, yield different verdicts on the rationality of different kinds of Pascalian wager.

That rationality is *relative* has been denied on the grounds that (A″) is a universally valid absolute. Nothing substantive hinges on my terminology, however, as its intended sense should be clear. Besides, my usage actually does comport with patterns of established English use. For example, according to one leading paradigm of relativity, a predicate is relative if it allows qualification by "for": if proposition P were true *for S_1* but not for S_2, truth would be relative, despite any purported universal norm such as "ω is true for S iff ω works for S." The same goes for rationality.

1.2 Multiple Wagers

Schema (A)–(C) covers two separate lines of argument that I call the *metaphysical wager* and the *mundane wager*. Metaphysical wagers focus on the potential costs and benefits of heaven, hell, nirvana, reincarnation, and so forth while mundane wagers focus on the empirical consequences of religious faith for our present lives here on earth. The dominating-expectations version of the metaphysical wager assumes a decision matrix where the exact finite values do not matter (see Table 9.1).[2]

---

[2] This has been called "the Canonical Wager" (Jordan, 2006), which seems misleading. A variety of wagers appear in the *Pensées*, and in that sense they are all canonical.

Table 9.1 Metaphysical Wager

	God Does Not Exist	God Exists
Atheistic payoff:	Finite	Finite at best
Theistic payoff:	Finite	∞

Objections to the metaphysical wager, and replies, are summarized in Saka (2002). For instance, the many-gods objection observes that Pascal's matrix oversimplistically leaves out relevant possibilities. We don't know whether God will reward the devout, or instead favor the honest skeptics who apportion their beliefs to the available evidence. We don't know what prizes, penalties, or indifference may await the followers of competing religions. (For developments of this objection, see Saka (2001) and Duncan (Chapter 7 of this volume) and Oppy (Chapter 13); for replies, see Bartha (Chapter 12), Franklin (Chapter 1), Jordan (Chapter 5), and Pasternack (Chapter 8).)

Because the values in the last column of Table 9.1 are too contentious or too speculative to posit, the mundane wager turns to worldly benefits. Religion, it is widely believed, is a great good. According to Pascal, theism will make one "faithful, honest, humble, grateful, full of good works, a true friend, ... I will tell you that you will gain even in this life ... " (L418/S680). According to William James, religious belief produces "a new zest which adds itself like a gift to life, and takes the form either of lyrical enchantment or of appeal to earnestness and heroism ... an assurance of safety and a temper of peace and, in relation to others, a preponderance of loving affections" – whereas "sadness lives at the heart of every merely positivistic, agnostic, and naturalistic scheme of philosophy" (1936 [1902], p. 140). In short, we get the values listed in Table 9.2.

Of course, the theist's life on earth is not guaranteed to be better than the atheist's. Revising what Pascal and James actually say, the outcomes schematized above are merely what's most likely, and it is for this reason that adopting theism counts as gambling or wagering. The statistical nature of my opening argument is explicit in contemporary renditions of the mundane wager, including Schlesinger (1994), Betty (2001), Fagan (2006), Jordan (2006, this volume), Casey (2009), Myers (2000), McBrayer (2014), Rota (2016a), and above all an enormous empirical literature including influential meta-analyses by Koenig et al. (2001, 2010).

Metaphysical and mundane considerations can be combined into one master wager (see Table 9.3).

Table 9.2 Mundane Wager

	God Does Not Exist	God Exists
Atheistic payoff:	Finite n	n + some metaphysical unknown, to be ignored
Theistic payoff:	Finite n + x	n + x + some metaphysical unknown, to be ignored

Table 9.3 Hybrid Wager

	God Does Not Exist	God Exists
Atheistic payoff:	Finite x	Finite at best
Theistic payoff:	Finite x + n	∞

Although this argument from strong dominance might be what Pascal and James had in mind, my present concern is less biographical, historical, or exegetical than philosophical. Philosophically speaking the hybrid wager is basically a conjunction of the metaphysical and mundane wagers, which puts it at a dialectical disadvantage: if either component wager fails then the hybrid fails. The wise Pascalian therefore asserts a *disjunction* of the metaphysical and mundane wagers so that if either component succeeds then theism is rational.

2 The Agent-Relativity of Belief

Decision theorists recognize that p is a function of S, but the importance of this relativity is exaggerated by some Pascalians and under-appreciated by others.

2.1 The Metaphysical Wager

Apparently conceding that the many-gods objection disarms the metaphysical wager for today's multicultural sophisticates, some scholars make a point of relativizing the Wager to Pascal's contemporary audience, for whom it is said Catholicism and atheism were the only possibilities (Rescher, 1985; Franklin, 1998, Chapter 1 of this volume; Nehr, 2012; cf. Wetsel, 1994, p. 177; Wood, Chapter 2 of this volume). This doesn't wash, however. Pascal's peers knew of Greco-Roman paganism, Judaism, Islam, new-world paganism, and multiple brands of Protestantism; they knew of alleged Satanism, from witchcraft trials and stories of the devil-worshiping Templars; and they knew, from their acquaintance with the foregoing, that still other religions could readily be hypothesized.[3] Indeed, almost everyone who has ever at any time or place heard of Pascal's Wager has surely heard of multiple religions, and any would-be wagerer – any agnostic seeking belief according to its profit – is likely to recognize as epistemic possibilities more than just two or three theologies. Even though the rationality of the metaphysical wager is relative in principle, therefore, in practice the relativizations hardly matter. (Compare: because all human beings move through space at about the same speed, they experience roughly the same time dilation; thus, even though time is theoretically relative, any choice of anthropocentric reference frame would carry nearly the same implications as any other.)

Because relativizations hardly matter, if the many-gods objection renders the Wager irrational for a modern audience then it renders the Wager irrational for Pascal's contemporaries; and if defenses against the many-gods objection are valid for some then they will tend toward validity for all.

2.2 The Mundane Wager

The mundane literature, in contrast, largely fails to acknowledge that rationality is relative. Absolutism is implied when the agent is omitted, as for instance in Ovid's *Ars Amatoria*: "It is convenient [to believe] that there be gods, and, as it is convenient, let us believe there are." (To those who object that Ovid obviously meant "it is convenient *for us* to believe ...," I reply that this hardly helps. Did Ovid intend to include just himself and his reader, or the larger circle of all educated upper-class urban Romans, or all free citizens of the empire; or did he assume, and mean to allude to, a cross-cultural human condition? Because Ovid's intention is not at all obvious, his message implies that the identity of "us" does not matter – that the convenience of theism possesses a universal or absolute validity.) Absolutism is explicit when Jordan writes, "*anyone* who has at least as much evidence" for theism as for atheism "is rational to cultivate theistic belief" (2006, p. 3, italics added), it is implicit when Jordan writes that a prudential argument may be sound even if its premises are not accepted (p. 210), and it is conveyed in wagers formulated with generic pronouns.

---

[3] Rescher's relativization is especially inexplicable because, for him, it is theologically important that Pascal's Wager extend to all human beings.

What's missed is that populations are variable. "Religiosity is more strongly tied to happiness among women (relative to men), among African-Americans (relative to whites), among older people (relative to younger ones), and among North Americans (relative to Europeans)" (Lyubomirsky, 2007, p. 234). A retired black American female and a young white German male may reasonably have conflicting beliefs about the value of attending church. Rationality is guided by belief, and belief varies with demographic status.

Belief (or knowledge) is most notably a product, in part, of one's sociopolitical environment. Romans under Nero knew that Christians faced death, medieval Europeans knew that Christians enjoyed totalitarian validation, Soviets knew that Christians suffered professionally, and today the informed know that states the world over support religions, directly or indirectly (Hamilton, 2005; Lynn, 2015; IHEU, 2016). For Third World denizens who stand to reap the benefits of missionary charity, feigning or adopting belief in God is presumably rational. Children with religious parents may "go along in order to get along," and even adult agnostics are prudent to seek faith when threatened with ostracism from their families (as some of my students have reported to me). In short, for most times in history, at most loci in the world, one's religion has a definite impact on one's well-being.

That said, prudence favors religion more often than not. To begin with, theists enjoy a numerical advantage. Since they constitute the vast majority of citizens in the vast majority of societies, and since in-group bias is part of human nature (Hewstone et al., 2002), in most societies atheists are stigmatized (Edgell et al., 2006; Wing, 2017); indeed, in several nations atheism is a capital crime. On top of that, qualitative differences put the two groups in asymmetric positions, as prejudice is, on average, stronger in theists than atheists (Hall et al., 2010; Galen and Beahan, 2013). Though the non-religious are disadvantaged in theistic societies like the United States, the religious are not disadvantaged in atheistic societies like Sweden (Saroglou et al., 2004; Snoep, 2008; Diener et al., 2011; Gebauer et al., 2012; Lun and Bond, 2013; Stavrova et al., 2013).

At a first pass, whether the mundane wager is rational varies according to one's circumstances, as mediated by beliefs. In most cases theism, regardless of its epistemic status, expresses a deep, perceptive understanding of what is in the interests of those surrounded by religious majorities.

3 The Agent-Relativity of Value

In toting the costs and benefits of religious belief, one must make value judgments in order to decide what counts as cost and what counts as benefit.

To some extent moral judgments are shared, but variation is greater than usually appreciated. The human mind, susceptible to a cognitive illusion known as the false-consensus bias, systematically imagines that agreement obtains where it does not (Marks and Miller, 1987). Disagreement about values, in turn, has consequences for both wagers.

3.1 The Metaphysical Wager

According to Pascalians, the outcome of wagering for God, if God exists, is eternal life and happiness. Yet happiness is not a universal end: in some cultures it is regarded as sinful, or feared or scorned for its "shallowness, foolishness, and vulgarity" (Joshanloo and Weijers, 2014; cf. Oishi and Koo, 2008; Suh and Koo, 2008). Nor is immortality universally valued, with plenty expressing some sort of aversion to its perceived tedium (e.g., Williams, 1973b). Even when it is preferred, subjective preferences for it are not always strong enough to fuel the Wager. To see that this is so, consider the following *limit argument*.

Future utilities tend to have discounted value, and increasingly future utilities tend to have increasingly discounted value. If the discount is steep enough, the sum of heavenly utilities will converge on a limit. Subjective discounts are plausibly steep indeed, as they could help to explain why many religious believers are nonetheless willing to commit mortal sins; why few believers are willing to expend their every resource in attaining unimpeachable holiness; and why many skeptics remain unmoved by Pascal's Wager quite apart from the many-gods objection. Heaven will also tend to have finite appeal for those who grok what literal eternity actually entails: if not the unremitting persistence of a frozen mental state then, for any finite mind, the repetition of some experiences in the exact same sequences, over and over and over, without end.

Because infinite rewards stretched into the infinite future may have finite desirability in the here and now, merely postulating the existence of an infinitely rewarding heaven is insufficient to secure an absolutely valid wager. The Pascalian needs to argue that the prospect of heaven truly trumps all else, and not silently assume it. Otherwise, for the dominance argument to work, it must be established for any given subject that the value of an atheistic lifestyle is inferior to a theistic lifestyle, a matter which will vary from individual to individual; and for the EU argument to work, it must be established for any given subject that the probability of God's existence is high enough to compensate for heaven's limited value, which again will vary from individual to individual.

The limit argument assumes the traditional view that heaven is an unending sequence of positive finite goodness, but of course other views are possible too (for Pascal, heaven is "an infinity of infinite happiness," an infinite sequence of infinite utilities). Positing unbiblical views of heaven, however, undercuts the role that tradition plays for most wagerers, thereby adding to the force of the many-gods objection. More to the point, to acknowledge diversity amongst evaluations of heaven is to acknowledge that the rationality of the metaphysical wager must be relativized to the agent's utility function. The lure of heaven, in short, raises not only mathematical issues with infinity (Oppy, Chapter 13 of this volume; Wenmackers, Chapter 15), but an axiological concern heretofore unrecognized in the Pascalian literature.

3.2 The Mundane Wager

Pascalians variously credit faith with promoting humility and charitable giving, and with reducing suicide rates, divorce rates, use of recreational drugs, and premarital sex, but these implied valuations are tendentious. Although religious doctrine teaches that suicide is among the few unpardonable sins, secular utilitarians see voluntary euthanasia as good (least-bad) for those who suffer from painful terminal diseases. Christianity regards marriage as a sacred union, made in the eyes of God, yet for many others a series of intimate pair-bondings, in the context of a modern lifespan in a modern economy, maximizes EU. While Christianity teaches that all pride is sinful because everyone is fallen and no one achieves anything without the grace of God, the ethos of pagan humanism valorizes correctly proportioned pride (with American educators going so far as to exalt false pride). Where Christians see private charity as a virtue, social Darwinists see support for the weak as bringing the human race down. On the opposite end of the political spectrum, my own view is that religious charity undermines the public will to support robust social welfare programs. Evidence for this is that Christian political parties overwhelmingly tilt right.[4]

There are *multiple* ways to assign values, and some align with religious outlooks. The moral and metaphysical strands of a worldview quite often

---

[4] The pattern extends to religious parties in general. My survey of the world's top fifty nations, measured by GDP (thus indirectly factoring in population size, modern industrial condition, and global impact), yielded ninety-eight current political parties characterized by Wikipedia in religious terms, with seventy-five of these classified as left/right: of these, eight are on the left and sixty-seven are on the right (en.wikipedia.org/wiki/List_of_countries_by_GDP_(nominal); en.wikipedia.org/wiki/List_of_ political_parties_by_region).

correlate statistically with each other, and can reinforce each other, both epistemically and pragmatically (Weeden and Kurzban, 2014, ch. 4, report that young adults *first* establish a lifestyle – either conservative or hedonistic – and then adopt religious practice to suit). Thus theists plausibly maximize the good as they see it by maintaining their theism, and likewise for atheists. In this case, the mundane wager rationally justifies theists in their belief, and atheists in their nonbelief. As Alexei Panshin says in his novel *Masque World*: "Monism promises you only one thing: to make you very happy. There is a catch, of course. To be happy as a Monist, you must accept Monist definitions of happiness."

Another catch is that "those who pursue religion for intrinsic reasons ... are happier than those who pursue it as a means to an end" (Lyubomirsky, 2007, p. 234). Since pursuing religion for Pascalian reasons is pursuing it as a means to an end, the specter is raised that (on average) religion is of net benefit only to non-wagerers.

The foregoing considerations give independent motivation for thinking that certain worldviews are like mathematical attractors (q.v. Sober, this volume), and in "stable equilibrium" (Bartha, Chapter 12). In line with this idea, philosophers endorse the rationality of cognitive conservatism, all else being equal (Quine and Ullian, 1978; Harman, 1986), while empirical studies reveal that human beings are prone to cognitive homeostasis (Edwards, 1982), to effort- and choice-justification (Qin et al., 2011), and to ownership and endowment effects (Morewedge and Giblin, 2015). Extrapolating from the examples in the studies just cited, it would seem that mere possession of a belief-state (be it theism or atheism) in itself makes it valuable to a subject, and that performing a chosen action (be it attending church or going fishing) increases the preference for that action. Thus, one's prior religious convictions are relevant. If you are already a theist, you may find religious devotions to be rewarding, while abandoning theism may incur apostatic costs (Maslen, 2016). On the other hand, if you are an atheist persuaded by pragmatic reasoning to convert then, following Pascal's prescriptions, you may participate in rituals that you find to be tedious or alienating, you may lose time reading Scripture, and in general you will pay conversion and opportunity costs. As a result, it may be that theists and atheists are both rational.

The attractor thesis is also supported by empirical studies that show a curvilinear relation between religiosity and self-reported happiness: the especially religious are the happiest, and then come affirmative atheists, with the modestly religious and unsure agnostics at the bottom (Mochon et al., 2011; Newport et al., 2012; Yeniaras and Akarsu, 2016). The devout will be motivated toward greater devotion while many of the marginally religious will be drawn further from faith, assuming that changes are incremental and rational. Since the extremely religious and the extremely atheistic are at local peaks, each will think, from their own experience, that they are more rational than the other. In this scenario, however, the religious are the most rational while atheists are stuck in a mental trap.

4 Perspectival Relativity

The rationality of an action can be judged from a number of perspectives. Perspectives are quasi-indexical, and to capture them I deploy here the analytic framework developed in my semantics research program (2007):

(R) S's action A is rational, from Σ's perspective, iff Σ believes that A promotes S's good.

One salient perspective is that of the agentive subject S: when $\Sigma = S$, (R) describes internal rationality. Another kind of salient perspective is that of any third-party judge of S: when $\Sigma \neq S$, (R) describes a potentially stronger external rationality. Among the external cases, we get increasingly powerful rationalities as we strip away Σ's biases and ignorance. At the limit Σ is an ideal observer, a perfectly informed judge whose beliefs match the actual facts exactly. In that case perspective needn't even be mentioned, with (R) effectively collapsing back to (A), thus describing objective rationality.[5]

When individual subjective rationalities coincide, they yield intersubjective rationality. As intersubjective rationalities grow in size and density, we get socio-politically important kinds of rationality. To the extent that Σ subsumes all or almost all members of a given community, (R) describes consensus rationality. Because two heads are better than one, consensus rationality carries greater normative weight than any one individual's rationality. (For a Bayesian account linking rationality to consensus, see Lehrer and Wagner, 1981; for Kant's account, see Pasternack, 2014.)

Decision theory's perspective is largely internal, building S's point of view into credence and utility functions p and u.[6] This perspective may be appropriate when we want to *explain* S's behavior, but what we want is some concept of external rationality for *evaluating* S's behavior (Nathanson, 1994,

---

[5] Foley (1994) posits an "objective" rationality, but conflates it with external subjective rationality, and fails to consider the sort of objection to be given below by my United example.

[6] Not entirely internal, however: while p and u depend on S in (A″), the mathematical operations used in combining them do not.

ch. 8). Compare: when we want to explain S's logical inferences, we cite what (we think) S believes about validity, but when we want to assess those inferences we use what *we* believe about validity. When we want to describe and explain S's perceptual beliefs, we cite what (we think) S's experiences are like, but when we want to assess them, we invoke what we ourselves take to be in S's environment. By the same token, when we want to assess S's rationality we use our own understanding of what's probable and what's valuable. For example, suppose that someone switches from Baptism to Episcopalianism because he thinks it will open business doors and make him a multimillionaire. If we regard his belief as groundlessly delusional then our external perspective would judge his conversion to be irrational despite its internal rationality. (Qualifications need to be added, perhaps with modal *woulds* and *oughts*, which go beyond the scope of this chapter.) Since external standards are generally relevant to assessing prudential rationality, yet the Wager is normally cast in terms of standard decision theory's internal rationality, prevailing formulations of the Wager are incomplete at best.

S's action best promotes the good, from a transcendentally objective point of view, only if it successfully achieves the best outcome that is potentially attainable. This gives us a concept of rationality that is either controversial or useless. For suppose that the criteria of ordinary rationality – factoring in wanderlust, price, convenience, etc. – lead S to book United flight 1010. Now suppose that, unbeknownst to anyone, mechanical fatigue destines 1010 for mid-air disintegration. Then objective rationality claims that S would be irrational to take 1010, which doesn't seem right. In contrast, to modify the example, suppose that T plans to destroy 1010, and warns S: "I wouldn't take that flight if I were you. I can't say anything more at this time, but it wouldn't be a smart move." T realizes that taking the flight may remain internally rational for S, but from T's external vantage it is not the "smart" thing to do, it is not rational, it is not what S *should* do. Moreover, T's judgment as to what's rational is objectively better than S's.[7]

To summarize, I want to orient our discussion to a perspective that is more rational than S's, without going so far as to appeal to unattainable transcendental objectivity. To do that I would invoke widespread intersubjectively acceptable standards. One such standard is that wagerers must take into account information already in their possession; otherwise their matrices are irrational. Another standard is that wagerers must take initiative in

---

[7] Σ and S may differ not only over how to promote S's good, but over what counts as S's good – whence the notion of irrational desire. Cf. Rescher on "perverse" desires (1988, ch. 6), Nathanson on self-destructive desires (1994, ch. 10), Wood on Pascal's "depraved" goods (Ch. 2 of this volume).

acquiring new information, expending effort that is proportionate to reasonably projected gains/losses. The hedged conclusions of the previous sections, therefore, call for reconsideration.

4.1 The Metaphysical Wager

Given the infinite stakes involved, wagers must (if rational) be based on decision matrices that meet our highest standards. If you were building a nuclear power plant, you would conduct a risk assessment that would admittedly not consider every conceivable contingency; there is an unending number of them and, besides, most will seem to have negligible likelihoods or incalculable probabilities. Yet considering only those few contingencies that viscerally move you, or that are "intuitive" or "simple," would be inexcusable. In real life you would invest time and money to hire consultants to examine at least those dangers which the opponents of nuclear power publicly worry about. By the same token, metaphysical wagerers would be correct if they merely rejected the demand to consider *every* rival religious hypothesis. However, many disregard relevant religious views simply because they come from exotic cultures, and they disregard metaphysical hypotheses that their professional colleagues have asked them to consider (James dismisses possibilities that are not "live"; Schlesinger dismisses those that are not subjectively "simple"; Jordan dismisses those that are novel to him, "bizarre," or lacking "intuitive appeal"; and others flat ignore epistemically possible hypotheses). If wagerers truly think that every soul faces possible infinite reward/punishment then they will hold that the question of theism is more significant than the question of whether to build some power plant. Since the latter calls for open-minded research, the former should too. Instead of thinking off the top of one's head, or nakedly emoting, one should at least look in readily available reference books on world religions. It takes very little drawing-room anthropological study to reveal rival religious hypotheses that undermine decision-theoretic reasons for believing in any one of them.[8]

If S has no access to a library, and is illiterate to boot, an oversimplistic decision matrix may be internally rational for S. For the purposes of evaluating whether theism is truly warranted, however, we need to judge from *our* perspective. We can't judge from higher perspectives, for they are inaccessible to us, and we shouldn't judge from lower ones, because that would pointlessly

---

[8] Franklin (Ch. 1 of this volume) coincidentally uses a nuclear power plant analogy, to ends opposite mine.

handicap us. To return to the nuclear analogy, suppose that the power company commissions two independent risk assessments, each of which is perfectly reasonable relative to the knowledge and resources of its author. From the company's perspective, which sees that assessment #1 takes into account everything that is in #2, plus more, the rational action to take is the one grounded in #1. The more complete the matrix, the more rational the resulting decision.

4.2 The Mundane Wager

The stakes are not as high for mundane wagerers, but still substantial. The impact of religious faith on life satisfaction may rival the impact of one's career. Since one undergoes years of training for the sake of a career, wagerers can be expected to spend considerable time and effort, if they are to be rational, researching the pros and cons of joining the denominations and cults that are open to them. Few do the work, however, that sincere commitment to instrumental rationality entails.

Adopting a minority religion comes with special costs and so, all else being equal, the dominant religion is certainly to be preferred. All else is not always equal, however. A few hours' googling reveals that Hindus are the happiest religious group in the UK (UK Office of National Statistics, 2016), and a Buddhist monk living part-time in the West is the happiest man known to neuroscience (Shontell, 2016). More fundamentally, an authentic desire to maximize EU does not start by reflecting on the benefits of one's own religion; it starts by asking what activities are known to best promote happiness, and turns to those. Given the attested value of exercise (Lyubomirsky, 2007, p. 245), perhaps true wagerers should spend time at the gym instead of church.

To clarify, I do not suggest that *theists* have a duty to investigate religious hypotheses where atheists do not. I am saying that Pascalian *wagerers* – those who decide to adopt or reject religion on the basis of its prudential value – have a duty, by their own standards, to investigate their options while non-wagerers, by their own standards, do not. For non-wagerers to be rational, evidentialism must be acceptable, the evidence must shape belief, and that's it. For wagerers to be rational, evidentialism must be dismissible, calculations of utility must shape belief, *and enormous calculations must be performed* when enormous utilities are credibly at stake. If this burden weighs more heavily on theists than atheists, it is because many theists justify their belief on the basis of instrumental considerations, while atheists for the most part consider themselves evidentialists.

5 Decision-Rule Relativity: The Wagerer's Dilemma

The mundane wager hinges on the empirical premise that believing in God promotes some given mundane good. This premise is widely taken for granted, but what does the evidence actually say?

Studies tend to show a modest, curvilinear correlation between religion and purportedly positive indicators at the individual level, and a more robust *inverse* linear correlation at population levels (on average the least religious nations are the best educated, the most prosperous, the healthiest and, on self-report surveys, the happiest; and likewise for states within the United States). David Myers, a renowned specialist on well-being and an evangelist who likes to emphasize religion's benefits, puts it this way:

> Curiously, irreligious *places* (nations, states) and highly religious *individuals* tend to exhibit high levels of health, well-being, and prosociality. Religious engagement correlates negatively with prosociality and well-being across aggregate levels (countries and American states), and positively across individuals. [2012; cf. 2013]

These correlations are sometimes ascribed causal significance:

(α) Religion confers net benefits at the individual level.
(β) Religion confers net harms at the societal level.

Claims (α) and (β) admit competing responses. *Skeptics* reject both, the *"straight-forward"* interpretation (q.v. Newport et al., 2012) accepts both, and *discriminators* favor one over the other.

Skepticism is reasonable: mere correlation does not spell causation, and moreover current research is methodologically deficient – not just a little, but systematically and shockingly so (inter alia Sloan, 2006; Haybron, 2008). The straight position too is reasonable: although correlation does not demonstrate relevant causation, it does raise its Bayesian likelihood, especially when plausible mechanisms are postulated and especially given that bidirectional causality is possible (x's contributing to y does not rule out y's contributing to x). As for methodological shoddiness, mortals have not the luxury of waiting until unimpeachable experiments are designed and run; the data are what they are, for the time being, and provisional acceptance is better than paralysis. What's not reasonable, on the present evidence, is discrimination. At some future date better evidence may confirm either (α) or (β) while discrediting the other, but *for now* they enjoy epistemological parity, and to selectively cherry-pick between them (as, e.g., Jordan, 2006, does) is irrational.

Unless we are to be skeptics, then, we are left with a puzzle. On the one hand the straight interpretation holds that religion promotes well-being *within* countries, and on the other hand it holds that religion depresses well-being *across* countries.[9]

These two propositions can be reconciled if we assume that secularism is beneficial to everyone, while religiosity benefits adherents and harms others, for instance by externalizing its costs. As a result, it is possible for massively secular societies to be better off on the whole than massively religious societies, while religious individuals in any given society are better off than their secular compatriots. The logic here is illustrated by the Prisoner's Dilemma, or the Tragedy of the Commons. For instance, consider a common resource such as the sea. If you comply with the laws that govern fishing in season, you can catch a certain amount per year; but if you fish year round, you can catch twice as much. However, if everyone breaks the law, the fish population will collapse (see Table 9.4). Regardless of what everyone else does, you are better off defecting. But if everyone else reasons the same way then everyone defects, which is suboptimal for everyone. Assuming the Tragedy of the Commons as our model of religious outcomes, we get a matrix something like that laid out in Table 9.5.

Cell (b) is set at an arbitrary baseline. Cell (a) is set higher to represent the difference in quality of life between secular and religious societies. Row (c/d) is set higher than (a/b) to represent the benefits of religion as reported in the pro-wager literature: Jordan (2006, p. 91) favors the coefficient of determination r^2 at just 0.5 percent; Rota (2016a, p. 37) quotes a range of 2–6 percent; and the literature at large typically reports a "small" or "modest" effect size. These numbers pull S to religion (because 115 > 110 and 105 > 100), and everyone else too, even though everyone would do better by rejecting religion.

The DILEMMA matrix is suggested by the uncontroversial correlation measures reported by David Myers and others. Additional grounds for taking it

Table 9.4 Tragedy of the Commons

	Everyone else cooperates	Everyone else defects
S cooperates:	(a) 10	(b) 2
S defects:	(c) 20	(d) 4

---

[9] Fagan (2006: 18) and Casey (2009: 7) assert that the truth of (α) entails the falsehood of (β), but as we shall see this is a logical fallacy.

Table 9.5 Wagerer's Dilemma

	Everyone else predominantly rejects religion	Everyone else predominantly chooses religion
S rejects religion:	(a) 110	(b) 100
S chooses religion:	(c) 115	(d) 105

quite seriously are sketched in section 5.1, and their implications for rationality are explored in section 5.2. Namely, I distinguish between classical rationality, which validates the Wager, and cooperative rationality, which validates an opposing anti-Wager. I also consider reasons for preferring cooperative rationality, and conclude my survey with compromise positions that variously straddle and stand between the Wager and anti-Wager.

5.1 Mechanisms of Happiness

Proposition (α) is very popular in our religious culture, and in my experience it needs very little explaining or defending. In contrast, (β) evokes both head-scratching from those who cannot imagine how it might be true, and passionate rebuttals that set a higher evidential standard for (β) than for (α). Accordingly, I will briefly motivate (α) and then linger over (β). In each case my aim is to address the skeptic's worry that correlation does not spell causation, and to document robust correlations. (Exceptions to generalizations, I must emphasize, do not refute statistical tendencies.)

As for (α), it is easy to speculate about causal mediators, between faith and well-being, at the individual level. Theism plausibly manages the terror of mortality, sustains optimism that justice will prevail, and adds meaningfulness to life; and organized religion plausibly builds social capital, promoting friendships, support networks, and a sense of belonging. What's more, some of these ideas enjoy empirical support, with some reputable studies going so far as to claim that *all* of religion's benefits accrue from the social solidarity it fosters (Diener and Seligman, 2002; Putnam and Lim, 2010). In short, few question assigning higher values for theism than for atheism.

As for (β), it is equally easy to speculate about causal mediators between faith and unhappiness at the collective level. For instance, religion seems to inculcate an us-versus-them mentality that contributes to wars and ethnic strife. The history of religious warfare is too well known to need rehearsal, and on top of that the empirical evidence shows a stark correlation between religiosity, on the one hand, and racism and ethnocentric prejudice on the other (see citations in section 2).

As another example, religion plausibly promotes traditional sex roles and suppresses family planning, while gender inequity diminishes happiness for all (Lahn, 2015) and uncontrolled birth rates plausibly contribute to social ills. In one notorious study, the legalization of abortion in the United States was tied to dramatically reduced crime rates in later decades, the allegation being that unwanted children tend to grow up to be delinquents (Levitt and Dubner, 2005, ch. 4; on the inevitable ensuing controversy, see Klick, 2004). Be that as it may, societies would fare better if their population sizes matched their natural resources. Yet religion seems to inspire indifference to the environment, at least in the United States where atheists are substantially more concerned about climate change than Christians, mainline Christians are substantially more concerned than fundamentalists and evangelicals, and degree of religiosity within a denomination correlates with unconcern (Mooney, 2015; cf. Pew, 2015).

Such indifference is understandable given that theism plausibly encourages complacent passivity: if God ultimately guarantees justice, and piety earns you eternal bliss, there is hardly reason to worry about building and maintaining cultural and material infrastructure. That more than 40 percent of Americans expect the Second Coming of Jesus by the year 2050 (Pew, 2010), and that this rate is sometimes even higher among Muslims (Pew, 2012), would help to explain their respective societies' lackluster performances regarding not only environmental conservation but public welfare and intellectual investment. My speculations are one-sided, needless to say, but recall that the point of the present exercise is to appreciate "straightforward" perspective (β), not to prove it. Besides, my speculations *are* backed by a certain amount of evidence.

First, religion correlates with political affiliation. Among US evangelicals, 28 percent lean Democrat (center-left) and 56 percent lean Republican (center-right), while the ratio among atheists is 69 to 15 percent (Pew, 2014). The same qualitative pattern obtains in the UK and Canada (Clements and Spencer, 2014; Mang, 2009); and, globally, religious political parties are predominantly on the right (fn. 5, above). These facts are relevant insofar as political affiliation translates into societal well-being. Democrat governance is said to be better for the US economy than Republican governance, according to both the expert opinions of economists and actual performance measures (Anon., 2015). Some 19 percent of Democrats, and 58 percent of Republicans, believe that higher education harms society (Pew, 2017). Democrat policies purportedly work better for society all around than Republican policies:

among scientists, who are comparatively data-driven and able to see through rhetoric, 81 percent lean Democratic while 12 percent lean Republican (Pew, 2009). Internationally, according to the conservative magazine *Forbes*, welfare states are the happiest (Helman, 2011; cf. Rothstein, 2010; Radcliff, 2013).

Second, religion has a rich long history of discouraging the life of the mind – learnedness, reason, inquiry, science, and education. To be sure, Christianity has contributed *enormously* to literacy, schooling, and dialectic, and there is much to be said for being skeptical of (β). But again, the present topic is not to prove (β) but to hear it out. From Christianity's sacred Scripture we have: "with much wisdom comes much sorrow, with much knowledge much grief" (Ecclesiastes 1:18; cf. Matthew 23:10). From Christianity's early theoreticians and prominent apologists, we have: "there is no need of any wish to learn, except of Jesus Christ" (Tertullian, third century); "the learning and study of these matters [natural phenomena] is impious" (Eusebius, fourth century); "The disease of curiosity ... drives us to try and discover the secrets of nature, ... which man should not wish to learn" (Augustine, fifth century). From Catholics we have suppression of popular access to the Bible (e.g., the Wycliffe and Tyndale affairs), and from Protestants we have "reason is the greatest enemy that faith has" (Martin Luther). In every century, religious forces have banned and burned books. If religion *on balance* valued knowledge, it would be hard to explain why freethinkers disproportionately contribute to science (Larson and Witham, 1998; Pew, 2009; Ecklund et al., 2016), and why 25 percent of US Christians have a college degree, compared to 42 percent of atheists (Pew, 2014). (My focus has been on Christianity because it's more intellectual than Islam, by far, and together the two constitute well over half the world's population.)

To summarize, there are plenty of hypotheses to explain why religious faith benefits the individual believer, and at least one, pertaining to the social nature of religious practice, is supported by the evidence. At the same time there are plenty of reasons to suspect that religious faith harms society. Available data demonstrate connections linking religiosity to ethnic conflict, uncontrolled population growth, ecological shortsightedness, contempt for education and science, and conservative government (cf. Duncan, 2013, p. 394). Whether these qualify as harms is at least partly a matter of opinion, and direction of causality is not always clear, but the reported correlational figures are incontestably what they are, and mechanisms that could plausibly explain them have been postulated.

Of course, religious sects are not all alike, and some are decidedly beneficial for all of society. Yet irreligious populations are not all alike either, and it is important not to compare the very best theists with average atheists, or average atheists with the very worst theists. At any rate, until proponents of the mundane wager slice ideologies and practices more finely, I follow the available anglophone literature in focusing primarily on theism, Christianity, and atheism as such, i.e., in the aggregates.

5.2 Superrationality and the Anti-Wager

The Tragedy of the Commons, which is clearly not hypothetical, demonstrates that *if everyone is rational* in the atomistic manner prescribed by standard decision theory *then everyone is worse off*. It suggests that we should follow instead a cooperative kind of rationality consistent with what Hofstadter (1983) calls superrationality. This cooperative rationality can be distinguished from the classical rationality of defectors ("defective rationality"?) by means of the following rules:

- (R_{cl}) To be rational, select the action that maximizes EU.
- (R_{co}) To be rational, select the best action to secure Pareto optimality, at which no one can improve without making another worse off. (This rule prescribes cell [a] in the cases of TRAGEDY OF THE COMMONS and WAGERER'S DILEMMA; its workings are illustrated below.)

The difference between classical and cooperative rationality can be seen as a matter of perspectival relativity. When Σ applies rule (R_{cl}), formula (R) from section 4 describes classical rationality; when Σ applies (R_{co}), (R) describes cooperative rationality.

The two rationalities converge when one-on-one game-playing is indefinitely iterated, or whenever cooperation can be rewarded and defection punished. (In those cases, the players' actions lose their probabilistic independence, a condition assumed in the rest of my discussion.) Furthermore, even without sanctions, standard decision theory can emulate cooperative rationality by building into everyone's utility function a preference for acting cooperatively. Yet classical rationality remains inadequate, for in some sense cooperating seems to be the rational thing to do regardless of sanctions or the satisfaction of acting atomistically (assuming the outcomes of the TRAGEDY/DILEMMA matrices).

To make cooperative rationality more appealing, I begin by rebutting the charge that it is oxymoronic. Perhaps *morality* requires acting cooperatively, it has been said, but *rationality* does not. This conviction, I believe, is induced merely by familiarity with (R_{cl}), and by the shaping of our minds by the social environment we have grown up in. To appreciate that (R_{cl}) is not analytically valid, note that it is rejected in forager cultures (Henrich et al., 2005); cooperation is prized over atomism in Asian cultures (Nisbett, 2003); and rival rationalities have been advocated by Western scholars (Kant, 1997 [1785]; Hofstadter, 1983; Rescher, 1985; Gauthier, 1986; Anderson, 1993).

Moreover, cooperative reasoning must not be confused with altruistic or collectivist reasoning. To see the differences, and to see how (R_{co}) works, consider the matrix laid out in Table 9.6. In each cell, the first number represents the outcome value for agentive subject S; the second, for T. When S and T share the information in this matrix as common knowledge, different decision rules yield different results.

- If everyone is classically rational, cell B2 will be selected. S reasons that action B is the best option under each condition (1, 2, 3). T reasons that action 2 is the best option under condition A, and ties for best under B and C.
- If everyone is altruistic, cell C3 will be selected. S reasons that row C is the best for T, and T reasons that column 3 is the best for S.
- If everyone is utilitarian, cell A3 will be selected. S and T both see that its sum is greater than the sum in any other cell, and act accordingly.
- If everyone is cooperative, everyone will "select the best action to secure Pareto optimality" according to a hierarchy of criteria. (i) The best action for securing Pareto optimality is, foremost, one that guarantees it. For example, in THE COMMONS matrix the optima are found in cells (a), (b), and (c). If S cooperates then an optimal outcome is assured regardless of what everyone else does, but if S defects then S risks losing an optimal outcome. The same reasoning applies to everyone else, and so the "best" reasoning – the safest – yields outcome (a). (ii) If no action uniquely guarantees an optimal outcome then it would seem reasonable to revert at this stage to classical reasoning. In the COMPARISON matrix, the competing Pareto optima are A1, A3, and C1. Of these, A1 dominates the others and is

Table 9.6 Strategies Compared

	T's action 1:	T's action 2:	T's action 3:
S's action A:	7, 7	1, 8	9, 3
S's action B:	10, 2	2, 2	10, 1
S's action C:	6, 9	1, 9	8, 4

selected. (iii) In cases where the preceding criteria fail to determine a decision, principles of parity, utilitarianism, turn-taking, etc. arguably should come into play.

Cooperative rationality thus does not sacrifice self-interest for the greater good, nor does it demand totalitarian uniformity of beliefs and values on the part of the players. It aims at maximizing the good of the agent just as much as (R_{cl}) does, but with a different strategy.

Decision strategies can be mixed, and descriptively speaking often are: it is just a fact that in some populations some actions will be altruistic, some utilitarian, and so forth. For normative purposes, however, we seek to find consistent general *rules*. I therefore focus on cases where everyone follows the same pure strategy. Thus, S is rational if S does S's part in creating the best outcome S can expect, given that others are doing their part in trying to create the best outcome they can expect.

As a final defense of (R_{co}), note that the prisoner's *dilemma* is so-called for a reason. The prisoner's thinking is pulled in two directions at once, which indicates that rationality is not univocal. In view of the fact that if everyone follows (R_{cl}) then everyone can be worse off, and if everyone follows (R_{co}) then everyone can be better off, cooperative rationality is at least as good a norm as classical rationality.

5.3 Bottom Lines

So far I have entertained two propositions: (α) religion confers net benefits at the individual level, and (β) religion confers net harms at the societal level. Collectively these constitute the wagerer's dilemma, assuming that cell (a) > (d). One response to the dilemma is to accept (α), (β), and (R_{cl}). From this perspective *the mundane wager is sound*, and belief in God is rational. The opposite response is to accept (α), (β), and (R_{co}). From this perspective *the mundane wager is unsound*. The latter perspective does not merely reject Pascal's Wager but turns it completely around. It yields a version of what I call the anti-Wager:

(A) Action A is rational iff A promotes the good.
(B*) Believing in God demotes the good.
(C*) Therefore believing in God is irrational.

I myself disavow the first premise, as I favor epistemic rationality over instrumental rationality in the matter of religion. For those who follow expedience in shaping their religious faith, however, the anti-Wager should be taken much more seriously than it is.

Other responses to the wagerer's dilemma are possible too. One is to accept, along with (α) and (β), the validity of both (R_{cl}) and (R_{co}). From this perspective *the Wager and anti-Wager alike are sound*. This gives us another route to the pluralist position, floated in section 3, whereby theism and atheism are both rational. Alternatively, one might be torn between (R_{cl}) and (R_{co}). Unable to decide which decision rule to follow, the conflicted follows neither. From this perspective (R_{cl}) and (R_{co}) both lie inert, and *the Wager and anti-Wager alike are unsound*. Finally, it is arguably rational to follow the skeptic in rejecting both (α) and (β). From that perspective, too, *the Wager and anti-Wager alike are unsound*.

My conclusions carry substantial caveats. First, our empirical knowledge of correlates and causes will almost certainly develop as time goes by, and as it does it may render one or more of the foregoing perspectives untenable. Second, from any given perspective there will surely be identifiable demographic subgroups that make exceptions to any universal rationality-claim; and in that way, as well, rationality is relative.

10 The Role of Pascal's Wager in Authentic Religious Commitment

Joshua Golding

1 Introduction

Pascal's Wager argues that belief in God is rational based on pragmatic considerations. The argument claims that even if from a cognitive or evidential point of view one cannot determine whether God exists, it is still rational to believe in God based on the premise that one has more to gain than to lose by believing in God. Pascal's argument has raised numerous objections. One objection is that a belief in God based on such an argument would be insincere or inauthentic. More broadly, one may wonder what role the Wager can play in the life of a sincere religious person. If indeed a wager-based belief is plagued by some form of inauthenticity, the question arises as to whether the Wager can be remedied in such a way at to legitimize its role in a devout religious life.

This chapter aims to address the following two questions. First, is a belief in God that is based on a wager-style argument authentic? To consider this question, we shall need to discuss the notion of an authentic belief. There may be some reason to think that, in a religious context especially, such a belief is not authentic. But our discussion here will also take us somewhat outside the religious context, as we shall consider whether any belief that is based on pragmatic considerations is authentic. Second, what role can the Wager play in a sincere or devout religious life? The only way to address this question is to consider the issue from the point of view of a specific religion. As the religion with which I am most familiar is Traditional Judaism,[1] we shall consider this question from that perspective. This may also offer some readers a fresh

---

[1] For the purpose of this chapter, Traditional Judaism teaches that there is a supreme God, who created the world and gave the Torah to Israel. The Torah includes divine commandments (*mitzvot*). By fulfilling the divine commandments, it is possible to live a good and holy life, and attain a good relationship with God, both in this world, and the next. Needless to say, there are many forms of Judaism; this chapter analyzes the Wager from one Jewish perspective.

perspective on the Wager, as it is usually studied within a Christian framework.

Before proceeding to our main discussion, a few preliminaries are in order. Some readers may be surprised at the prospect of analyzing Pascal's Wager from a Jewish point of view. In its original context, Pascal's Wager seems to be an argument for *Christian* belief. However, one can also read Pascal as arguing for belief in God, whilst leaving open the question of what version of monotheism one should adopt. To sidestep this controversy, here we shall make use of the notion of a "wager-style" argument. A wager-style argument is any argument that claims one should believe that God exists based on the premise that one has more to gain than to lose by being a believer. Defined in this way, a Jew (or Muslim or Christian) might have a belief in God that is based on a wager-style argument. For example, suppose a person is faced with the question about whether to believe in the God of Israel, who gave the Torah and the commandments, and who holds the promise of some great good for those who are faithful. Suppose he finds himself in the position of cognitive doubt about whether the God of Israel exists. Now suppose this person accepts the premise that he has more to gain than lose by being a Jewish believer, and he decides to believe on this basis. The question of this chapter is whether such a belief in God is authentic, and, more broadly, what role can the Wager play in the life of a devout religious Jew. I venture to say that many (though perhaps not all) of my conclusions will find agreement among Christians and Muslims as well. Readers from those traditions or others may be stimulated by this chapter to reflect on the extent to which the conclusions of this chapter apply to their tradition.

It is important to distinguish the main questions of this chapter from other issues that are, so to speak, in the neighborhood. First, the topic of this chapter is *not* to investigate the general soundness of the Wager argument. It may turn out that Pascal assumes certain premises that are unfounded, or that his premises are true but the conclusion does not follow from the premises. In general, a wager-style argument relies on the premise that one stands to gain some great value only if God exists and one believes in him; if this premise falls, the argument falls. It is not our purpose here to worry about this issue. For the purpose of this chapter, suppose that a person finds the premise to be true, and thinks that the argument is convincing. The question is whether a belief in God based on such an argument is authentic, and what role such an argument might play in the life of an active religious believer. Even if the belief is authentic, that does not entail that the argument is free from other flaws.

Second, it is not our purpose to investigate the whether it is *morally permissible* or *ethically upright* to believe in God based on a wager-style argument. Many philosophers (such as W. K. Clifford) have questioned whether it is morally permissible to form a belief based on pragmatic considerations. Our concern is whether there is some kind of *inauthenticity* involved in having such a belief. These issues are related, for if the belief is inauthentic, that might be one reason for thinking that such a belief is morally questionable. But these issues are not identical, as (a) even if the belief is authentic, there might be other reasons for holding that it is immoral to form such a belief, and (b) there may be cases where inauthentic belief is not necessarily immoral. For the purpose of this chapter, we shall set aside the moral question, and focus on whether such a belief is authentic.

A brief overview of the main claims of this chapter may be helpful. I shall argue that in many (though perhaps not all) cases, a belief based on a wager-style argument will turn out to be *inauthentic*. However, I shall also suggest a remedy – namely, that the argument may be revised in such a way as to dispense with *belief* and instead be construed as supporting a *pragmatic assumption* that God exists. As I shall explain, a person who makes such a pragmatic assumption based on a wager-style argument is not engaged in an inauthentic commitment. Furthermore, a pragmatic assumption that God exists can play a strong role in a very active Jewish religious life. Nevertheless, we shall also find that, at least from a Jewish perspective, there are certain limitations to the religious life of a person who makes only a pragmatic assumption that God exists, and who does not have a genuine belief in God. In the final analysis, I shall suggest programmatically that a wager-style argument may play a wholesome role in a religious life if it is coupled with a cognitive argument for knowledge of God based on religious experience. Whether or not such an argument can be successfully constructed is beyond the scope of this chapter.

2 Is a Wager-Style Belief that God Exists Inauthentic

In this section, we shall examine three reasons for thinking that a belief in God based on a wager-style argument is inauthentic. The first involves a concern that, specifically, a belief in God that is based on pragmatic considerations is somehow impious and therefore *religiously inauthentic*. I shall argue that this concern is rather easily dismissed. The next two reasons have nothing in particular to do with religion, but rather with any belief on any topic that is generated from pragmatic considerations. In effect, the worry here will be that such a belief is *epistemically inauthentic*. I shall argue that these reasons are not easily dismissed, and that, for many people, a belief in God based on pragmatic considerations would be epistemically inauthentic.

The first reason for thinking that a belief in God that is based on pragmatic considerations is inauthentic comes from a religious point of view. It seems that a belief in God that is based on a wager-style argument is motivated by self-interest. Surely, God (if he exists) does not approve of someone's believing in him for the purpose of self-gain. If someone believes in God on this basis, his belief is *religiously authentic* in the sense that it is not formed in a religiously appropriate way.

In my view, this consideration is dismissible, for two reasons. First, from whose religious point of view is the judgment being made that one should not believe in God based on self-interest? Most, if not all, of the major religions teach that at least one reason, if not the only reason for worship of God is self-interest, either in this world, or the next. The Bible is filled with the teaching that worship of God and faith in God is ultimately in the best interest of the worshipper.[2] Second, the Wager can easily be construed as arguing that the value at stake in being a believer is not only a matter of self-interest, but also a matter of what is objectively best for a person. Pascal does speak of "an infinity of infinitely happy life to be won." But it is obvious that Pascal himself thinks that the attainment of heaven or eternal life with God is not only in one's own self-interest; it is also the best possible condition for a human being. A wager-style argument need not appeal only to self-interest; it may also appeal to what is best for the human being. Hence, it is not religiously inauthentic for a person to believe in God on the basis of the Wager.

Let us move on to other, more epistemic worries. A second consideration runs as follows. Obviously, just because a person professes to have a certain belief does not mean that he really does have that belief. He could be lying. More subtly, a person could be in the position of lying to himself. He could tell himself that he has a certain belief, but not actually have that belief. Alternatively, a person could act as if he has a certain belief, but not actually have that belief. In this sense, a belief is *authentic* only if it is really held by that person; an inauthentic belief is not really a belief at all. It is a belief that a person claims or professes to have, but he doesn't in fact really have it.

Note that the issue of whether a belief is authentic does not depend on whether the belief happens to be true or false. It is possible for someone to

---

[2] See Deuteronomy, 6:24, 10:13. The book of Psalms is laden with this teaching. See Psalm 1, 23, and *passim*.

mistakenly, yet authentically, believe something; it is also possible for someone to believe something inauthentically that happens to be true. Clearly, the issue of whether a belief is authentic is relative to a given person.

Some philosophers might propose that a belief is authentic for a given person only if that person bases his belief on cognitive or evidential reasons. However, without further substantiation, this proposal would be question-begging in our context. A proponent of the Wager might simply reject this proposal and insist that a belief could be authentic even if it is not based on cognitive or evidential reasons. How then may we understand the notion of an authentic belief, in such a way that is does not beg the question against those who are inclined to accept a wager-style argument for belief in God?

I suggest the following. Given that p is some proposition, imagine that a person were to say, "I believe p, but I don't really think p is true." Surely, this would elicit a puzzled look, if not a jeer. An appropriate response would be, "If you don't think p is true, you don't really believe p!" If such a person were to insist that he does believe p, even though he thinks p is not true, we could say that his belief in p is not authentic. Formally stated, a belief p is authentic for a person, S, only if "S thinks p is true." This does not mean that S must be *certain* of p, nor does it mean that he has cognitive or evidential reasons for it. Someone may genuinely think that p is true, even though he is not certain of p and even though he cannot give any reason to substantiate p. Notice that we have not *defined* what it means for a belief to be authentic; we have only stated one necessary condition for an authentic belief.

What do I mean by the phrase "S thinks p is true"? Admittedly, this is hard to define, and I shall not even try to do so. But, we can say the following. Generally, when a person is asked the question "Do you think p is true?" – the answer should be immediate and come from the heart (or the mind). The answer to such a question is either "yes," or "no." If the answer is "no," then the person does *not* think p is true, and if he were to say that he believes p, he would be insincere. If this is correct, then, the belief that God exists is authentic for a person only if he or she thinks it is true that God exists.

We may now consider whether a belief in God that is based on a wager-style argument is authentic in the sense described. For some people, it may happen that they find the argument convincing, and they also happen to think God exists independently of the Wager. If so, their belief is not solely based on the Wager after all. We may set that case aside, for our concern is with those persons who base their belief on the Wager. By hypothesis, such persons were in the position of *not* believing that God exists before they considered the

Wager argument. Before considering the Wager, the response of such persons to the question, "Do you think there's a God?" would be "No." Furthermore, even after hearing the Wager and finding it rationally persuasive, it is unlikely that they would immediately start thinking that God exists. It is possible that this could happen in some cases, but in many cases, that simply will not happen. In those cases where it does happen, the belief is authentic, for the person actually thinks there is a God. But for most people, if they were to start professing a belief in God immediately after hearing the Wager, their belief is inauthentic. They simply don't have the belief at all, even though they may say that they do.

We now move on to a third consideration. In his original argument, Pascal himself realizes that just because a person finds the Wager persuasive does not mean that he will suddenly start believing in God. For such persons, Pascal prescribes that one should act like a believer (pray, take holy water, etc.) in order to inculcate the belief in oneself. On the surface, it seems that here Pascal recommends that a person convince himself that God exists by repeatedly saying things like, "I believe in God." This seems like a strategy of self-inducement or self-brainwashing. Now, for some persons, such a procedure may not even work. If it doesn't work and they still profess a belief in God, their belief is inauthentic because of the consideration raised above. On the other hand, suppose this strategy does succeed in getting a person to think and believe that God exists. The problem is that the belief is inauthentic because it was not generated in a proper manner. It was generated because of a kind of self-brainwashing. I hasten to point out that saying this does not entail that the only proper way that a belief may come about is through cognitive or evidential reasons. All we are saying is that a belief that is generated by self-inducements of the sort Pascal describes seems like a disingenuous belief.

Toward the end of this chapter, I shall suggest an alternative way of understanding Pascal's concluding recommendation. At this point, it seems there are two good reasons for thinking that, for many persons, a belief in God that is based on a wager-style argument would be inauthentic.

3 Pragmatic Assumption (instead of Belief) that God Exists

I have argued that for many persons, a belief in God based on a wager-style argument would be inauthentic. I now wish to propose a remedy. The argument may be revised in order to escape this problem. Instead of arguing for a *belief* that God exists, the Wager may be revised as an argument in favor of a *pragmatic assumption*[3] that God exists, for the purpose of attaining a good relationship with God. This requires explanation and defense.

What is meant by a *pragmatic assumption*? A pragmatic assumption that p is not the same thing as a belief that p. A person who makes a pragmatic assumption that p guides (at least some of) his actions in such a way as if p were true, for the sake of achieving some goal. Such a person need not think that p is true. However, he cannot be in the position of thinking that p is definitely false. For example, a person who wishes to find a friend lost at sea may – indeed must – assume that his friend is still alive for the sake of attaining the goal of saving him. Since the goal of saving the friend is only achievable if the friend is actually alive, the pursuit of that goal entails that, at least for this practical purpose, the person must act as if the friend is alive. Such an assumption may make sense, but it won't make sense if he definitely thinks his friend is dead. Similarly, a person who applies for a job may – indeed must – assume that the job has not already been given to an insider. Other cases in real life abound. We often make assumptions for practical purposes even when we know that our assumptions are likely to be false. We do so precisely when some gain is at stake that makes it worthwhile to make some pragmatic assumption, and so long as we think there is some live possibility that the assumption is true.

It is also important to the note that a pragmatic assumption that p is not the same thing as a belief that there is some small probability that p. A person could believe that there is some small probability that p, but still not act on the assumption that p. For example, a non-religious person may believe there is some small probability that God exists, but for all practical purposes, he does not act on the assumption that there is a God. On the other hand, a person who makes a pragmatic assumption that God exists will tend to act in very different ways from a person who merely believes that there is a probability that God exists.

Now let us return to the case of the Wager. Suppose a person is in the position of cognitive doubt about whether God exists. If asked whether he thinks that God exists, he would say, "No." Now suppose this person hears and finds convincing the following wager-style argument: A person can attain some great gain (heaven, bliss, eternal life) only if God exists and one has a good relationship with God. The only way or the best way to attain that relationship is to adhere to some religious path. (Whether these premises are true is not our topic here.) Based on these premises, this person concludes that it is

---

[3] For more on this notion, see Golding (1990).

pragmatically rational to pursue some religious path, even in the face of doubt about whether God exists. It follows that it is pragmatically rational for him to assume that God exists, for the purpose of pursuing a good relationship with God. In other words, he should now guide his actions as if God exists, for the purpose of attaining that goal.

Incidentally, this approach has another advantage over Pascal's original argument. Let's assume that one can gain infinite eternal bliss only if God exists. (Some critics of the Wager reject even this assumption, but let's assume it is true here. This assumption is reasonable, especially if infinite eternal bliss is understood as partaking in the infinite eternal goodness of God.[4]) Yet, Pascal further assumes that one may access the great gain at stake in being religious only if one *believes* in God. From the point of view of Pascal's skeptical interlocutor, this assumption is (or should be) questionable. Perhaps belief in God *per se* is neither necessary nor sufficient for gaining eternal bliss. Perhaps there are *other* things a person needs to *do* (or *avoid*) in order to access eternal bliss. The argument of the Wager should be that one should engage in whatever actions will promote the good relationship with God, even in the face of doubt about whether the conditions are such that one will succeed.[5] A consequence of this is that it is rational to make a pragmatic assumption that God exists, in order to guide one's actions toward that great end.

Let us return to the main thread and summarize what we have said thus far. In many cases, a person who forms a belief in God's existence based on a wager-style argument will have an inauthentic belief. But there is nothing inauthentic about making a *pragmatic assumption* that p even where one doesn't think p is true. The only case where it would not make sense is if the person thinks that p is definitely *not* true. Even then, such a commitment would not be *inauthentic*; it would just be silly. In any event, by hypothesis, a person who accepts the Wager argument already thinks there is a live possibility that God exists. A person who makes a pragmatic assumption that God exists based on the Wager argument is not engaged in an inauthentic commitment.

4 The (Limited) Role of Pragmatic Assumption in Religious Life

Our next task is to consider how far such a pragmatic assumption can take a person in an active religious life. To consider this question is to consider what role the Wager argument can play in the life of a religious person. While

---

[4] For example, see Moshe Chaim Luzzatto (1978, pp. 37ff.).
[5] For a sustained effort to work out such an argument, see Golding (2003).

the remedy I have suggested dissolves the problem of authenticity, it comes at a cost. As we shall see, the cost is that without belief in God, the life of a religious person will be rather vexed, to say the least.

We shall use Traditional Judaism as our framework to consider this question. Suppose that a person is convinced by the wager-style argument above that it makes sense to pursue a good relationship with the God of Israel, even in the face of doubt about whether such a God exists. Suppose also that, having read the first few sections of this chapter, this person wishes to avoid an inauthentic belief in God's existence. Instead, he adopts a pragmatic assumption that the God of Israel exists for the purpose of pursuing a good relationship with Him. Given that such a person does not have a *belief* that God exists, what kind of religious life can such a person lead? For the sake of brevity, in the following discussion we shall refer to this person as the "faithful Jewish pragmatist."

The answer is not simple. On the one hand, a pragmatic assumption that God of Israel exists can play a strong role in a very active religious life. Nevertheless, there are some severe limitations to the devotional life of the faithful Jewish pragmatist.

Let's state the positive side first. If the God of Israel exists, then the Torah represents the divine commandments (*mitzvot*) that are binding upon the Jewish people, and it is through keeping the Torah that the Jew attains a good relationship with God. Much of the religious life of the Jew includes following or keeping the positive and negative commandments. The faithful pragmatist will have much to do, and much to avoid doing. Many of the commandments involve moral and socially productive behavior, so he will be committed to those commandments. Many of the commandments involve diet. Just to name a few examples, such a person will refrain from eating non-kosher food, refrain from eating bread or bread products on Passover, and he will fast on Yom Kippur. Other commandments involve the observance of special days, such as refraining from work on the Sabbath and on religious holidays. He will not only refrain from violating the negative commandments but also keep many of the positive commandments. He may hear the blast of the Shofar on Rosh Hashanah, eat meals in a *sukkah* (festive hut) on the Feast of Tabernacles, and light the Menorah on the festival of Hanukkah. All of these activities are part of this person's quest to attain a good relationship with God, even in the face of doubt as to whether or not God exists.[6]

---

[6] While many of the commandments are bodily, some are cognitive. Moses Maimonides held that it is a positive commandment to *know* that God exists. See Maimonides, *Mishneh Torah (Code of Jewish Law): Laws of the Foundations of the Torah*, I:1. The faithful pragmatist is not in the position to fulfill this commandment.

Aside from bodily commandments, another important aspect of Jewish religious life is study of Torah. The faithful Jewish pragmatist may be deeply involved in the study of Torah. He may engage in this study not merely as an intellectual exercise, but as an attempt to understand what he must do in order to relate to God well – should it be the case that God actually exists. The addition of that last modifying phrase makes the enterprise of learning Torah sound somewhat peculiar. But many Jews (and non-Jews) have studied Torah under a condition of doubt whether the Torah is genuinely divine; some later come to accept the Torah as divine and some do not. In any case, if by "learning" one simply means "understanding the words" of the Torah, both the person who actually believes in God and the faithful pragmatist can "learn" Torah.

The situation of the faithful pragmatist becomes thornier, when we consider other aspects of Jewish religious life, such as Jewish *liturgy*. One aspect of Jewish liturgy is *prayer*. Jewish thinkers have noted that prayer involves two aspects, namely, *petitionary prayer* (*bakashah*) and *worship* or *service of the heart* (*avodah she-belev*). In petitionary prayer, a person asks God for his needs, both material and spiritual. If a person makes only a pragmatic assumption that God exists, can he sincerely engage in petitionary prayer? The situation of such a person reminds us of the amusing prayer penned by Renan on behalf of the agnostic: "Oh God – if you exist, save my soul – if I have one!" Nevertheless, despite its peculiarity, such a prayer and indeed any petitionary prayer could be said by the faithful pragmatist with great sincerity.

On the other hand, the prospect of engaging in genuine *worship* of God seems remote. Can I genuinely "serve" God if I don't fully believe that God exists? Can I be a devoted servant of a Master whose existence I doubt? A conditional devotion does not seem like a genuine devotion after all. At best, this seems like a rather vexed devotion indeed.

Another aspect of Jewish liturgy is thanksgiving and blessings. Can the faithful pragmatist sincerely engage in thanking and/or blessing God's name? As with petitionary prayer, these activities could be carried out in a conditional manner. The faithful pragmatist can utter words of thanksgiving and praise, whilst thinking in the back of his mind that his thanks and praise may after all be a waste of time and breath. I leave it to the reader to consider whether such a thanksgiving counts as a genuine thanksgiving. However, even if the *act* of thanksgiving doesn't quite make sense for the faithful pragmatist, it seems that the faithful pragmatist can cultivate a thankful disposition about his natural gifts.

Another aspect of Jewish liturgy is affirmation. Two of the most important aspects of daily Jewish liturgy are the *Shema* and the *Kedushah*.[7] The *Shema* is typically understood as an affirmation that there is only one God. In the Talmud, the *Shema* is regarded as the acceptance of the kingdom of heaven upon oneself, and the acceptance of the commandments.[8] If the *Shema* is an affirmation of creed, it would seem disingenuous for the faithful pragmatist to assert it. If it is a statement of acceptance or commitment to God and to keep the commandments, it seems less problematic. On the other hand, the *Kedushah* is affirmation of the holiness or transcendence of God. If a person does not genuinely believe that the God of Israel exists, can he recite these passages with any sincerity? The notion of a conditional sanctification ("God, if you exist, you are holy!") seems vacuous.

Finally, Jewish religious life also consists in the cultivation of certain spiritual virtues (singular: *middah*; plural: *middot*). Again, here the situation of the faithful pragmatist is rather vexed. Certain religious emotions and certain spiritual virtues are available to him, but others are not. We have already noted above that the faithful pragmatist may cultivate the virtue of thankfulness, or at least, a thankful disposition. Another important spiritual virtue is humility (*anavah*), which we may understand to mean a sense of one's limitations and flaws. It seems that the faithful pragmatist can cultivate humility, as this particular virtue does not presuppose a belief in God. Another virtue is *yirah*, which is fear or reverence of God. Can a person who does not actually believe in God have fear of God? It seems that one can be in the position of fearing something or someone about whose existence one is dubious. If I even think there *might* be a hidden video in a room where I am taking a test, I might reasonably and genuinely be afraid to cheat. Similarly, if I think there *might* be an almighty God, I can genuinely cultivate a kind of fear of God. But can I cultivate *reverence* under those circumstances? That seems more dubious. Another important *middah* is *ahavah* or love of God. Can the faithful pragmatist genuinely have *love* of God? Can you love someone if you don't fully believe in his existence? A conditional love does not seem like much of a love at all. At best, it is surely a vexed form of love.

Ultimately, for Judaism, the best relationship with God consists in *devekut* or "bonding" with God. Many Jewish thinkers (philosophers and mystics alike) regard *devekut* as the pinnacle of the best relationship with God. In *devekut*, a person reaches a most intimate relationship where it is as if

---

[7] The Jewish prayer book is known as the *Siddur*. See Scherman (2001). The *Shema* is found on p. 90 and the *Kedushah* on p. 100.
[8] *Mishnah Brachot* (Blessings), ch. 2: Mishnah 2.

one partakes of the very divine nature of God.[9] The way to arrive at *devekut* is to keep the commandments, study Torah, and worship God with *yirah* and *ahavah*. For many Jewish thinkers, the great eternal reward in the next world is itself a form of *devekut*. Now it is possible for someone to *pursue* this goal whilst not having a firm belief in God. However, it does not seem possible for someone to *attain* this goal whilst not having a firm belief in God. Indeed, if one reaches *devekut*, one is in some sense bound together with God; in that state, one *knows* that there is a God, just as much as one knows that oneself exists. Knowledge of God is itself part of *devekut*. The faithful Jewish pragmatist cannot have *devekut* – unless he becomes no longer a pragmatist but a knower of God.

In sum, the life of the faithful Jewish pragmatist can be very busy with religious activity. He can vigorously keep many of the commandments and engage heavily in the study of Torah. He may sincerely engage in petitionary prayer and in verbal expression of his pragmatic commitment. There are also certain *middot* that he may pursue and cultivate. But in other respects his religious life will be stunted. There are some *middot* that he cannot attain unless he becomes a genuine believer in God's existence. Most importantly, he cannot attain the ultimate relationship (*devekut*) with God unless he somehow comes to *know* that God exists.

5 Conclusion

Our discussion has taken a significant turn. We just noted that the faithful Jewish pragmatist cannot be in the position of having *devekut*, since this requires knowledge of God. But, this limitation is not unique to someone whose commitment to Judaism is based on a wager-style argument. This limitation plagues any religious Jew who merely *believes* in God, but does not *know* God. The pursuit of knowledge of God is itself part of the religious life, not just for the Jewish pragmatist, but for any religious Jew. Until and unless one has such knowledge, one cannot attain the best relationship with God.

The question of whether knowledge of God is possible, and if so how, is beyond the scope of this chapter. Nevertheless, let us consider two very different options for the theologian on this question. One view might be that such knowledge is impossible until after death. On that view, the religious person spends his life pursuing a goal that can only take place, if at all, in the

---

[9] How *devekut* happens or can happen is not our subject here. However, see the discussion of *devekut* in Golding (2018).

next world. In a sense, the religious person is a kind of tragic hero. His entire way of life is based on a commitment whose significance remains in question throughout his life.

A more optimistic, if also more ambitious view is that there is a way of attaining knowledge of God in this life, and that the way to do so is by living a devout religious life and coming to know God through religious experience.[10] This suggestion takes us back to Pascal's concluding recommendation toward the end of the Wager. Perhaps we misunderstood Pascal's recommendation as a form of self-brainwashing. Rather, perhaps Pascal intends that by living a religious life, a person may cultivate genuine religious experience of God, to the point where one comes to know God. A religious Jew would do so by keeping the commandments and studying Torah, and engaging in prayer and meditation. The goal would be to attain knowledge of God and *devekut* in this life.

In summary, the main claims of this chapter are as follows. For most persons, a belief in God that is based solely on a wager-style argument would be epistemically inauthentic. However, a pragmatic assumption that God exists based on a wager-style argument is not inauthentic. Such an assumption can form the basis of a very active religious Jewish life. It can also fuel a spiritual journey in which a person cultivates certain *middot*, and progresses toward *devekut*. One of the goals of this journey would be to replace his pragmatic assumption that God exists with a cognitive experience that brings about knowledge of God. During periods of cognitive doubt, the religious person may fall back on the pragmatic assumption. Ultimately, a commitment based on a wager-style argument plays a wholesome role in a religious life only if it can be coupled with a cognitive argument based on religious experience that leads to knowledge of God. Whether such an argument can be successfully constructed is beyond the scope of this chapter.

---

[10] The argument from religious experience is considered by William James in his classic, *Varieties of Religious Experience*. More recent exponents of this argument include Richard Swinburne and William Alston. See works cited in the bibliography.

Part III
Extensions

11   The Arbitrary Prudentialism of Pascal's Wager and How to Overcome It by Using Game Theory

Elliott Sober[*]

1 Distinguishing Two Objections to Pascal's Wager

In his Wager argument, Pascal[1] makes assumptions of two kinds. He assumes some facts about what God would be like if such a being existed. He also assumes some facts about what human lives would be like if God existed and what those lives would be like if God did not. I'll call these two classes of conditional assumptions Pascal's "theology."[2] Pascal then asks you to decide, based on prudential considerations (the desire for heaven, the fear of hell, etc.), whether to believe in God. Pascal claims to show that even if your evidence leads you to assign God's existence a very low probability, there is a compelling prudential reason for you to believe in God. The many-gods objection to the Wager argument questions the correctness of Pascal's theology, noting that there are alternative theologies that issue in prudential decisions about whether to believe in God that differ from the one that Pascal recommends. In our article "Betting against Pascal's Wager," Gregory Mougin and I claimed to have found a different objection. Pascal assumes a theology (let's call it "P theology") and then subjects belief in God to prudential scrutiny, but why not do the reverse? Why not believe what the evidence tells you about the existence of God, and then do a prudential adjustment of your theology? Rather than retain Pascal's theology and change from atheism (or agnosticism) to theism, why not retain your atheism

---

[*] I am grateful to Emily Barrett, Paul Bartha, Branden Fitelson, Daniel Hausman, Gürol Irzik, Gregory Mougin, Gregory Nirshberg, Clinton Packman, Lawrence Pasternack, Alan Sidelle, and Michael Titelbaum for their help.

[1] Here and in what follows, I'm not discussing what the historical Pascal thought and wrote, but about the decision-theoretic argument that now bears his name.

[2] This definition of "theology" is stipulative. The Pascalian claim that you won't go to heaven or hell if there is no God is part of what I'm calling a theology, and this is something with which many atheists would agree (though they might be dismayed to hear this called a theology). Since I want the existence of God to be logically independent of the theologies I'll consider in this chapter, I have a second stipulation. By "God" I'll mean a being that has a mind and created the universe.

Table 11.1 Wager Assuming P-Theology

Wager 1: Deciding whether to believe in God (when you assume P theology)			
		States of the World	
		God exists	God does not exist
Actions	Believe in God	+1000	−2
	Don't believe in God	−1000	+2

Table 11.2 Wager Assuming X-Theology

Wager 2: Deciding whether to believe in God (when you assume X theology)			
		States of the World	
		God exists	God does not exist
Actions	Believe in God	−1000	−1000
	Don't believe in God	+1000	+1000

Table 11.3 Wager Assuming God Does Not Exist

Wager 3: Deciding whether to believe a theology (when you assume that God does not exist)			
		States of the World	
		P theology is true	X theology is true
Actions	Believe P theology	+2	+995
	Believe X theology	+7	+1000

(or agnosticism), and replace Pascal's theology with another? For example, why not embrace X theology, which says that atheists and agnostics go to heaven and theists go to hell, regardless of whether God exists? The point of these questions is not to endorse the alternative argument that Pascal ignores, but to suggest that Pascal's argument is no better than the prudential argument that Mougin and I described (Mougin and Sober, 1994).

The differences between the many-gods objection to Pascal's Wager and the one that Mougin and I formulated are described in the three accompanying tables. Table 11.1 describes the payoffs that accrue from your decision about whether to believe that God exists when you assume that P theology is true. Table 11.2 describes the payoffs that accrue from the same decision if you assume that X theology is true. Table 11.3 involves using prudential considerations to choose a theology when you assume that God does not exist.

The many-gods objection involves comparing Wagers 1 and 2; the objection that Mougin and I constructed involves comparing Wagers 1 and 3.

What justifies the assignments of payoffs in these three wagers? In Wager 1, it is assumed that if God exists, the relevant payoffs are heaven and hell – your decision about whether to believe in God makes a huge difference in how well off you'll be. It also is assumed that if God does not exist, you're better off not believing in God, but only by a little. This small difference is due to the fact that the religious observances you'll embrace if you believe in God will be less gratifying than the alternative activities you'll indulge in if you don't believe. In Wager 2, the assumption that X theology is true yields the result that believing that there is no God dominates believing in God; regardless of whether God exists, you do better by being an atheist. In Wager 3, the assumption that there is no God entails that it doesn't much matter which theology you believe,[3] but the choice does matter a little; if you assume that there is no God, then, believing P theology may induce occasional fear and trembling, whereas believing X theology will lead you to happily anticipate a rosy futurity, and this is true regardless of whether P theology or X theology is true.

2 Clarifying the Charge of Arbitrary Prudentialism

I hope the comparison of Wager 1 and Wager 3 makes it clear what I mean by "arbitrary" in the title I've given to this chapter. The prudentialism that Pascal deploys in Wager 1 is arbitrary in the sense that it arbitrarily singles out belief in God for prudential scrutiny while exempting his P theology from that assessment. Wager 3 is also arbitrary; it arbitrarily singles out theologies for prudential scrutiny while neglecting to do the same for the agent's conviction that God does not exist. This arbitrariness is a defect in both arguments; if you're going to be a prudentialist, you should eat the whole enchilada! A better wager would involve four choices and four states of the world; the states are G&P, G&X, notG&P, and notG&X (where G = God exists), and the four possible choices are B(G&P), B(G&X), B(notG&P), and B(notG&X), where "B(π)" means believing proposition π.[4] I say that this wager is "better" than Pascal's, not that it is perfect. After all, there are other possible theologies beyond P and X, and there are actions additional to believing this or that conjunction (e.g., suspending judgment).

---

[3] Notice that the upper-left entry for Wager 3 is the same as the lower-right entry for Wager 1.
[4] Here and in what follows I assume that $B(\varphi \& \psi)$ is the same as $B(\varphi) \& B(\psi)$.

Table 11.4 Wager with No Constraints on Prudentialism

Wager 4: Both belief in God and belief in a theology are subject to prudential assessment.

		States of the World			
		G&P	notG&P	notG&X	G&X
Actions	B(G&P)	heaven and feel good	rot and feel good	hell and feel good	hell and feel good
	B(notG&P)	hell and feel bad	rot and feel bad	heaven and feel bad	heaven and feel bad
	B(notG&X)	hell and feel good	rot and feel good	heaven and feel good	heaven and feel good
	B(G&X)	heaven and feel bad	rot and feel bad	hell and feel bad	hell and feel bad

The new and better 4 × 4 problem is depicted in Wager 4 (see Table 11.4). Each cell contains a conjunction; the first conjunct describes the payoff after death (rot, heaven, or hell); the second conjunct describes the payoff before. The choices considered in Wager 1 are located in the four upper-left cells;[5] those considered in Wager 3 are located in the center four. Notice that if Pr(notG&P) = 1, then B(G&P) and B(notG&X) are equally good, and both are better than B(notG&P) and B(G&X). Mougin and I say that being certain that notG&P is *prudentially unstable*; if you are certain, based on your evidence, that this conjunction is true, prudential considerations will lead you to switch to another conjunction. Pascal described one such shift, but neglected the other.

Mougin and I also say that if most, but not all, of the probability is concentrated on notG&P, then what you ought to do will depend on how the remaining probability is distributed. If all of that remainder is assigned to G&P, then B(G&P) is prudentially best. If more is assigned to X theology than

---

[5] Mougin and I made a small modification in the standard modern formulation of Pascal's Wager shown in Wager 1. As mentioned, the standard formulation assumes that if there is no God, then not believing in God has a slightly higher payoff than believing, since not believing frees you from the tedium of religious observances. There are two problems with this assumption. First, it confuses belief in God with embracing some religion and its practices. Second (and more importantly), if you fully believe Pascal's theology, being an atheist will torment you while you live, and being a theist will bring relief, whether or not God in fact exists. This change makes a difference in what you need to assume about Pr(G) for Pascal's Wager to reach the conclusion that belief in God is prudentially justified. The standard version (Wager 1) has the consequence that Pr(G) must be greater than some positive threshold whose value can be calculated from the payoffs (so long as heaven and hell involve finite payoffs). In the version of the Wager that Mougin and I prefer, no such constraint is placed on the value of Pr(G); belief in God is the prudential thing to do, no matter what value Pr(G) has. The overall analysis that Mougin and I defended then, and what I will say in what follows, are not affected by this detail.

to G&P, then B(notG&X) is best. This shows that Pascal's preferred revision conclusion isn't justified by the assumption that P-theology is very probably true.

3 Tinkering and Generalizing

In the remainder of this chapter, I will tinker and generalize. Tinkering is needed because I want to use Bayesian ideas to think about belief revision. Pascal's Wager and its cousins begin with probabilities assigned to states of the world, but then conclude in a dichotomous choice between believing a proposition and not believing it, or between believing it and believing its negation. Dichotomous belief, though familiar from our everyday thought and speech, does not sit well with the Bayesian idea that belief is a matter of degree. We therefore need to describe how prudential considerations can lead agents to change their *degrees of* belief. They start with initial assignments based solely on the evidence at hand; then prudence takes charge and agents change their minds. So, I'll continue to talk about proposition G in what follows, but I need to replace the question of whether to believe G with the question of what your degree of belief in G should be. As for theologies, recall that Pascal's P theology is a function from dichotomous beliefs to payoffs. Once degrees of belief replace dichotomous beliefs, theologies must be adjusted accordingly. This means that P and X will eventually be left behind.

A second type of tinkering is needed, though it is less central. The standard conception of Bayesian updating (updating by strict conditionalization) entails that probabilities of 1 and 0 are *sticky*.[6] That is, if you now are certain that a proposition is true, or that it is false, then you are stuck with that conviction forever after. This tells against the description that Mougin and I gave of what happens when an agent begins deliberating with the certainty that notG and P are both true. Bayesians need to consider instead the case in which the agent starts off *almost* certain that notG and P are true.

4 Skyrms on the Dynamics of Rational Deliberation

So much for the tinkering. The generalizing that is needed involves providing a fuller picture of how agents who begin with different evidence-based credences will change their minds as prudential considerations are taken

---

[6] See Titelbaum (2013) for a framework in which this stickiness no longer obtains.

into account. It isn't enough to say what agents should do if they are very certain that G is false and are very certain that some theology is true.

Brian Skyrms has laid the groundwork for the work that needs to be done. In his 1990 book *The Dynamics of Rational Deliberation*, he describes how game theory can be used to model the deliberations of two or more rational agents in cases where the utility of each agent's chosen action depends on what other agents are going to do. In Skyrms's framework, an agent's process of deliberation comes in many small adjustments; it isn't done in a single step. In many such cases, the agents eventually converge on an equilibrium (where no agent can do better by changing unilaterally) and prudential revision ceases. A simple and suggestive game that Skyrms describes (1990, p. 54) gives the flavor of his framework. It is called *The Winding Road*:

> Two cars approach a blind curve from opposite directions. Each would prefer that they are both driving on the left or both driving on the right. There are two pure equilibria, equally attractive, but if Row goes for one of them and Column goes for the other, they will end up in trouble. If, however, Row believes Column expects him to drive on the left and believes that Column believes him to believe this, and so on, and Column believes likewise about Row, and each believes that the other is rational and that the other believes that he is, then they each have good reason to drive on the left.

Table 11.5 lays out the payoffs that characterize the two players. The point values of the payoffs don't matter; all that matters are equalities and inequalities among the numbers. For example, you could put (500, 500) in the upper-left and lower-right cells and the analysis would be the same.

The game begins with the two players each having their own probabilities for driving on the left and driving on the right. All values are possible (though not 0 and 1, for the reason mentioned earlier). These values are common knowledge and each player takes that information into account in deciding whether and how to change his or her initial assignment. Each player decides what to do by using the principle of maximizing expected utility, their new

Table 11.5 Winding Road

		Column	
		Left	right
Row	Left	1,1	0,0
	Right	0,0	1,1

The Arbitrary Prudentialism of Pascal's Wager   231

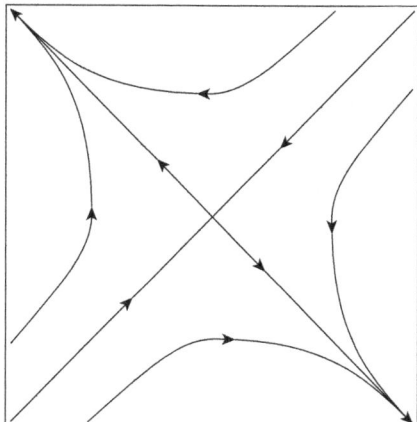

Figure 11.1 Winding Road: In the winding road game, two agents who each deliberate about whether to drive on the left or drive on the right will, given the common knowledge assumption, converge on the certainty that both will drive on the left or on the certainty that both will drive on the right. This figure is reprinted from *The Dynamics of Rational Deliberation* by Brian Skyrms, Cambridge, MA: Harvard University Press, Copyright © 1990 by the President and Fellows of Harvard College, p. 55.

probability values then become common knowledge, and the game goes into its second round of deliberation and revision, then into its third, and so on. In each round, the agents make a small adjustment in their probabilities; they don't jump to certainty or near certainty about what they'll do all at once. The result is that the agents' repeated revisions will converge on both assigning a probability of 1 to left or on both assigning a probability of 1 to right, depending on where they begin.[7] These trajectories are shown in Figure 11.1. The x-axis represents the probability that Column chooses right and the y-axis represents the probability that Row chooses left.

The players' process of revising their probability assignments takes the form of hill climbing; this is because the players always choose to maximize their expected payoffs until they are at a peak and can't go up any more. There are two peaks and two valleys on this surface as shown in Figure 11.2. Notice that half of the unit square in Figure 11.1 leads to the equilibrium (Pr(Row chooses left) = 1, Pr(Column chooses left) = 1) and half leads to (Pr(Row chooses right) = 1, Pr(Column chooses right) = 1). Well, *almost* half; the point in the center of the square is an unstable equilibrium. If the two agents initially

---

[7] "Converge" doesn't mean that the players reach these limit states.

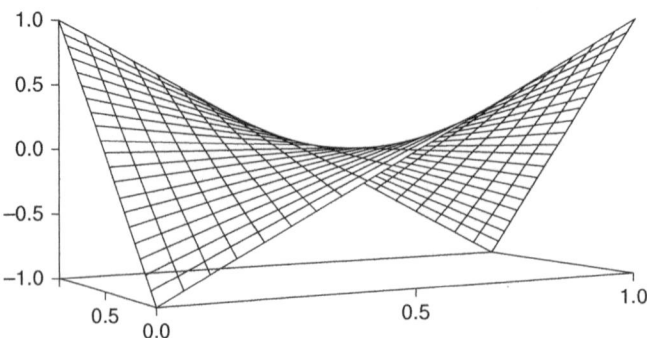

Figure 11.2 Hill Climbing: If believing that God exists and believing P* theology are simultaneously subjected to prudential scrutiny, a single agent will revise her probabilities for both so as to converge on assigning a probability of 1 to both or a probability of 0 to both. This figure is reprinted from *The Dynamics of Rational Deliberation* by Brian Skyrms, Cambridge, MA: Harvard University Press, Copyright © 1990 by the President and Fellows of Harvard College, p. 17.

perch there, they are indifferent about whether to go Northwest or Southeast. Each resembles Buridan's ass.

5 From the Winding Road to Pascal's Wager

In the passage I quoted from Skyrms, you'll notice the reliance on a very substantial common-knowledge assumption. Drivers on real-world highways will mostly lack this kind of knowledge unless they live in cultures in which conventions about driving on the right or left are already in place.[8] However, applying Skyrms's treatment of the Winding Road to Pascal's Wager encounters no such difficulty. Instead of having *two* agents, we consider a *single* agent. Applying this format to the propositions that Mougin and I considered, we obtain Wager 5, whose payoff matrix is shown in Table 11.6. For convenience, I now use payoffs of 1 and 0. Notice that cells no longer contain pairs of payoffs, and columns now represent belief states, not states of the world. The Winding Road is a game in the technical sense used in game theory, since the payoff to each player depends on what the other player does; the Pascalian deliberation problem I now am describing isn't a game in that sense, which is not to deny that it is like Solitaire.

[8] Skyrms sees his analysis of this and other games as contributing to Lewis's (1969) project of using game theory to explain how conventions can arise without agents needing to make explicit agreements with each other.

Table 11.6 Conjunctive Wager

Wager 5: payoffs for four conjunctive beliefs		
	B(P)	B(X)
B(G)	1	0
B(notG)	0	1

Wager 5 describes payoffs that attach only to full-on beliefs. To map how an agent gradually changes his or her degrees of belief by moving in the direction of higher payoffs, we need to say what payoffs attach to different degrees of belief. An agent begins with evidence-based probabilities for God's existing and for one or more theologies; these together determine what the agent's expected payoff is. The agent then modifies those two degrees of belief for prudential reasons, and his or her new probabilities then determine a new expected payoff. As in Skyrms's treatment of the Winding Road, the modifications involve a series of small changes. As noted earlier, the theologies P and X need to be replaced, since they don't discuss degrees of belief. The replacements I'll consider are these:

(P*) $\Pr(\text{God exists}) = \Pr(\text{You go to heaven})$
(X*) $\Pr(\text{God exists}) = 1 - \Pr(\text{You go to heaven})$

For the sake of simplicity, I assume that P* and X* are the only possible theologies. Each says that an agent's expected payoff depends on the agent's *probability* that God exists, but not on whether there is a God in fact. This marks a departure from what P says.

If $\Pr(P^*) = p$ and $\Pr(\text{God exists}) = g$, then $\Pr(\text{You go to heaven}) = pg + (1-p)(1-g)$. With payoffs of 1 for heaven and 0 for hell, this means that the expected utility $EU(G = g \ \& \ P = p) = pg + (1-p)(1-g)$. The expected utility is maximal when p and g are both 1 or both 0; it is minimal when one probability is 1 and the other is 0. Notice that $\Pr(P^*)$ is a second-order probability.

This new wager can be located in the unit square of Figure 11.1. Now the x-axis represents the agent's probability that God exists and the y-axis represents his or her probability that X* theology is true. Pascal addressed his Wager to atheists who use the evidence at hand to assign a small probability to God's existing. Figure 11.1 shows that these atheists must assign the P* theology a very high probability if they are to be in the zone of attraction that will lead them to converge on $\Pr(G) = 1$. In fact, the symmetry of the figure allows one to say something more precise. If an agent initially thinks that $\Pr(\text{God exists}) = e$, then his or her initial value for $\Pr(P^*)$ must be greater than $1-e$ for prudential deliberation to lead the agent to converge on $\Pr(G) = 1$. This point applies to theists as well as to atheists. In general, the higher your initial, evidence-based probability is for God's existing, the lower the threshold is for how probable P* must be for the dynamics to have you converge on full-on theism. Theists get to be initially more skeptical about P* theology than atheists are able to be.

This new wager is less arbitrary than both Wager 1 and Wager 3, but it does nothing to show that Pascal can achieve what he wants. Pascal wants to convince atheists that they should do their best to become theists. For Figure 11.1 to lead to this conclusion, the atheists in question must think that the P* theology has a very high probability. How many atheists do you know who think that this theology is almost certainly true? The atheists I know believe nothing of the kind. I doubt that this difficulty is going to dissolve by making the decision problem more complicated. Pascal assumed that there is just one theology to consider; Mougin and I assumed that there are two, and the present chapter doesn't push that number up. My suspicion is that bringing in more possible theologies is not going to help.

I have framed the separate roles of evidential and prudential considerations – with the first preceding the other – in deference to the usual way that Pascal's Wager is formulated, but there is no need to think that evidence ever suffices to determine rational degrees of belief. This is important for Bayesians, since posterior probabilities cannot be deduced from observations, but require prior probabilities. So, the two-step process considered here involves evidence-plus-priors first, after which prudential considerations are brought to bear. The required separation is clear enough, provided that the priors involved are not influenced by the prudential considerations later adduced.

6 The Scope of Prudential Scrutiny

Pascal's prudential scrutiny of a single proposition and the expansion of that scrutiny described here — from scrutinizing one proposition to scrutinizing two — raises a broader question. Is there any limit to prudential scrutiny's proper scope? For example, does the idea of *unbridled* prudentialism make sense, wherein the degree of belief that a rational agent has in any and all propositions should be adjusted up or down in light of prudential considerations? Even if it makes sense to prudentially scrutinize *each* proposition, it is hard to see what it would be like to scrutinize *all* of them simultaneously.[9] This is because decision-theoretic reasoning always depends on assumptions

---

[9] There is an analogy here with Quine's (1953) thesis that every belief is rationally revisable; see Sober (1981) for discussion.

that are not themselves put under the prudentialist's lens. In particular, when you compute the expected utility of an action, you seek to determine what that expected utility is *in fact*, not what you *want* it to be.[10]

Even if unbridled prudentialism is untenable, there may be reason to embrace a constrained prudentialism. It is the conventionalism that Carnap (1950a) advocated in his essay "Empiricism, Semantics, and Ontology" that inspires this suggestion. The Carnapian idea is that the proposition *electrons exist* and the proposition *physical objects exist* differ epistemically. The former can be justified by observations, but the latter cannot be. This point about physical objects suggests that we might want to adopt a convention; we may want to accept the proposition that physical objects exist because it is useful to think in those terms, even though no evidence could ever justify the proposition.[11]

I prefer to modify this Carnapian idea, rephrasing it so that it addresses the problem of discriminating between two or more contrasting propositions. The following two pairs of propositions have something important in common:

(O1) Physical objects exist.
(O2) Physical objects do not exist, but all the experiences we'll ever have will be as if physical objects exist.
(E1) Electrons exist.
(E2) Electrons do not exist, but all the experiences we'll ever have will be as if electrons exist.

The two O's are empirically equivalent, and so are the two E's. The Carnapian thought outlined above says that proposition O1 differs epistemologically from proposition E1; my present point is that experience cannot discriminate between the E's any more than it can discriminate between the O's (Sober, 1990, 2008).

The constrained prudentialism I have in mind asserts that if two propositions are empirically equivalent but incompatible, it is permissible to accept one of them and reject the other based on prudential considerations. If "God exists" and "God does not exist" were empirically equivalent, Carnapian conventionalism and Pascalian prudentialism would be on the same page. However, that's a very big *if*, since atheists and theists often insist that there is evidence that discriminates between the two propositions.

---

[10] Here there is a point of contact with Quine's (1976 [1936]) claim that unbridled conventionalism is untenable because extra-conventional logical principles are needed to see what a convention entails; Quine makes this point specifically for the case in which there are infinitely many sentences that the conventionalist claims are true by convention.

[11] Carnap thought that adopting the belief that physical objects exist is to adopt a convention concerning what language to speak. Quine (1976 [1963]) criticized Carnap on that score. I think that Carnap's linguistic formulation is extraneous to his epistemological point.

12 Pascal's Wager and the Dynamics of Rational Deliberation

Paul Bartha[*]

> You want to find faith and you do not know the road. You want to be cured of unbelief and you ask for the remedy: learn from those who were once bound like you and who now wager all they have. These are people who know the road you wish to follow, who have been cured of the affliction of which you wish to be cured: follow the way by which they began ... I tell you that you will gain even in this life, and that at every step you take along this road you will see that your gain is so certain and your risk so negligible that in the end you will realize that you have wagered on something certain and infinite for which you have paid nothing.
>
> (L418/S680)

1 Introduction

Pascal's Wager is widely understood by philosophers as a decision-theoretic argument. The most familiar version is represented by Table 12.1.

The rows represent the actions available to me: *Wager for God, Wager against God*. The columns represent the two possibilities: *God exists, God does not exist*. The dots represent the finite payoffs attached by Pascal to three of the four outcomes. The fourth outcome, salvation, has infinite value. I assign a real number $p > 0$ as the probability that God exists. I then calculate that the expected utility of wagering for God is infinite, while that of wagering against God is merely finite. My conclusion is that I am rationally required to wager for God. I should act in ways that foster religious belief and which, presumably, will increase my degree of belief in God.[1] Following (Jordan, 2006), let us refer to this version of Pascal's argument as the Canonical Wager.[2] Like all

---

[*] Thanks to Graham Oppy and Chris Stephens for helpful comments and suggestions.
[1] Note that it is now standard to characterize the two options in this way (Jordan, 2006; Hájek, 2003) rather than as "Believe" and "Don't believe," which are not actions under the agent's control.
[2] Hacking (1972) identifies three versions in the *Pensées*. Additional versions are discussed by philosophers: the Jamesian wager (James, 1956 [1896]; Jordan, 2006 and Ch. 5 of this volume),

Table 12.1 Pascal's Wager

	(p) God exists	$(1-p)$ God does not exist
Wager for God	∞	.
Wager against God	.	.

versions of the Wager, it offers an answer to a normative question: What should a rational agent do in the face of uncertainty about the existence of God?

There is a related but distinct explanatory question: How can we explain the evolution and stability of an individual agent's religious beliefs? Many people believe (to a high degree) in God, many reject religious belief, and many are agnostic. Individual religious beliefs have a complex history. Some people might actually arrive at religious belief after following a self-imposed regime such as the one recommended by Pascal. Although we cannot hope to explain individual details, perhaps we can account for the evolution and persistence of distinct forms of religious belief.[3]

This chapter argues that Pascal's Wager offers us tools to answer not just the normative but also the explanatory question. Actually, the two questions are closely related. Although Pascal writes that "reason impels you to believe," the result (even if you find the Wager persuasive) is not instant conversion but rather a decision to take "steps along the road." Suppose the decision-theoretic argument leads you to take the first step. What if doubt remains? Should you run through the argument again? This question motivates a dynamical model: the Wager leads to a process of belief revision in which subjective probabilities evolve over time in response to well-defined pressures.

Pascal's Wager can be interpreted as offering a rule for the revision of your credence (subjective probability) that God exists, in the absence of decisive evidence. Your credence, p, reflects your best current judgment of the probability that God exists, and serves (along with your utilities) as the basis for the expected utility calculation that drives Pascal's argument. But your credence also has a second role: it reflects the strength or "weight" of your inclination to wager for God. The version of Pascal's Wager in Table 12.1 may be read as an argument that if $0 < p < 1$, your credences are dynamically unstable: the calculation of expected utilities shows that you have reason to act in ways that

---

formulations involving large finite utilities (Mougin and Sober, 1994; Jordan, 2006; Duncan, Ch. 7 of this volume), formulations using infinitesimal or vague probabilities (see Wenmackers, Ch. 15 of this volume and Rinard, Ch. 14), and others.

[3] For a different decision-theoretic analysis of faith, see Buchak (2012).

will increase the value of *p*. We can explain stable religious beliefs as evolutionary equilibria. In anachronistic terms, Pascal's principle states: *revise your credences until you achieve equilibrium*. Let us call this *Pascalian belief revision*, as it clearly differs from evidence-based belief revision.

There are at least three motivations for developing a dynamical model for Pascalian belief revision. First, such an approach is suggested by Pascal himself. Pascal first likens the Wager to a single cast of the dice: "there is no room for hesitation, you must give everything." However, he notes that our passions hold us back: "if you are unable to believe, it is because of your passions, since reason impels you to believe and yet you cannot do so." In the passage cited at the outset of this chapter, Pascal outlines the steps needed to diminish this opposing passion and bring about belief. So, the development of a model for this interplay of reason and passion is, in a sense, a contribution to our understanding of the *Pensées*.

Second, over the years, numerous difficulties have emerged with the Canonical Wager. What if my credence that God exists is 0, infinitesimally small, or vague? What if the infinite utility that appears in the decision table makes no sense? What if it is impious or epistemically impermissible to alter one's beliefs on pragmatic grounds, rather than on the basis of evidence? What if adopting a mixed strategy, such as flipping a coin to decide whether to wager for or against God, is just as good as wagering for God (since the mixed strategy also has infinite expected utility)? Finally, there is the many-gods objection: What if we assign positive credence to the existence of several different gods, each of whom offers an infinite reward? Some of these problems appear to be unsolvable for the Canonical Wager. As we shall see, however, we can offer answers when we turn to a dynamical model.

In particular, a dynamical model offers scope for exploring the many-gods objection. Here is a glimpse of how this works. One version of the many-gods objection asks us to assign positive credence not just to a traditional god, but also to a skeptic-loving deity who rewards agnostics and atheists with infinite utility. *Wager-against* then has infinite expected utility, as does *Wager-for*, so there appears to be no basis for choosing between these options. So far, this is a strong objection. Now let's introduce the dynamical perspective. If you decide to wager against all gods, what happens next? You take steps to reduce your credence in all gods. That includes your credence in the skeptic-loving god. In lowering your credence in the skeptic-loving deity, however, you undercut the core assumption that leads you to wager against all. Of course, the details of how your credences change are important, and that is why we need a clear model.

Finally, there is a third reason for exploring a dynamical approach to Pascal's Wager: an analogy with the many successful applications of evolutionary game theory to problems of social choice. Thanks to the work of Skyrms (1996) and others, we know that evolutionary models of interaction can shed light on long-standing puzzles about rationality and cooperation. Stable behavior can emerge and persist even if it violates classical norms of rationality. In the same way, it may turn out that a dynamical approach to Pascal's Wager can explain the emergence and stability of forms of religious belief.

Some of these ideas are developed elsewhere (Bartha, 2012, 2016). My purpose in this chapter is to motivate the value of a dynamical approach to Pascal's Wager by drawing on earlier results, suggesting some extensions, and showing how a dynamical model allows us to raise and explore a large number of interesting and important questions.

The chapter proceeds in three stages. Section 2 presents what I call *relative utilities* and *relative decision matrices*, which allow us to answer a few simple objections to Pascal's Wager. Section 3 outlines the basic *many-wagers* model, which enables us to respond to some forms of the many-gods objection. Section 4 develops the dynamical model further. This allows us to consider more difficult versions of the many-gods objection, and to explore a wide variety of puzzles about religious belief.

2 Relative Utilities and Relative Decision Matrices

In this section, I begin by reviewing two objections to the Canonical Wager alluded to above: the mixed-strategy objection and a simple version of the many-gods objection. It will become clear that these two objections are closely related. They all come down to the problem of how to represent the infinite utility of salvation. I introduce what I call relative utilities and relative decision matrices as tools for representing infinite utility and for answering both objections. I will use these devices to develop a dynamical approach to Pascal's Wager in section 3.

The mixed-strategy objection, suggested by Jeffrey (1983a) and Duff (1986), and fully elaborated by Hájek (2003), runs as follows: any mixed strategy between *Wager-for* and *Wager-against* that assigns a positive probability to *Wager-for* has infinite expected value.[4] Getting *Heads* on a coin toss, rolling

---

[4] If a mixed strategy assigns probability $q > 0$ to *Wager-for*, then its expected utility is $q \cdot p \cdot \infty$ (plus a finite term), which equals ∞ regardless of the value of q.

Table 12.2 Many Jealous Gods

	god_1	god_2	...	god_n	No god exists
Wager for god_1	∞	.	...	.	.
Wager for god_2	.	∞	...	.	.
...					
Wager for god_n	.	.	...	∞	.
Wager against all	.	.	...	.	.

double-six on a pair of dice, or waiting for lightning to strike nearby as preconditions for wagering: all of these strategies have infinite expectation and are just as good as the pure act of wagering described by Pascal. Hence, there is no imperative to cultivate religious faith.

The simple many-gods objection drops the unfounded assumption that we need only consider one possible deity seriously. Suppose that there are n possible deities to whom we assign positive probability. Suppose that each is a *jealous* god, offering the same infinite reward to its followers. We have the decision table shown in Table 12.2. As before, the dots represent finite payoffs. A straightforward calculation yields infinite expected utility on each row apart from the last. Each of the n wagers appears equally good. By the symmetry of the situation, we have no basis for settling upon any of the wagers (although we may rule out *wager against all*).

These two objections have a common structure. We can see very well that we ought to prefer the pure strategy to any mixed strategy, since it gives us the highest probability of the infinite outcome. Who would hesitate to choose a gamble that gives 99 percent probability of an infinite reward over a gamble that gives 1 percent probability of the same infinite reward? We can see very well that we ought to wager for the deity in Table 12.2 with the highest subjective probability. But given the way that we have represented infinite utility and given our naïve calculations of infinite expected utility, we have no apparent way to defend these obvious-sounding positions.

The most promising line is the one taken by Schlesinger, who offers the following principle:

> In cases where the mathematical expectations are infinite, the criterion for choosing the outcome to bet on is its probability. [Schlesinger, 1994, p. 90]

Applied to the mixed-strategy objection, Schlesinger's Principle tells us to choose the pure strategy. Applied to the many-gods objection of Table 12.2,

Schlesinger's Principle tells us to wager for the deity that has highest subjective probability.[5] These are very plausible recommendations.[6]

Unfortunately, there are two problems. First: the principle may give us the wrong result if we allow different magnitudes of infinite reward. Suppose we change the example in Table 12.2 so that the reward offered by god_1 is infinitely better than the reward offered by god_2, even though both are infinitely good. In this case, we might prefer to wager for god_1 over god_2, even if the probability of god_1 is a little lower than that of god_2. Second: Schlesinger's Principle as stated here is an "add-on" to standard decision theory that might lead to inconsistency. Indeed, infinite utility is inconsistent with standard decision theory (McClennen, 1994). We should only embrace Schlesinger's Principle in the context of a rigorous non-standard decision theory that allows for infinite utilities.

The key point of departure from standard decision theory, as pointed out by McClennen (1994), is that we need to drop the assumption of *Continuity*.

Continuity. Whenever an agent prefers A to B to Z, there is some gamble between A and Z such that the agent is indifferent between B and that gamble:

$$B \sim [p_A, (1-p)Z].$$

Here, $[p_A, (1-p)Z]$ represents a gamble with probability p of yielding A and $(1-p)$ of yielding Z. *Continuity* is incompatible with any set of preferences where one outcome is infinitely better or worse than another. Whether *Continuity* is a requirement of rationality is much debated. For present purposes, I note simply that we can motivate dropping *Continuity* with examples much more straightforward than Pascal's Wager. Dreier (1996) offers a case along the following lines. Suppose you are a financial manager who prefers A, *avoiding financial loss*, to B, *financial loss*, to Z, *committing a sin* (such as defrauding the investors whom you represent). You have such a horror of Z that you prefer B to any gamble $[p_A, (1-p)Z]$ with $(1-p) > 0$, i.e., any gamble that might result in the outcome Z. In such a case, your preferences violate *Continuity*, but they are not obviously irrational.

---

[5] In the case of a tie there is again no basis for decision, but that is no more problematic than deciding between two actions with equal finite expected utility. They are equally good.

[6] Indeed, Schlesinger's Principle as applied here conforms to a principle of decision theory sometimes referred to as the Better-Chances Condition (Resnik, 1987): in comparing gambles between two identical outcomes, prefer the gamble that offers the better chance of the better outcome.

There are a number of ways to develop a decision theory that accommodates violations of *Continuity*, and hence allows infinite utilities. One approach (Hausner, 1954) uses *lexicographic utilities*: vector-valued utilities where different components of the vector represent "dimensions of value," some of which are valued infinitely more than others. I have proposed an alternative, *relative utility theory*, that is well suited to the development of a dynamical approach to Pascal's Wager.[7] Here is a brief outline of the theory.

The basic idea is that relative utilities are like generalized ratios of (non-negative) utility intervals. Let $\preccurlyeq$ be a preference ordering: $B \preccurlyeq A$ means that you prefer A to B. If $Z \preccurlyeq B$ and $Z \preccurlyeq A$, then the relative utility

$$U(A, B; Z)$$

is the utility of A relative to B with base-point Z. Roughly speaking, it is the ratio of the "length" of the utility interval B–A over the "length" of the utility interval Z–A. The base-point Z functions as the zero or reference point for the intervals. Interval lengths can be 0, positive or infinite, so the function always takes values from 0 to ∞.

In this chapter, I will only need three special cases:

1. *Infinite relative utility*.

$$U(A, B; Z) = \infty \leftrightarrow B \preccurlyeq [p_A, (1-p)Z] \text{ for } 0 < p \leq 1.$$

Any gamble that offers even a slight chance of A is preferred to B.

2. *Zero relative utility*.

$$U(A, B; Z) = 0 \leftrightarrow A \preccurlyeq [pB, (1-p)Z] \text{ for } 0 < p \leq 1.$$

Any gamble that offers even a slight chance of B is preferred to A.

3. *Equal relative utility*.

$$U(A, B; Z) = 1 \leftrightarrow [p_A, (1-p)Z] \preccurlyeq B$$

and

$$[pB, (1-p)Z] \preccurlyeq A, \text{ for } 0 \leq p < 1.$$

You prefer the pure outcome B to any non-trivial gamble between A and Z, and you prefer the pure outcome A to any non-trivial gamble between B and Z.

---

[7] Hausner (1954) shows that we can drop *Continuity* and still retain a version of expected utility maximization. Bartha (2007) shows the same thing for relative utility theory. See Bartha (2012) and Bartha and DesRoches (2016) for further details and comparisons between relative and lexicographic utilities.

Figure 12.1 Case 1: Infinite Relative Utility $U(B, A; Z) = \infty$ and Case 2: Zero Relative Utility $U(A, B; Z) = 0$

Figure 12.2 Case 3: Equal Relative Utility $U(A, B; Z) = 1$

The crucial point here is that we don't attempt to define absolutely infinite utility. Instead, in special case 1, we define relatively infinite utility: one outcome, A, is infinitely better than another, B, relative to Z. The definition of this concept can be given entirely in terms of ordinary, well-defined preferences. Although I have only sketched the concept here, it should be clear that relative infinite utilities escape many of the objections that could be raised against absolutely infinite utility. They can be meaningfully defined and they can represent the preferences of ordinary, finite agents.

Some pictures may be helpful for the three special cases (see Figures 12.1 and 12.2). In each picture, a set of outcomes is represented as an interval with the least-preferred outcome, the base-point Z, on the left. The zig-zag line represents a discontinuity: outcomes to the right are, roughly speaking, infinitely preferred relative to those on the left. Note that for Case 1 and Case 2 the picture is the same. For Case 3, the story is as follows: even though B is strictly better than A, you are unwilling to trade the pure outcome A for any gamble that carries a positive chance of ending up with Z.

In short, relative to the base-point Z, the three cases characterize situations where B is infinitely better than A, where A is infinitely worse than B, and where A and B are both infinitely better than Z and we don't want to risk losing either.[8]

---

[8] What if there is another discontinuity between A and B? As discussed in Bartha (2007), we have a choice between $U(B, A; Z) = \infty$ and $U(B, A; Z) = 1$, depending on our willingness to take the risk of ending up with Z.

Table 12.3 Canonical Wager, Relative Decision Table

	(p) God exists	(1- p) God does not exist	
Wager for God	1	0	$U = p$
Wager against God	0	0	$U = 0$

The three values $U(B, A; Z) = \infty$, 0 and 1 will be the only values we need in what follows.

A relative decision matrix is a table of relative utilities, computed relative to a fixed optimal outcome ("Best") and a fixed base-point ("Worst"). Relative decision theory tells us that we can calculate relative expected utilities in the same way as ordinary expected utilities. The guiding principle is to act so as to maximize relative expected utility. To see how this works, let us represent the decision problems in Tables 12.1 and 12.2 using relative decision matrices as laid out in Table 12.3.

We compute relative expected utility U in the same way as we compute expected utility in ordinary decision theory.[9] Since $U(Wager\text{-}for) = p \cdot 1 + (1 - p) \cdot 0 = p$, and $U(Wager\text{-}against) = 0$, we should select *Wager-for*. Notice also that the mixed strategy q *Wager-for* + $(1 - q)$ *Wager-against* has relative expected utility qp, which is strictly less than that of the pure wager. Relative decision theory provides a vindication of Schlesinger's Principle and a straightforward way to respond to the mixed-strategies objection.

Similarly, we can respond easily to the version of the many-gods problem stated above. Table 12.4 is the relative decision table. The relative expected utilities (relative to best and worst outcomes) of each possible act are easily computed and they validate the intuitive solution: we should wager for the god with the highest subjective probability.

In general, to convert from a "naïve" decision matrix containing ∞ to a relative decision matrix, the procedure is simple:

- Replace each occurrence of ∞ with 1.
- Replace each finite value ($\cdot$) with 0.

If you are prepared to accept, at least provisionally, the claim that relative utilities are a good way to represent (relatively) infinite utility, and that relative decision matrices provide a rigorous way to solve decision problems

---

[9] I write $U(Wager\text{-}for) = p$ as shorthand for the three-place expected relative utility $U(Wager\text{-}for, B; Z)$ of the act relative to the best (B) and worst (Z) outcomes in the decision table. It's usually clear what the best and worst outcomes are. This convention will be adopted throughout the chapter.

Table 12.4 Many Jealous Gods, Relative Decision Table

	(p_1) god$_1$	(p_2) god$_2$	...	(p_n) god$_n$	(q) No god exists	
Wager for god$_1$	1	0	...	0	0	$U = p_1$
Wager for god$_2$	0	1	...	0	0	$U = p_2$
			...			
Wager for god$_n$	0	0	...	1	0	$U = p_n$
Wager against all	0	0	...	0	0	$U = 0$

involving (relatively) infinite utilities, then we have answers to both problems raised at the start of this section. And we are ready to move on to dynamical models of Pascal's Wager.

One final point: What about a case where there are different levels of infinite reward? Suppose that the infinite reward offered by god$_1$ is so much better than the other rewards that we would willingly trade any of them even for a tiny positive chance of receiving the reward from god$_1$. Relative decision tables can accommodate this modification. Remember: all values in the table are computed relative to the best and worst outcomes. So, we need only replace each "1" with "0" in rows 2 through n. We are left with a single "1" in row 1. In this case, $U(Wager\ for\ god_1) = p_1$ while U is 0 on all other rows; hence, we should wager for god$_1$.

I discuss relative utilities more thoroughly elsewhere.[10] It is time to return to the development of a dynamical model for Pascal's Wager.

3 Dynamics, Part I: The Many-Wagers Model

We need a dynamical model to think about sophisticated versions of the many-gods objection. These versions ask us to consider gods who reward people other than their followers, gods who reward skeptics, and so on. To keep things manageable, we define a set of theological possibilities: the *Pantheon*. Here is a fairly comprehensive set, defined in terms of which groups receive the infinite reward; see Table 12.5.[11]

---

[10] See Bartha (2007, 2012, 2016). For a critique, see Colyvan and Hájek (2016). One important restriction, noted by Oppy (Ch. 13 of this volume), is that the theory cannot handle infinitesimal probabilities. Finally, an important clarification: although differences in utility among finite outcomes are not represented in the decision tables shown here, they are not ignored by relative decision theory. See Bartha (2007, 2012).

[11] We could expand the possibilities by considering additional types of gods; see Example 10. Another idea is to consider alternative theologies (Mougin and Sober 1994; Sober, Ch. 11 of this volume); this is briefly considered in Example 6.

Table 12.5 The Pantheon

Jealous gods	Reward only their followers.
Nice gods	Reward their followers plus at least one other group. (*Very nice* (universalist) gods are a special case – see below.)
Grouchy gods	Don't reward their followers, but reward at least one other group. (*Very grouchy* gods reward everybody except their followers.)
Detached gods	Reward independently of wagering behavior. (Special cases: the *universalist* god rewards everybody, the *pointless god* rewards nobody.)

Table 12.6 Jealous and Grouchy

	(p_J) J (Jealous exists)	(p_G) G (Grouchy exists)	$(1 - p_J - p_G)$ No god exists	
Wager for jealous	1	0	0	$U = p_J$
Wager for grouchy	0	0	0	$U = 0$
Wager against all	0	1	0	$U = p_G$

Consider the scenario alluded to near the end of section 1: you assign positive credence both to a traditional jealous god and to a grouchy skeptic-loving god who rewards only non-believers. Let's go straight to the relative decision matrix (Table 12.6).

When you compute relative expected utilities, you find that you should *wager for jealous* if $p_J > p_G$, and *wager against all* if $p_G > p_J$. In the latter case, what happens next? To *wager against all* is to take steps to reduce your tendency to believe in any deity – i.e., to lower both p_J and p_G. But *wager against all* is only attractive if you assign appreciable credence to the grouchy god. Thus, you are taking steps that lower your credence in the grouchy god. Given the above decision table, *wager against all* appears to be self-undermining when viewed as a long-term project, rather than as a single decision.

These considerations motivate a shift to a dynamical view of Pascal's Wager. On this approach, you treat your credences both as probabilistic inputs for calculations of (relative) expected utility and as inclinations to make the corresponding wagers. This leads to three ideas.

3.1 Many-Wagers Model: Qualitative Statement

- *Rather than choosing one wager, choose them all "to a degree."* Assign each possible wager a weight that matches the corresponding current subjective probability for that deity.

- *Use relative expected utilities to update your credences.* Start with an initial set of credences. Update by increasing your credences for wagers with above-average relative expected utility, and decreasing your credences for wagers with below-average relative expected utility.
- *A viable set of credences must be stable under updating.* For a stable set of beliefs, there should be no upward or downward pressure on any of your credences.

We need an explicit updating rule. The many-wagers model proposed in Bartha (2012) uses the Replicator Dynamics, adapted from evolutionary game theory. In evolutionary game theory, we begin with distinct population groups, each defined by a fixed strategy. Based on assumptions about starting proportions and frequency of encounters, we calculate the relative fitness of each strategy. Groups with above-average relative fitness increase in proportion; those with below-average fitness shrink. At equilibrium, each group has the same relative fitness. For the many-wagers model, we make the following adaptations. Theological possibilities are the analogue of population groups, and initial credences correspond to starting proportions. Wagers are the analogue of strategies, and the relative expected utility of a wager corresponds to the fitness of a strategy. The basic theory is encapsulated in the following updating rule, the definition of equilibrium, and the definition of a stable equilibrium – and in each case, the technical definition is followed by an informal characterization. These ideas will be illustrated in the examples that follow.

3.2 Many-Wagers Model: Precise Statement

3.2.1 Updating Rule

Suppose there are finitely many theological possibilities $S_1, \ldots, S_n$ with corresponding wagers $W_1, \ldots, W_n$. Suppose that the relative decision matrix is A, and the initial subjective probabilities are given by $\boldsymbol{p} = (p_1, \ldots, p_n)$, where p_i is the initial subjective probability for S_i.

Let $U(W_i) = (A\boldsymbol{p})_i = A_{i1} p_1 + \ldots + A_{in} p_n$ represent the relative expected utility of wager W_i. Let $\overline{U} = p_1 U(W_1) + \ldots + p_n U(W_n)$ represent the average relative expected utility. The updated subjective probability p_i' for S_i is given by

$$p_i' = [U(W_i)/\overline{U}] \, p_i.$$

Informally: shift support for S_i up or down depending on whether the relative expected utility for W_i is above or below average.

3.2.2 Equilibrium Distribution

A probability distribution $\{p_i\}$ over possibilities $S_1, \ldots, S_n$ counts as an equilibrium distribution if $p_i' = p_i$ for all i.

In other words: at an equilibrium, the updating rule produces no changes.

3.2.3 Stable Equilibrium Distribution

An equilibrium distribution $\{p_i\}$ over possibilities $S_1, \ldots, S_n$ is stable if for any small (and mathematically admissible) set of changes Δp_i to the probabilities, repeated application of the updating rule to the distribution $\{p_i + \Delta p_i\}$ leads to reconvergence to $\{p_i\}$.

In other words: at a stable equilibrium, your beliefs are robust. In the event of small changes to your probabilities (slight doubts or shifts in belief), the updating rule carries you back to the original probability values.

3.3 Examples

Example 1: (Original Wager)

The relative decision table is Table 12.3. There are two states (*God exists, God does not exist*) with initial probabilities p and $1 - p$, and two corresponding wagers (*Wager-for, Wager-against*). The relative utilities are $U(\text{Wager-for}) = p$ and $U(\text{Wager-against}) = 0$. The average relative expected utility is easily computed:

$$\overline{U} = pU(\text{Wager-for}) + (1-p)U(\text{Wager-against})$$
$$= pp + (1-p)0$$
$$= p^2$$

If p' is the revised credence that *God exists*, then by the updating rule:

$$p' = [p/p^2]p$$
$$= 1$$

Provided $p > 0$, $p' = 1$ after a single iteration. The only equilibria are $p = 0$ or $p = 1$, and only $p = 1$ is a stable equilibrium.

Starting from any value $p > 0$, our revision rule leads immediately to $p = 1$. In fact, any updating rule consistent with the "qualitative statement" above has a single stable equilibrium at $p = 1$.

Example 2: (Many Jealous Gods)

The relative decision table is Table 12.4. The relative utilities are $U(Wager\ for\ god_n) = p_n$, and $U(Wager\ against\ all) = 0$. Application of the many-wagers model shows that if one deity has highest initial subjective probability, that probability converges to 1.

What about ties? Suppose that there is a two-way tie for highest subjective probability: $p_1 = p_2$, and all other probabilities are lower. The many-wagers model carries you to an equilibrium where $p_1' = p_2' = 0.5$. However, it is not a stable equilibrium. Suppose there is a slight perturbation, an increase in your probability for god_1. Application of the updating rule leads to $p_1' = 1$. The conclusion is that the only stable equilibria are those in which there is full belief (with credence 1) in a single jealous god.

Example 3: (Jealous and Grouchy)

The relative decision table is Table 12.6. The relative utilities are $U(Wager\ for\ jealous) = p_J$, $U(Wager\ for\ grouchy) = 0$, and $U(Wager\ against\ all) = p_G$. The average relative expected utility is $\overline{U} = p_J^2 + p_G(1 - p_J - p_G)$. If we suppose that $p_J > 0$, $p_G > 0$, and $p_A > 0$ initially, then $p_G' = 0$ after one iteration and both p_G'' and p_A'' drop to 0 after two iterations of the updating rule. The only equilibrium values for (p_J, p_G, p_A) are $(1, 0, 0)$, $(0, 0, 1)$ and $(0, 1, 0)$, and only the first of these is a stable equilibrium. So long as $p_J > 0$, the initial support for *Wager against all* disappears quickly. The many-wagers model provides an effective response to this version of the many-gods objection.

Example 4: (Jealous and Universalist)

What does the many-wagers model say about belief in a universalist deity who rewards everyone, regardless of beliefs? The short answer is that wagering for such a deity is pointless. To see this, start with positive credences for both a traditional jealous god J and a universalist god N. The relative decision matrix is shown in Table 12.7.

Table 12.7 Jealous and Universalist

	(p_J) Jealous exists	(p_N) Universalist exists	$(1 - p_J - p_N)$ No god exists
Wager for jealous	1	1	0
Wager for universalist	0	1	0
Wager against all	0	1	0

For the relative utilities, we have $U(J) = p_J + p_N$, $U(N) = p_N$ and $U(A) = p_N$, where we let J represent wagering for the jealous god, and so forth. So long as $p_J > 0$ initially, only J has above-average relative expected utility. There is no stable equilibrium except $p_J = 1$.

Referring back to our definition of the Pantheon: any detached deity, whose practice of rewarding or not rewarding is independent of wagering behavior, appears to be doomed by an evolutionary model. Such a deity cannot be part of a stable equilibrium if there is a credible alternative involving a jealous god. The relative utility of wagering for the jealous god will always be larger than that of wagering for the detached god. This observation knocks out not just the universalist god, but also the pointless god who rewards nobody.

4 Dynamics, Part II: Challenges

The many-wagers model extends the basic idea of Pascal's Wager to a dynamical model that shows how an agent's credences might change over time. Credences rise or fall, based on relative expected utilities, until a stable equilibrium is achieved. As we have seen, this model lets us respond effectively to some versions of the many-gods objection. At the same time, however, some interesting new puzzles and challenges emerge. I briefly review a few of them here. My point is not to provide a definitive resolution for these problems, but rather to illustrate the explanatory potential of the many-wagers framework.

Example 5: (The Happy Universalist)

Example 4 raises an interesting puzzle. If something like the many-wagers model is right, how can we explain the popularity and stability of belief in a universalist god who rewards everybody? One simple answer is to assign zero credence to any god who rewards followers on the basis of religious belief, perhaps because you consider such a god implausible or immoral. Since this answer takes us outside the whole Pascalian framework, let us set it aside and ask: Does the many-wagers model itself provide a route to universalism?

Here is the sketch of an answer. Wagering for a jealous god is riskier than wagering for a god who rewards everyone. Jealous gods are fussy that their believers get things exactly right.[12] By contrast, a universalist god is reassuringly non-fussy.

---

[12] Oppy (1991) exploits this point in developing an argument that there is at least a countable infinity of possible jealous gods, and each should receive only infinitesimal credence.

Table 12.8 Jealous and Nice Cartel

	J_1	J_2	N_1	N_2	N_3	No god exists
Wager for J_1	1	0	0	0	0	0
Wager for J_2	0	1	0	0	0	0
Wager for N_1	0	0	**1**	**1**	**1**	0
Wager for N_2	0	0	**1**	**1**	**1**	0
Wager for N_3	0	0	**1**	**1**	**1**	0
Wager against all	0	0	?	?	?	0

Here is a way to implement that reasoning in the many-wagers model. Suppose your relative decision matrix includes a few jealous gods (J_1 and J_2) and a *nice cartel*: a large block of unfussy gods (N_1 to N_3) who reward each other's followers. Table 12.8 sets out the picture, with the rewards of the nice cartel shown in bold type.

If you start with a perfectly symmetrical set of credences (equal for each possible god), then wagering for any of the gods in the cartel has three times the relative expected utility of wagering for a jealous god. The many-wagers dynamics will carry you to full belief in the nice cartel, equally distributed over the three participating deities. The jealous gods drop out and the remaining deities are universalist in a qualified sense: they reward all wagers still in play.

Of course, the deities in the nice cartel are not *strictly* universalist: they don't reward followers of the jealous gods. Furthermore, although we have an equilibrium, it is not a *stable* equilibrium.[13] So, although this qualified form of universalism might arise on the many-wagers model, it is not really an adequate response to the rejection of universalism in Example 4.

Example 6: (The Happy Atheist/Agnostic)

A second puzzle: How should we explain the popularity and stability of a strongly agnostic position that assigns credence 1, or nearly 1, to *No god exists*?[14] Here are three straightforward answers. First, you could be a strict evidentialist and reject non-epistemic reasons for belief change. But then you must also reject the Pascalian framework. Second, you might have infinitesimal credence that God exists. Unfortunately, you must then reject the many-

---

[13] If there is a slight increase in credence in favor of one of the members of the cartel, there is no restoring pressure that takes you back to equality. Every distribution over the members of a nice cartel is an (unstable) equilibrium.

[14] I'm not relying on sharp definitions of atheism or agnosticism here. The agent just has a stable and very high credence that *No god exists*.

Table 12.9 Finite Utilities

	p God exists	$1 - p$ God does not exist
Wager-for	1	f_1
Wager-against	f_3	f_2

wagers model.[15] Third, you might assign a large finite value as the relative utility of salvation. The relative decision matrix is laid out in Table 12.9.

If $f_2 - f_1$ is sufficiently large relative to the probability p ($p/(1 - p) < f_2 - f_1$ is sufficient), then *Wager-against* has higher relative expected utility than *Wager-for*, and the updating rule leads to $p = 0$ as a stable equilibrium. This solution, however, takes us away from the Canonical Wager, which insists on representing the utility of salvation as infinite.

Does the many-wagers model provide a path to atheism? It might. Perhaps we can refine the model to include epistemic constraints. The many-wagers model almost entirely ignores evidence[16] in order to propose a mechanism for updating credences on pragmatic grounds. It might be possible to develop a model that combines pragmatic and epistemic considerations. On such a model, an agnostic might concede a small positive (non-infinitesimal) probability that God exists, but resist the upward pressure coming from comparisons of relative expected utility.[17]

Indeed, the many-wagers model is open to the charge of arbitrary prudentialism (Mougin and Sober, 1994; Sober, Chapter 11 in this volume) because it allows prudential considerations to influence our credences about gods, but not about other things. Sober (Chapter 11 in this volume) shows that if we allow pragmatic considerations to influence our credences in background theological beliefs, we can perhaps be brought to an equilibrium where atheism is prudent. I will not assess Sober's model here, but I will briefly discuss one important idea.

Sober suggests that an agent might take seriously a theology (Sober calls it X-theology) on which atheists receive an infinite reward and theists do not, independently of which god (or whether any god) exists. Suppose ϵ represents the probability that X-theology is true, while traditional theology has probability $1 - \epsilon$. The relative decision matrix of Table 12.3 must be modified.

---

[15] The relative utilities framework cannot accommodate infinitesimal credences. Relative utilities are defined using preferences over ordinary gambles (with standard probability values).
[16] It does not ignore evidence about how to maximize expected relative utility.
[17] See Example 10 for one simple way to achieve this result.

Table 12.10 X-Theology

	p God exists	$1-p$ God does not exist
Wager-for	$1-\epsilon$	0
Wager-against	ϵ	ϵ

The new table (Table 12.10) is shown here, with the relative utility attached to each of the four possible outcomes.

In this table, for the sake of simplicity, we hold fixed the probability, ϵ, of X-theology (Sober allows it to vary). This limited incorporation of Sober's idea opens an avenue to stable atheism. The relative expected utility of *Wager-for* is now $p(1-\epsilon)$, while the relative expected utility of *Wager-against* is ϵ. If $p > \epsilon/(1-\epsilon)$ initially, the dynamics carries you toward stable full belief, while if $p < \epsilon/(1-\epsilon)$, the dynamics carries you toward stable atheism.

Example 7: (A Grouchy Cartel)

Although the many-wagers model plausibly disposes of some versions of the many-gods objection, it introduces some surprising new many-gods problems. Consider a *grouchy cartel* of two possible gods (G_1 and G_2), each of whom rewards the other's followers (see Table 12.11).

Initially, you have non-zero probabilities p_1, p_2, and p_3. It is easy to see that the relative expected utilities are $U(\text{Wager for } G_1) = p_2$, $U(\text{Wager for } G_2) = p_1$ and $U(\text{Wager against all}) = 0$. A quick calculation shows that $\overline{U} = 2p_1 p_2$. There is only one stable equilibrium: $p_1 = \frac{1}{2}$, $p_2 = \frac{1}{2}$, and $p_3 = 0$. At this unique stable equilibrium, you are equally disposed to wager for each grouchy god. This idea extends to larger cartels of grouchy gods: the stable equilibria assign equal credence to each god in the cartel (if the pattern of rewards is symmetrical).

Cartels are clearly an important feature of the many-wagers model. They deserve a clear definition.

4.1 Cartels

Let G be the set of all deities $g_1, \ldots, g_n$ in a decision table.

- g_1 *supports* g_2 if g_1 rewards those who wager for g_2.[18]
- g_1 and g_2 are *connected* if either supports the other.

---

[18] The g_i represent all of the columns, including "no god exists."

Table 12.11 Grouchy Cartel

	(p_1) G_1	(p_2) G_2	$(p_3 = 1 - p_1 - p_2)$ No god exists	
Wager for G_1	0	1	0	$U = p_2$
Wager for G_2	1	0	0	$U = p_1$
Wager against all	0	0	0	$U = 0$

- A *cartel* K is a subset of G that is *closed under connection*: if g is connected to any member of K, then g is a member of K.

The set consisting of a single jealous god is a cartel. So is a set of mutually supporting nice gods (Example 5), and so is a set of mutually supporting grouchy gods (Example 7). In general, we can organize a relative decision matrix so that each cartel forms a "block" that is not connected to the other blocks.

Here is the most basic result about cartels: *in a stable equilibrium, there is exactly one cartel*. By way of illustration, consider the following example, pitting a jealous god against a cartel of two grouchy gods.

Example 8: (Two Cartels)

There are two stable equilibria here in Table 12.12: (1, 0, 0, 0) and (0, ½, ½, 0). You can give full credence to J, or credence 0.5 to both G_1 and G_2. There is a third equilibrium that gives equal credence to all three gods: (⅓, ⅓, ⅓, 0). This is an unstable equilibrium: a slight boost to your credence in J leads to the disappearance of the grouchy cartel, and a slight drop in your credence for J leads to the disappearance of J. The same point applies to any relative decision matrix: in a stable equilibrium, only one cartel can survive. We saw a special case of this result in Table 12.4: in an array of jealous gods, only one can survive. The generalization shows that each cartel behaves like a single jealous god.

Example 9: (Grouchy and Nice)

Here, in Table 12.13, we have a cartel consisting of a grouchy god who rewards skeptics and a nice (universalist) god who rewards everyone. There is a problem for the many-wagers model. Every starting configuration with positive probability for all three options leads to convergence to full belief that *no god exists*. The only stable equilibrium is (0, 0, 1).

Table 12.12 Jealous and Grouchy Cartel

	J (jealous)	G_1	G_2	No god exists
Wager for J	1	0	0	0
Wager for G_1	0	0	1	0
Wager for G_2	0	1	0	0
Wager against all	0	0	0	0

Table 12.13 Grouchy and Nice

	G (grouchy)	N (nice)	No god exists
Wager for G	0	1	0
Wager for N	0	1	0
Wager against all	1	1	0

In Example 7 and Example 9, we have some very strange scenarios involving cartels. In both cases, the many-wagers model leads to stable beliefs that nobody would take seriously. That's a problem for the model! I close this section by sketching three possible defences.

The first and simplest response to a case like Example 7 is the observation that grouchy cartels have small "basins of attraction" compared to jealous gods. What initial set of credences would lead to the equilibrium in Example 7? Suppose that we initially started with the possibilities in Example 8: the grouchy cartel, one jealous god, or no gods at all. A quick calculation shows that you'd need your initial credence in the grouchy cartel (the sum of your credences for G_1 and G_2) to be double the value of your initial credence in the jealous god, J. For a grouchy cartel of three gods, the initial credence in the grouchy cartel would need to be three times the initial credence in the jealous god, and the pattern continues for larger cartels. In short, a grouchy cartel is like an inefficient version of a jealous god. The upshot is that given most initial sets of credences, grouchy cartels will not survive.

The second line of defence, suggested in Bartha (2012), is to move to a more sophisticated notion of stability. In evolutionary game theory, the notion of stability takes into account the possible emergence of *mutations*, variations on existing strategies. The analogous concept in the many-wagers model is the emergence of new theological possibilities, with corresponding options for wagering.

4.2 Strongly Stable Equilibria

An equilibrium distribution $\{p_i\}$, $1 \leq i \leq n$, over possibilities $S_1, \ldots, S_n$ is *strongly stable* if for any new possibility S_{n+1}, and any (mathematically admissible) set of changes Δp_i to the probabilities, $1 \leq i \leq n + 1$, application of the updating rule to the distribution $\{p_i + \Delta p_i\}$ leads to convergence to $\{p_i\}$ (and to $p_{n+1} = 0$).

An equilibrium is strongly stable if the updating rule restores the equilibrium even if you introduce a new theological possibility (with small adjustments in your set of probabilities).

If we impose the requirement that a viable equilibrium must be strongly stable, then we can reject the solution of Example 9. The reasoning is as follows. While at or close to the equilibrium in which your credence that no god exists is nearly 1, suppose that you entertain the slight possibility that a jealous god exists after all. You add a row and a column to your decision table, which becomes:

Example 9A: (Grouchy and Nice Plus Jealous God)

See Table 12.14. It's crucial here that you assume N, the nice deity, would reward belief in the jealous god: N rewards all wagers. The introduction of this new option destabilizes the belief that no god exists and the result is convergence to belief in the jealous god.

It must be acknowledged that the strengthened stability requirement does not help us directly to deal with the grouchy cartel of Example 6, where you assign credence ½ to each grouchy god. That equilibrium is strongly stable; new theological possibilities cannot gain a foothold. However, the general idea of introducing "mutants" (novel theological possibilities) can still help with this example, provided you apply it to your initial set of credences. This is essentially a version of the "basins of attraction" response: from a sufficiently diverse initial set of credences, you are unlikely to arrive at a grouchy cartel.

Table 12.14 Grouchy, Nice, and Jealous

	G (grouchy)	N (nice)	J (jealous)	No god exists
Wager for G	0	1	0	0
Wager for N	0	1	0	0
Wager for J	0	1	1	0
Wager against all	1	1	0	0

Example 10: (Oppy's Virtuous God)

In our examples thus far, rewards have always been linked to wagering behavior. As our final example, we consider a possibility raised by Oppy (Chapter 13 in this volume). Suppose you assign positive credence to a *virtuous god* V who rewards evidentialists – i.e., those who revise credences solely on the basis of evidence. Let's suppose that the only other possibilities you take seriously are *a jealous god exists* or *no god exists*. The situation is represented in Table 12.15.

Notice: your probability p that the jealous god exists appears in the bottom-left corner of the table. That is because if you refuse to wager, you stick with your initial probabilities: p for jealous god and q for the virtuous god. Therefore, the relative utility of "Refuse to wager" and J is p: you still have a chance of the reward from the jealous god. Notice also: we assume (with Oppy) that if we refuse to wager, the virtuous god will not count this decision as a form of wagering.

This example raises an interesting challenge for the many-wagers model. The decision table favors "Refuse to wager" if $p^2 + q > p$: for example, if $p = ½$ and $q = ⅓$. But, if you refuse to wager, then your credences remain at their initial values. In this case, your credences and your decision not to wager are stable. It seems that with sufficiently high credence in V relative to J, the updating process of the many-wagers model never gets started. In my view, this is not a problem for the many-wagers model. Suppose that this analysis is right, and that your initial credences are such that you refuse to wager. Then the many-wagers model has the resources to put a cap on the use of pragmatic reasoning to modify credences. Far from counting as an objection, this result is arguably a boon for the theory. For example, it gives us a way to explain the stability of agnosticism (see Example 6).

Table 12.15 Jealous and Virtuous

	(p) J (*jealous*)	(q) V (*virtuous*)	$(1 - p - q)$ No god exists	
Wager for J	1	0	0	$U = p$
Wager for V	0	0	0	$U = 0$
Wager against all	0	0	0	$U = 0$
Refuse to wager	p	1	0	$U = p^2 + q$

There is, however, another interesting response to this challenging case that rejects the preceding analysis. Rather than use the table to make a single decision, we should use it to revise our credences repeatedly, in the spirit of the many-wagers model. The only two acts in the table that receive non-zero relative utility both give support to the jealous god. Specifically, the initial weight attached to "Wager for J" is p, while the weight attached to "Refuse to wager" is q. The other two wagers have zero weight.[19] The average relative expected utility is thus

$$\overline{U} = p^2 + q(p^2 + q).$$

The calculation of your new credences is somewhat more complicated than usual. If we apply the usual updating rule, we have

$$p' = \frac{p^2}{p^2 + q(p^2 + q)} \quad \text{and} \quad q' = \frac{q(p^2 + q)}{p^2 + q(p^2 + q)}.$$

Here, p' represents the new weight of "Wager for J," while q' represents the new weight of "Refuse to wager," which commits you to sticking with your original credences (with probability p of believing in J). Since both acts contribute to your new credence in J, we combine them for the following results:

$$\text{New credence in } \mathbf{J} = p' + q'p$$
$$= \frac{p + p^2q + q^2}{p^2 + q(p^2 + q)} p$$

$$\text{New credence in } \mathbf{V} = q'q$$
$$= \frac{p^2 + q}{p^2 + q(p^2 + q)} q^2$$

So long as $p > 0$, these results indicate that your credence in V will decrease and your credence in J will increase.[20] Less formally: your new probability for J is a mixture of your old credence p (which you'd keep if you refuse to wager) and your boosted credence if you wager for J. Inevitably, your probability for

---

[19] Your inclination to believe in V translates into an inclination to *refuse to wager*.
[20] This follows because $\frac{p^2+q}{p^2+q(p^2+q)} q^2 < q$ if and only if $q(p^2 + q) < p^2 + q(p^2 + q)$, which holds so long as $p > 0$.

J will rise. The only stable equilibrium, once again, is where you have full credence in the jealous god.

For the second analysis to work, we must assume that the updating process does not itself count as wagering. But we had to make a very similar assumption for the first analysis, namely, that using the decision table to decide to refuse to wager did not count as wagering.

5 Conclusion

Rational choice theorists have embraced the idea that decision theory and game theory are enriched by shifting to an evolutionary perspective. Within such an approach, the key idea is to look for evolutionarily stable strategies. The evolutionary framework offers a powerful set of tools: basins of attraction, mutations, and other ideas. This approach allows us to explain the emergence and stability of a wide variety of social behaviors.

The exploration of Pascal's Wager is similarly enriched by shifting from the perspective of a single argument to a dynamics of Pascalian belief revision: the many-wagers model. The idea that we might need multiple steps was suggested by Pascal, though its real interest becomes apparent only in the context of many-gods problems. In order to develop a plausible dynamics, we first need a finite representation for the infinite utilities associated with wagering. That is provided by relative utilities. We then exploit the analogy with the Replicator Dynamics.

The examples in this chapter illustrate three levels of complexity in approaching challenges to Pascal's Wager. First: relative utilities alone can handle some of the simplest issues, such as the mixed-strategies objection. Second: the basic many-wagers model allows us to respond to simple versions of the many-gods objection. Finally, more complicated many-gods problems can be explored using tools such as mutation and sophisticated versions of the many-wagers model. These ideas take us well beyond Pascal's original argument into new terrain, allowing us to explore diverse scenarios for religious belief.

13 Infinity in Pascal's Wager

Graham Oppy

1 Wagering with Infinite Utilities

The standard formulation of Pascal's Wager casts it in terms of absolute utilities: we use "∞" to represent the absolute utility of wagering for God if God exists. The standard formulation supposes that the decision set-up is represented by Table 13.1.

There are two available courses of action: wager for God, or fail to wager for God. There are two relevant conceivable states of the world: God exists, or God does not exist. The credence that is assigned to the claim that God exists is p, whence the credence assigned to the claim that God does not exist is 1 – p. The utility of wagering for God, given that God exists, is infinite; all of the other utilities are finite. The expected utility of wagering for God is p.∞ + (1 – p).A = ∞. The expected utility of failing to wager for God is p.B + (1 – p).C = a finite value. Given that one ought always to act so as to maximize expected utility, one ought to wager for God.

One might worry that the mathematics is suspect: "∞" is not a standard number. For now, let's suppose that there is no difficulty involved in revising the standard numbers N_S to the extended numbers N_S' via the addition of "∞," subject to the following rules:

(1) $\forall x \in N_S': \infty + x = \infty$
(2) $\forall x \neq 0 \in N_S': x.\infty = \infty$
(3) $\forall x \in N_S: \infty - x = \infty$
(4) $\infty - \infty$ is undefined
(5) $0.\infty = 0$

If you think that there are some rules missing from this list, don't worry: we'll have more to say about the rules governing "∞" later on. So far, I have just considered rules that might be required for the calculation of expected utilities in variations of Pascal's Wager, on the assumption that probabilities only take

Table 13.1 Pascal's Wager

	God Exists Pr(God Exists) = p, 0 < p < 1	God does not Exist Pr(God does not Exist) = 1 − p
Wager for God	∞	A, where A is finite
Fail to Wager for God	B, where B is finite	C, where C is finite

standard values. Also, note that I have not yet specified *which* standard number system is designated by "N_S."

There are many concerns that have been raised about Pascal's Wager in this standard formulation.

First, there are worries about the assignment of a credence to the claim that God exists. Does it make sense to suppose that we do or can assign credences to that claim? And, if it does make sense to suppose that we do or can assign credences to that claim, are we justified in supposing that the assigned credence can or must be strictly greater than zero and strictly less than 1?

Second, there are worries about the assignment of infinite utility to wagering on God's existence given that God exists. Even if we can construct a consistent mathematical formalism, we might think that it makes no sense to suppose that we can have infinite utilities. And, even if we suppose that it does make sense to suppose that we can have infinite utilities, we might think that it is a mistake to suppose that the utility of wagering on God's existence if God exists is infinite.

Third, there are worries about exactly what "wagering for God" amounts to. It cannot be "believing in God," because believing in God is not an action that you can just choose to perform. But, on Pascal's theology, it is not clear that anything less than believing in God will secure "an infinity of infinite happiness." It is simply not obvious that there *is* an action that you can choose to perform that has infinite utility if God exists.

Fourth, there are worries about the suggestion that it can be proper to adjust one's credences in the light of one's utilities. However, exactly, the Wager argument is supposed to work, the overarching idea is that your utilities alone give you a reason to revise up the credence that you give to the claim that God exists. But it seems questionable whether your utilities alone ever can give you an adequate reason to revise your credences.

Fifth, there are worries about the soteriological import of Pascal's Wager. The construction of the Wager seems to suggest that you may obtain "an infinity of infinite happiness" merely by acting on a dominant desire to acquire, for yourself, an infinity of infinite happiness. Experience teaches that, in this life, merely acting on a dominant desire to acquire happiness for oneself typically does not lead to happiness. Is it really credible to suppose that things stand differently with the next life?

Beyond these worries, there are three objections to the standard formulation of Pascal's Wager that seem particularly formidable.

First, there are many different actions that one might take that are bundled together under "fail to wager for God." For example, I might toss a coin in order to decide whether to wager. The expected utility of doing this is ∞. More generally, there is a vast range of "mixed strategies" that I could pursue, all of which have infinite expected utility. The assumption, in the standard formulation of the Wager argument, that there are just two available courses of action that need to be considered is false; and, when we correct this assumption, we no longer get out the conclusion that, in order to maximize expected utility, we must wager for God. (For more detailed development of this objection, see Duff, 1986 and Hájek, 2003.)

Second, there are many different conceivable ways that the world might be that are bundled together under "God does not exist." In particular, there are conceivable sources of infinite utility that we might obtain by performing actions other than wagering for God. Consider, for example:

(a) Very Nice Gods who reward everyone regardless of whether and how they wager
(b) Nice Gods, each of whom rewards those who wager for it, and all of whom reward wagers on some conceivable Gods while also not rewarding wagers on all of the other conceivable Gods
(c) Very Perverse Gods who reward everyone except those people who wager for them
(d) Perverse Gods, each of whom does not reward those who wager for it, and all of whom reward wagers on some conceivable Gods while also not rewarding wagers on all of the other conceivable Gods.

All of these conceivable Gods reward some kinds of wagering on Gods, but many reward different kinds of wagering on Gods. So, there are many different wagers all of which have infinite expected utility. The assumption, in the standard formulation of the Wager argument, that there are just two relevant conceivable ways that the world might be is false; and, when we correct this assumption, we no longer get out the conclusion that, in order to maximize expected utility, we must wager for God.

Third, given that we are countenancing infinite utilities – i.e., utilities with value ∞ – it seems that we should also countenance infinitesimal credences – i.e., credences with value ε. If ∞ is so large that it cannot be increased, then ε is so small that it cannot be properly decreased: if you remove a proper part of something of infinitesimal measure, you are left with something that is also of infinitesimal measure.

If we suppose that both ∞ and ε are added to N_S to form N_S', then it is, at least initially, plausible to suppose that we have the following rules (in addition to those given earlier):

(6) ∞/∞, n/∞, and ε/∞ are all undefined
(7) $\infty.\varepsilon$ is undefined
(8) $\varepsilon + \varepsilon = \varepsilon$
(9) $\forall n \neq 0 \in N_S$: $n.\varepsilon = \varepsilon$
(10) $\varepsilon - \varepsilon$ is undefined

If, in our standard formulation of Pascal's Wager, we set $p = \varepsilon$, then it turns out that the expected utility of wagering for God is undefined. And, in that case, the advice that we ought to maximize expected utility will not give us the conclusion that we ought to wager for God.

Perhaps it might be objected that we cannot make sense of infinitesimal credences. But it is hard to see a good reason for supposing that, while we can make sense of infinite utilities, we cannot make sense of infinitesimal probabilities. If we can make sense of the idea that a utility can be so large that it cannot be increased by adding to it, why can't we make sense of the idea that a credence can be so small that it cannot be properly decreased by subtracting from it? "Infinitesimal" is a natural dual to "infinite."

Even if we waive other objections, the system that we have to this point is not satisfactory. There is just too much that is undefined: for example, if we ask what is the sum of $(n + \varepsilon)$ and $(m - \varepsilon)$, we get back the answer that this sum is undefined. The best way to avoid this result that I have found so far is to suppose that the introduction of ε "perturbs" all of the standard numbers: instead of n, we have $n \pm \varepsilon$. We then have the following rules for the extended rationals (where n and m are standard rationals):

(11) $(n \pm \varepsilon) + (0 \pm \varepsilon) = n \pm \varepsilon$
(12) $(n \pm \varepsilon) + (m \pm \varepsilon) = (n + m) \pm \varepsilon$
(13) $\infty + (0 \pm \varepsilon) = \infty$
(14) $\infty + (n \pm \varepsilon) = \infty$
(15) $\infty + \infty = \infty$

(16) $(n \pm \varepsilon) - (0 \pm \varepsilon) = (n \pm \varepsilon)$
(17) $(n \pm \varepsilon) - (m \pm \varepsilon) = (n - m) \pm \varepsilon$
(18) $\infty - (0 \pm \varepsilon) = \infty$
(19) $\infty - (n \pm \varepsilon) = \infty$
(20) $\infty - \infty$ is undefined
(21) $(n \pm \varepsilon).(1 \pm \varepsilon) = (n \pm \varepsilon)$
(22) $(n \pm \varepsilon).(0 \pm \varepsilon) = (0 \pm \varepsilon)$
(23) $(n \pm \varepsilon).(m \pm \varepsilon) = (n.m) \pm \varepsilon$
(24) $\infty.(1 \pm \varepsilon) = \infty$
(25) $\infty.(0 \pm \varepsilon)$ is undefined
(26) $\infty.(n \pm \varepsilon) = \infty$
(27) $\infty.\infty = \infty$
(28) $(n \pm \varepsilon)/(1 \pm \varepsilon) = (n \pm \varepsilon)$
(29) $(n \pm \varepsilon)/(0 \pm \varepsilon)$ is undefined
(30) $(0 \pm \varepsilon)/(n \pm \varepsilon) = (0 \pm \varepsilon)$
(31) $(n \pm \varepsilon)/(m \pm \varepsilon) = (n/m) \pm \varepsilon$
(32) $\infty/(1 \pm \varepsilon) = \infty$
(33) $\infty/(0 \pm \varepsilon)$ is undefined
(34) $(n \pm \varepsilon)/\infty = (0 \pm \varepsilon)$
(35) $\infty/(n \pm \varepsilon) = \infty$
(36) ∞/∞ is undefined
(37) $(0 \pm \varepsilon)/(0 \pm \varepsilon)$ is undefined
(38) $(0 \pm \varepsilon)/\infty = (0 \pm \varepsilon)$

The interesting rule, given that we take this route, is that, despite gains in definition elsewhere, $\infty.(0 \pm \varepsilon)$ is undefined. The intuitive justification for this is that, even if we think that the product of the infinite and the infinitesimal should be finite, there is no satisfactory way of choosing any particular finite number to be that product. Speaking *very* loosely, it is quite intuitive to suppose that $n.(0 \pm \varepsilon) = (n.0) \pm (n.\varepsilon) = 0 \pm \varepsilon$. But, when we consider $\infty.(0 \pm \varepsilon) = (\infty.0) \pm (\infty.\varepsilon) = 0 \pm (\infty.\varepsilon)$, there is no way to sensibly assign a value to $\infty.\varepsilon$. It must be that ∞/∞ is not well-defined: if you divide an infinite set into infinitely many equinumerous subsets, the cardinality of those subsets could be any of 1, 2, 3, ... ∞. But, if ∞/∞ is not well-defined, then neither is $\infty.\varepsilon$, if ε is the multiplicative inverse of ∞. And $\infty.\varepsilon$ is also not well-defined if we suppose that, while ∞ is so big that addition makes no difference to it, ε is so small that subtraction of a proper part makes no difference to it.

What is our number system N_S? Strictly, I think, N_S is the rational numbers. What motivates the addition of "∞" to the rational numbers is the countable denumerability of the rationals: if we do not add "∞" to the rationals, there is a sense in which there are rational quantities that are not explicitly represented in the rational number system. Given that our concern is expressive completeness, if we want to extend the real numbers by including symbols for infinite numbers, then we cannot make do with just the pair of symbols: "∞" and "ε." On the assumption that the only infinite cardinal below the infinite cardinal that measures the real numbers is the infinite cardinal that measures the rational numbers, we should extend the reals by adding two infinite numbers – ∞_{little} and ∞_{big} – and two infinitesimals – ε_{little} and ε_{big}. (We should want to be able to say that there are ∞_{little} rational numbers and ∞_{big} real numbers, where $\infty_{little} < \infty_{big}$: there are fewer rational numbers than there are real numbers, even though there are infinitely many rational numbers.) I leave it as an exercise for the reader to think about the rules that should govern N_S', when N_S is the standard real numbers.

Of course, you might think that it is just a mistake to couch discussion of Pascal's Wager in terms of these extended number systems. You might well prefer to couch your discussion in terms of some other kind of non-standard number theory – Robinson's non-standard arithmetic, or Conway's surreal number system, or vectorial representations, or the like – or you might prefer to couch discussion of Pascal's Wager in terms of standard number systems and insist that the utility of wagering for God if God exists is very large but finite. But on all of these other approaches, the result of the calculation of the expected value of wagering for God turns out to be sensitive to the precise values that are attributed to the probability that God exists and the utility that is attributed to wagering for God if God exists, and there is nothing that privileges any particular assignment of those values over other assignments of those values. In order for it to have any bite, it seems that Pascal's Wager requires "∞"-valued utilities; and yet there seem to be very formidable obstacles that confront Pascal's Wager if it is formulated in terms of "∞." (For a more careful and detailed argument for this conclusion – paying consideration to all of the just-mentioned non-standard number theories – see Hájek, 2003. Perhaps some will find Hájek's arguments not quite so persuasive as I take them to be.)

2 Wagering with Relative Utilities

In some very interesting publications, Paul Bartha (2007, 2012) has explored ways of rehabilitating Pascal's Wager that are intended to avoid the bad consequences that follow from the formulation of the Wager in terms of "∞"-valued utilities, but without having recourse to some other kind of non-standard number theory. By working with "relative utilities," Bartha hopes to be able to show that Pascal's Wager can be formulated in a way that entirely avoids both the Duff/Hájek objection from "mixed strategies" and the many-gods objection. Bartha makes the following bold conjecture:

> If (1) we can assign positive probability to the existence of deities, (2) we can make sense of infinite utility, (3) we can justifiably revise our beliefs on pragmatic grounds, and (4) we can provide a valid formulation of Pascal's original argument, then the many gods objection poses no additional threat. [Bartha, 2012, p. 205]

Since my list of worries about Pascal's Wager is slightly more extensive, I shall interpret this conjecture in the following way: *if we meet all of the other objections to Pascal's Wager, then the many-gods objection is thereby already met.*

Bartha argues in detail that, if you think that the field of conceivable Gods is made up of Very Nice Gods, Nice Gods, Very Perverse Gods, and Perverse Gods, along with

(5) Jealous Gods, each of whom rewards all and only those who wager for it (of course, Pascal's God is a Jealous God) and
(6) Indifferent Gods who reward no one, no matter how they wager

then you should apportion all of your credence to the Jealous Gods. If he's right about that, then Very Nice Gods, Nice Gods, Very Perverse Gods, Perverse Gods, and Indifferent Gods simply do not make any difficulty for Pascal's Wager.

The *first stage* of Bartha's rehabilitation of Pascal's Wager is to recast it in terms of *relative utilities*. Let $A \ll B$ mean that B is preferred to A under the weak preference ordering $\ll$. If $Z \ll A$ and $Z \ll B$, then $U(A,B;Z)$ is the utility of A relative to B with base point Z. Let $[pA, (1-p)B]$ be a gamble that offers A with probability p, and B with probability $1-p$. We have the following three special cases:

(1) $U(A,B;Z) = \infty \leftrightarrow B \ll [pA, (1-p)Z]$ for $0 < p \leq 1$
(2) $U(A,B;Z) = 0 \leftrightarrow A \ll [pB, (1-p)Z]$ for $0 < p \leq 1$
(3) $U(A,B;Z) = 1 \leftrightarrow [pA, (1-p)Z] \ll B$ and $[pB, (1-p)Z] \ll A$ for $0 \leq p < 1$

A relative decision matrix is a table of relative utilities. We compute the matrix entries relative to the optimal outcome – i.e., in the case of Pascal's

Table 13.2 Relative Utilities

	God Exists Pr(God Exists) = p, 0 < p < 1	God Does Not Exist Pr(God Does Not Exist) = 1 − p
Wager for God	1	0
Fail to Wager for God	0	0

Table 13.3 Mixed Strategies

	God Exists and Coin Falls Heads Pr(God Exists and Coin Falls Heads) = p/2, 0 < p < 1	God Exists and Coin Falls Tails Pr(God Exists and Coin Falls Tails) = p/2	God Does Not Exist Pr(God Does Not Exist) = 1 − p
Outright Wager for God	1	1	0
Toss Fair Coin to Decide Whether to Wager for God	1	0	0
Neither of the Above	0	0	0

Wager, we compute U(A, salvation; Z). According to Bartha, this means that, when we are dealing with pure outcomes and simple cases, all values in our table are either 0 or 1. "1" indicates the best outcome: salvation; "0" represents all of the other outcomes. When we move to the framework of relative utilities, our initial decision matrix is transformed to appear as laid out in Table 13.2.

This recasting in terms of relative utilities immediately disposes of the standard 'mixed strategies' objection. If we represent the expanded decision matrix as laid out in Table 13.3, we get:

EU(Outright Wager for God) = p
EU(Toss Fair Coin to Decide Whether to Wager for God) = p/2
EU(Neither of the Above) = 0.

In order to maximize expected utility, one must wager for God.

It may be tempting to suggest that, if we countenance infinitesimal credences, and if we accept that, for infinitesimal p, $p + p = p$, then the "mixed strategies" objection survives in the case in which p is infinitesimal. Given that our representation in terms of relative utilities only does away with the infinite values, in the case in which credence for the existence of God is infinitesimal, we have:

> Pr(God Exists and Coin Falls Heads) = Pr(God Exists and Coin Falls Tails) = p
> EU(Outright Wager for God) = p + p + 0 = p
> EU(Toss Fair Coin to Decide Whether to Wager for God) = p + 0 + 0 = p
> EU(Neither of the Above) = 0.

The advice to maximize expected utility does not tell us what to do.

However, since the relative utility approach depends upon the assumption that we work only with standard real-valued probabilities, this criticism does not quite hit the mark. (We shall subsequently return to considerations about infinitesimal credences.)

The *second stage* in Bartha's rehabilitation of Pascal's Wager is to introduce some constraints on acceptable credences with respect to the many-gods objection. Given that the thrust of Pascal's Wager is that, in certain cases, you ought to update your credences in the light of your utilities, it may seem plausible that, for those cases, your credences ought to be stable equilibrium points under the updating in question.

The specific updating rule that Bartha proposes is as follows:

> Suppose that there are finitely many possibilities $S_1, \ldots, S_n$ with corresponding wagers $W_1, \ldots W_n$, a relative decision matrix A, and an initial subjective probability vector $\mathbf{p} = (p_1, \ldots, p_n)$, where p_i is the initial subjective probability for S_i. Let $U(W_i) = A(\mathbf{p})$ represent the expected relative utility of wager W_i. Let $\hat{U} = p_1 U(W_1) + \ldots + p_n U(W_n)$ represent the average (relative) expected utility. Then the updated subjective probability for S_i is $p_i' = p[U(W_i)/\hat{U}]$.

The consequent constraint that Bartha imposes on acceptable credences with respect to the many-gods objection is that a viable probability distribution for a Pascalian decision problem should be a stable equilibrium under this updating rule. (A probability distribution $[p_i]$ over possibilities $S_1 \ldots S_n$ is an *equilibrium distribution* if $p_i' = p_i$ for all i. An equilibrium distribution $[p_i]$ over possibilities $S_1 \ldots S_n$ is *stable* if for any small (and mathematically admissible) set of changes Δp_i, application of the updating rule to the distribution $[p_i + \Delta p_i]$ leads to convergence to $[p_i]$.)

Bartha observes that the precise details of the rule are not important; the rule is but one of a large family of evolutionarily robust updating rules that would deliver similar results across the kinds of cases in which we are interested.

Table 13.4 Jealous and Nice

	Jealous God Exists	Very Nice God Exists
Wager on Jealous God	1	1
Wager on Very Nice God	0	1

Table 13.5 Grouchy and Nice

	Grouchy God Exists	Nice God Exists	No God Exists
Wager on Grouchy God	0	0	0
Wager on Nice God	0	1	0
Wager on No God	1	1	0

What kinds of scenarios meet the requirement of being a stable equilibrium? As Bartha notes, one obvious scenario of this kind is one in which one gives full credence to a single Jealous God. However, as Bartha also notes, not all cases of stable equilibria consist of a single Jealous God.

Consider, first, a decision scenario in which you are deciding between wagering on a Jealous God and wagering on a Very Nice God (see Table 13.4). For simplicity, we ignore the wager on neither.

Suppose that the probability assigned to the Jealous God is p > 0, so that the probability assigned to the Very Nice God is (1 − p) < 1.

EU(Wager on Jealous God) = p.1 + (1 − p).1 = 1.
EU(Wager on Very Nice God) = p.0 + (1 − p).1 = 1 − p
Û = p.1 + (1 − p).(1 − p) = p + (1 − p)^2 = 1+ p^2 − p
P_J' = p. 1/(1 + p^2 − p) = P_J.[1/(1 − p + p^2)]
P_N' = (1 − p).(1 − p)/(1 − p + p^2) = P_N.[(1 − p)/(1 − p + p^2)]

Clearly, P_J' > P_J and P_N' < P_N. (Remember that 0 < p < 1, so 0 < p^2 < p, whence 1 − p + p^2 < 1.) If we iterate the process of redistribution of credence, P_J' goes to 1 and P_N' goes to 0. Moreover, as a special case, if we suppose that P_J' is 1-ε, and P_N' is ε, under redistribution, P_J' goes to 1 and P_N' goes to 0. So, the only equilibrium point is P_J = 1 and P_N = 0, and it is a stable equilibrium point. Given a choice between wagering on a Jealous God and wagering on a Very Nice God, Bartha's constraint on credences requires you to be giving all of your credence to the Jealous God.

Consider, second, a decision scenario in which you are deciding between a Grouchy God and a Nice God, but where there is also the 'atheist' option of wagering on neither of these Gods (see Table 13.5).

Suppose, again, that the probability assigned to the Jealous God is $p > 0$, the probability assigned to the Nice God is $q > 0$, and the probability assigned to there being no God is $(1 - [p + q]) < 1$.

$$EU(\text{Wager on Grouchy God}) = p.0 + q.0 + (1 - [p + q]).0 = 0$$
$$EU(\text{Wager on Nice God}) = p.0 + q.1 + (1 - [p + q]).0 = q$$
$$EU(\text{Wager on No God}) = p.1 + q.1 + (1 - [p + q]).0 = p + q$$
$$\hat{U} = p.0 + q.q + (1 - [p + q])(p + q) = (p + q - [p^2 + 2pq])$$
$$P_G' = p.0/(p + q - [p^2 + 2pq]) = 0/(p + q - [p^2 + 2pq]) = 0$$
$$P_N' = q.q/(p + q - [p^2 + 2pq]) = q^2/(p + q - [p^2 + 2pq])$$
$$P_O' = [1 - (p + q)].(p + q)/(p + q - [p^2 + 2pq])$$

Since $q^2 < q < p + q$, $P_G' < P_O'$ and $P_N' < P_O'$. So, in this case, the only stable equilibrium point is wagering on No God.

In order to get the result that the Pascalian wants, we need a further condition. Bartha opts for the following:

> A viable probability distribution for a Pascalian decision problem must be a strongly stable equilibrium. (An equilibrium distribution $[p_i]$, $1 \leq i \leq n$, over possibilities $S_1, \ldots, S_n$ is *strongly stable* if for any new possibility S_{n+1}, and any (mathematically admissible) set of changes Δp_i, $1 \leq i \leq n + 1$, application of the updating rule to the distribution $[p_i + \Delta p_i]$ leads to convergence to $[p_i]$, and, in particular to $p_{n+1} = 0$.)

The motivation for this proposal is the observation that, if we expand our decision scenario by adding in some additional Nice Gods and Grouchy Gods who reward one another but do not reward those who wager on No God, then it will cease to be the case that wagering on No God is a stable equilibrium.

Even if the condition of strong stability gives Bartha the result that he wants, I think that a slightly different condition may be mandated. When deciding what to do, you really should take all of the relevant possibilities into account. Given our account of the various Gods that are up for consideration, a *serious* decision scenario is one in which there are no asymmetries introduced in connection with the Nice Gods and the Perverse Gods. Consider, for example, the following, somewhat more complex, decision scenario in Table 13.6 (where wagering actions are specified in the left column, states of the world are specified in the top row, J is Jealous, I is indifferent, VN is Very Nice, VP is Very Perverse, the Ni are Nice, the Pi are Perverse, and 0 is the state in which there are no Gods).

Table 13.6 Jealous, Nice, Indifferent, and Perverse

	J	I	VN	VP	N1	N2	N3	N4	P1	P2	P3	P4	0
J	1	0	1	1	1	0	0	0	1	0	0	0	0
I	0	0	1	1	0	1	0	0	0	1	0	0	0
VN	0	0	1	1	0	0	1	0	0	0	1	0	0
VP	0	0	1	0	0	0	0	1	0	0	0	1	0
N1	0	0	1	1	1	0	0	0	1	0	0	0	0
N2	0	0	1	1	0	1	0	0	0	1	0	0	0
N3	0	0	1	1	0	0	1	0	0	0	1	0	0
N4	0	0	1	1	0	0	0	1	0	0	0	1	0
P1	0	0	1	1	1	0	0	0	0	0	0	0	0
P2	0	0	1	1	0	1	0	0	0	0	0	0	0
P3	0	0	1	1	0	0	1	0	0	0	0	0	0
P4	0	0	1	1	0	0	0	1	0	0	0	0	0
0	0	0	1	1	1	0	0	0	0	1	0	0	0

In this scenario, the only stable equilibrium position (given our assumption that each of the initial credences is some positive, finite value) is to give all of your credence to the Jealous God. Since this scenario has the kind of symmetry that is plausibly the target of the strong stability condition, it is plausible to conclude that the assessment of this scenario establishes that, if there is a choice between a Jealous God, an Indifferent God, a Very Nice God, a Very Perverse God, the full range of Nice Gods, the full range of Perverse Gods, and no God, you should wager on the Jealous God.

Does this mean that Bartha's conjecture is vindicated? Bartha himself urges caution:

> At the moment, I'm unsure whether or not other types of deity can participate in a strongly stable equilibrium. That leaves room for a remnant of the many-gods objection, and for doubts about the sufficiency of the requirement of strong stability. [Bartha, 2012, p. 204]

I am undecided about whether Bartha's caution is justified. Perhaps, if you accept the requirement of strong stability, you can conclude that the only stable position is to assign all of your credence to a Jealous God, or to a grouchy cartel (i.e., a bunch of grouchy Gods who only reward each other's followers). But, even in view of the considerations currently in play, it is not clear to me that we should accept the requirements of stability and strong stability. Consider, for example, the possibility of a jealous cartel: a group of Gods, each of whom rewards all and only those people who wager on one among that group of Gods. Let J be a regular Jealous God, let J1 and J2 form a

Table 13.7 Jealous, Nice, Perverse, and a Cartel

	J	J1	J2	P	N
J	1	0	0	1	1
J1	0	1	1	1	1
J2	0	1	1	1	1
P	0	0	0	0	1
N	0	0	0	1	1

jealous cartel, let P be a very Perverse God, and let N be a Very Nice God. If we are deciding just between these five, then Table 13.7 represents our decision problem.

In this decision problem, the updating rule "redistributes" probability away from J, P, and N toward J1 and J2. So, it is clear that no stable equilibrium distribution gives any credence to J, P, or N. However, because the updating rule does not make any redistribution between J1 and J2, it is also clear that there is no stable equilibrium distribution across J1 and J2. In short: any distribution of all of your credence over J1 and J2 is an equilibrium distribution; any distribution that gives some of your credence to any of J, P, and N is not an equilibrium distribution; and there is no stable equilibrium distribution. In this circumstance, it seems to me that the friend of Pascalian wagering who is attracted by equilibrium distributions ought to suppose that the right thing to do is to give all of their credence to J1 and/or J2. But, if that's right, then the friend of Pascalian wagering ought to insist that the requirement of strong stability is amended accordingly.

How should we think about a jealous cartel? Where a Jealous God says "You must believe in me (in order to obtain salvation)!" a God in a jealous cartel says "You must believe in someone who is enough like me (in order to obtain salvation): it needn't *be* me; near enough is good enough!"

If this result stands, it is bad news for Pascal and Bartha. Their God is a Jealous God; but if it is better to wager on a member of a jealous cartel than it is to wager on a single Jealous God, then – even if everything is otherwise in order – Pascal's Wager does not give the result that Pascal wants.

That's not to say that Pascal's Wager tells you to wager on a God who belongs to a jealous cartel. There are several problems here.

First, it might seem plausible to suppose that a bigger jealous cartel trumps a smaller jealous cartel. (The updating rule will "redistribute" probability away from the small jealous cartel to the bigger jealous cartel.) If so, how big should be the jealous cartel to which the God on which you wager belongs?

It looks as though the only acceptable answer to this question is: infinitely large! But, if that's right, then we are now looking at a decision problem involving at least infinitely many Gods that belong to jealous cartels, infinitely many Perverse Gods, and infinitely many Nice Gods. That means that we'll have an infinite number of occurrences of "1" in many of the rows in our table. But, if this is right, then our attempt to extricate ourselves from entanglement with infinities by moving to the framework of relative utilities appears to have foundered.

Second, I see no good reason to suppose that the list of kinds of conceivable Gods that we have been considering is complete. In particular, the Gods that we have considered so far distribute their rewards according to the Gods that are believed in by those who would like to have the rewards. But there are lots of conceivable Gods who, while represented as Indifferent on the table – because they do not distribute rewards according to the God-beliefs of those who would like to have the rewards – nonetheless do differentially bestow rewards. Consider, for example, a God who rewards only those who do not allow their credences to be affected by their utilities. More generally, consider the – plausibly infinite – class of conceivable Gods who will not reward acts ultimately founded on the aim of maximizing expected utility in the light of Pascalian calculation.

Perhaps it might be objected that, while it was fine to countenance a Very Nice God, a Very Perverse God, an Indifferent God, and a range of Nice Gods and Perverse Gods, it is not fine to countenance the greatly expanded range of Gods that have crept into my discussion. Bartha says the following, in the context of justifying the requirement of strong stability:

> It is important that [a] new theological possibility is "in the neighbourhood". From the bare fact that a deity appears to be logically possible, one need not – indeed, cannot always – infer positive probability. The thought here is that the many-gods objection rests on the view that it is not reasonable to assign positive probability only to one deity. I am generalising this point to the "pantheon of possibilities" [mentioned above]: anyone who assigns one of these gods a positive probability should be willing to entertain a tiny positive probability for the other types ... These are relevant possibilities for anyone who takes Pascal's argument and the many gods objection seriously. [Bartha, 2012, p. 202]

This might seem to open up the prospect of admitting into consideration Jealous, Very Nice, Very Perverse, Nice, Perverse, and Indifferent Gods, while not admitting into consideration jealous cartels and Gods who do not reward acts ultimately founded on the aim of maximizing expected utility in the light of Pascalian calculation. However, it seems to me that it would be very odd to admit Jealous, Very Nice, Very Perverse, Nice, Perverse, and Indifferent Gods for consideration while not admitting jealous cartels for consideration; and it seems to me that, if anything, it would be even odder to admit Jealous, Very Nice, Very Perverse, Nice, Perverse, and Indifferent Gods for consideration while not also admitting for consideration Gods who do not reward acts ultimately founded on the aim of maximizing expected utility in the light of Pascalian calculation. If we are prepared to countenance the rather hard to motivate behavior of the Very Perverse God and at least some of the Nice Gods and Perverse Gods, surely we ought to be prepared to countenance members of a jealous cartel whose behavior is plausibly motivated in much the same way that the behavior of Jealous Gods is motivated. And surely, too, we ought to be prepared to countenance Gods who do not reward acts ultimately founded on the aim of maximizing expected utility in the light of Pascalian calculation, since it seems readily intelligible that one might suppose that the behavior of such Gods is motivated by their respect for rationality and integrity.

Bartha justifies the claim that one cannot always "infer positive probability from logical possibility" by appeal to "Gale's denumerable infinity of sidewalk crack deities." Here is what Gale (1991, p. 350) says:

> From the fact that it is logically possible that God exists, it does not follow that the product of the probability of his existence and an infinite number is infinite. In a fair lottery with a denumerable infinity of tickets, for each ticket it is true that it is logically possible that it will win, but the probability of its doing so is infinitesimal, and the product of an infinite number and an infinitesimal is itself infinitesimal ... There is at least a denumerable infinity of logically possible deities who reward and punish believers ... For instance, there is the logically possible deity who rewards with infinite felicity all and only those who believe in him and step on only one sidewalk crack in the course of their life, as well as the two-crack deity, the three-crack deity, and so on, *ad infinitum*.

I do not think that we should be persuaded by Gale's argument. In N', it is not true that "the product of an infinite number and an infinitesimal is itself infinitesimal"; more generally, I do not think that there is any coherent theory of infinities and infinitesimals on which the product of an infinite number and an infinitesimal is always an infinitesimal. If N' is our background mathematical theory, then it is not true that it is logically possible for there to be a fair lottery with a denumerable infinity of tickets. The requirement that the lottery is fair means that each ticket has an equal chance of winning. But, in N′, there is no number x which satisfies $x.\infty = 1$. Moreover – though I admit that this is controversial – I do not think that there is any coherent theory of infinities and infinitesimals against which we can establish that it is logically possible that there is a fair lottery with a denumerable infinity of tickets. (See Oppy, 2006b, p. 188; and, for a dissenting view, Wenmackers and Horsten, 2013.)

Earlier, I gave a prima facie case for admitting infinitesimal credences: but that prima facie case involved uncountably many Gods. While, in a case involving denumerably many Gods, it is possible to give positive credence to all of them – for example, if, for all n, one gives probability $\frac{1}{2}^n$ to the n-crack deity, then one gives positive credence to all of Gale's sidewalk crack deities – there is no way that one can give positive credence to all of uncountably many Gods. If one is prepared to allow that there are uncountably many possible Gods, one can rightly insist that there is no legitimate inference of positive probability from logical possibility.

But, as I have already insisted, whether or not one is prepared to countenance uncountably many possible Gods, there is another option: one can allow that some of the Gods admitted for consideration in Pascal's Wager are given only infinitesimal credence. If one takes this option, then one will say that the many-gods objection rests on the view that it is not reasonable to assign *non-zero* probability only to one deity. When we "generalize to our pantheon of possibilities," what we say is that anyone who assigns one of these gods a positive probability should be willing to entertain a tiny – i.e., non-zero – probability for the other types.

3 Concluding Remarks

Time to take stock.

Bartha conjectured that, if we meet all of the other objections to Pascal's Wager, then the many-gods objection is already met. Moreover, he showed that, if all other objections to Pascal's Wager are already met, then, in a choice between a Jealous God, an Indifferent God, a Very Nice God, a Very Perverse God, the full range of Nice Gods, the full range of Perverse Gods, and no God, you should wager on the Jealous God. However, he worried that there might be other types of Gods that can participate in strongly stable equilibria – and, if that were so, then it would remain the case that, even if all other objections

Table 13.8 Jealous and Virtuous

	Jealous God Exists	Virtuous God Exists	No God Exists
Wager on Jealous God	1	0	0
Wager on Virtuous God	0	0	0
Virtuously Refuse to Wager	0	1	0

to Pascal's Wager were met, the many-gods objection would still be a significant objection to Pascal's Wager.

I have argued that the requirement of strongly stable equilibrium – and, indeed, the requirement of stable equilibrium – is not well-motivated. There are other types of Gods, no less worthy of consideration than those that figure in Bartha's deliberations, that are intuitively better wagers than the Jealous God. In particular, I have suggested that one does better to wager on a God that is a member of a jealous cartel than one does to wager on a Jealous God.

I have also argued that there are other types of Gods, no less worthy of consideration than those that figure in Bartha's deliberations, that make trouble for Pascal's Wager, but not because one would do better to wager on them rather than on a Jealous God. In particular, I have suggested that a "Virtuous" God, who does not reward acts ultimately founded on the aim of maximizing expected utility in the light of Pascalian calculation but who does reward properly motivated virtue, plausibly makes trouble for Pascalian wagering. Consider the decision scenario laid out in Table 13.8.

In this case, Bartha's updating rule does not apply, and there are no relevant notions of stable equilibrium and strongly stable equilibrium. Once we introduce Gods who reward non-wager actions, we introduce a barrier to the updating of credences in the light of utilities. And, once we've done that, we definitely do not get out the conclusion that one ought to wager on the Jealous God in preference to merely acting virtuously.

Finally, I have argued that there is an objection to Pascal's Wager that Bartha does not consider, but that interacts in interesting ways with Bartha's treatment of the many-gods objection. If we are prepared to countenance infinitesimal credences, then we should baulk at the move that recasts Pascal's Wager in terms of relative utilities. In the original formulation of Pascal's Wager, when infinite utility meets infinitesimal credence, we do not get well-defined results (and quite properly so). But, if we suppose that infinitesimal credences are in no worse standing than infinite utilities, then we cannot accept the assumption – built into the relative utilities framework – that there cannot be infinitesimal credences.

If I take seriously the idea that there are uncountably many possible Gods, and I understand the requirement of strongly stable equilibrium to require that all possible Gods are taken into account, then it will certainly be the case that there are undefined expected utilities in my wagering calculations. Even if Bartha's treatment of some cases involving finitely many Gods is quite compelling, it seems that the uncountably-many-gods objection remain a serious, independent objection to Pascal's Wager.

I think that there is a more general lesson here. I have focused on Bartha's approach because I think that it is the best extant treatment of infinity in Pascal's Wager. As I noted above, I agree with Hájek that, on all other extant approaches, the result of the calculation of the expected value of wagering for God turns out to be sensitive to the precise values that are attributed to the probability that God exists and the utility that is attributed to wagering for God if God exists, and there is nothing that privileges any particular assignment of those values over other assignments of those values. To date, there is no satisfactory treatment of infinity in Pascal's Wager.

14 Pascal's Wager and Imprecise Probability

Susanna Rinard

1 The Wager

A commonly discussed version of Pascal's Wager, and the one I'll be working with here, goes as follows. Either God exists or he does not; and you either believe that God exists, or you do not. There are, thus, four possibilities. The best is when God exists and you believe: here your reward is an eternity in heaven, an outcome of infinite value. (Here and elsewhere, Pascal assumes a traditional Christian theology.) The value of the possibility in which God exists and you don't believe is either finite, or negative infinity (call this h, for hell). The value of believing when God does not exist (call this b, for believer) and the value of not believing when God does not exist (call this a, for atheist or agnostic), are both finite. We make no assumptions about the comparative value of a and b; the argument is meant to go through regardless. These assumptions about utilities are summarized in Table 14.1.

Let's turn now to probabilities. It is commonly assumed that the agent has some precise, real-valued credence, between 0 and 1, in the proposition that God exists. Pascal's argument is meant to be persuasive even to those whose credence in God's existence is extremely low, but one crucial assumption is made: that your credence is not equal to 0 (and that it is real-valued, rather than infinitesimal).[1] On the standard Bayesian picture, one updates one's credences by conditionalizing on new evidence, and one bets in accordance with the odds set by one's credences. If so, then an agent with credence 0 that God exists would never increase their confidence in God's existence, no matter what evidence they might receive; and, they would not pay any price, no matter how small, for a bet that pays handsomely if God exists but nothing if God does not. But surely it is

---

[1] It is a very interesting and important issue whether one's credence that God exists might rationally be infinitesimal. However, for the purposes of this chapter I will set aside the possibility of infinitesimal-valued credences. I will also set aside the possibility of imprecise, or vague, credences that are imprecise or vague over some infinitesimal values.

Table 14.1 Pascal's Wager

	God exists	God doesn't exist
Believe	Infinity	b (finite)
Don't believe	h (finite or negative infinity)	a (finite)

not rational to take an attitude this extreme toward the proposition that God exists. Surely there is some possible evidence one might receive that ought to make one at least somewhat more confident that God exists. (For example, suppose a booming voice from the clouds repeatedly makes predictions that turn out to be accurate.) And surely one should be willing to pay some tiny price – a penny, say – for a bet that pays enormously – $10 billion, say – if God exists, but nothing if he does not. If so, then rationality requires one to have a non-zero credence in God's existence (though it may be tiny).

Now that we have made some assumptions about utilities and probabilities, we can calculate expected values. Let p be your credence that God exists. EV(believe) = (∞)(p) + (b)(1 − p). Infinity times any positive non-zero number is still infinity, and infinity plus or minus any finite number is still infinity. So, the expected value of believing is infinite. EV(Don't believe) = (h)(p) + (a)(1 − p). This will equal either some finite value, or negative infinity. Either way, the expected value of believing is higher than the expected value of not believing. So, Pascal concludes, we ought to believe in God.

Philosophers have raised numerous objections to this argument. If we can't believe voluntarily, and if ought implies can, then it's not true that we ought to believe, even if doing so has highest expected value. Even if we could choose to believe, is it really true that we ought to choose beliefs based on their expected value, rather than some other criterion (such as whether they are supported by our evidence)?[2] Also, one might worry that God would withhold the infinite reward from someone who chooses belief as a selfish gamble. Finally, there are possible divinities other than the Christian God, who, if they exist, would reward and punish different believers and non-believers, and Pascal's argument does not take these into account.

In my view, some of these objections (particularly the so-called many-gods objection) have considerable merit. It is not my aim here, though, to address

---

[2] I have argued elsewhere (Rinard, 2017) that the rationality of belief is to be determined in precisely the same way as the rationality of action – and so, if the rationality of an action is determined by its expected value, then the same is true of belief. So, in my view, Pascal is right that we ought to have whichever belief has highest expected value (on the assumption that this is true for action).

them. Rather, my focus is on the following question: Supposing that the argument, as stated, does go through, would it still go through if the agent's credence in God is not some precise real-valued number, but is rather *imprecise*?

2 Modelling Imprecise Credences

Traditional Bayesian approaches to modelling rational agents embody a number of idealizations. It is assumed that the agent is certain of all logical truths. It is assumed that they have thought of all possible theories before evidence starts rolling in, and that they never forget anything they have learned. It is assumed that they have a precise, real-valued credence between 0 and 1 in every possible proposition.

Plausibly, most real human beings lack all of these features. Sometimes we are uncertain of some logical truths. We didn't think of all possible theories before getting evidence – indeed, an important scientific skill is the creativity required to think up novel theories that might explain the evidence that we have. Sometimes we forget things that we once learned. And some of us lack precise credences in some propositions.

So, it seems, anyway. Of course, we might consider what it would take to defend the idea that, despite appearances, we really do have some of these features. For example, might it be that we are certain of the logical truth, and are uncertain only about which sentences express that truth? Might it be that, despite our inability to determine them by introspection, we actually do have real-valued credences after all, precise down to the millionth decimal point and beyond?

These are interesting possibilities that deserve to be investigated and taken seriously. However, it is not my aim to do so here. In particular, for the purposes of this chapter I will assume that, by and large, there are many propositions for which we lack perfectly precise credences.

The next natural question is whether this constitutes a rational defect. Does our lack of precise credences reflect a deficiency, or limitation, on our part? Would an ideally rational agent have perfectly precise credences?

Many philosophers have thought that the answer is *no*. Consider, for example, the proposition that humans will go extinct sometime in the next 6,000 years. Agent A1 has precise credence r in this proposition (P). A2 has precise credence r + .00000001 in P, and A3 has credence r − .00000001 in P. A4's credence in P is, however, vague over these (and other) values, not determinately equal to any of these three credences. Is A4 thereby irrational?

It's hard to see why they would be. Moreover, it may seem plausible that A4 is the only rational one of the bunch: assuming that these agents all have basically the same evidence as we do, A4's attitude may seem to better reflect that evidence than the extremely precise attitudes of A1, A2, and A3.

I will not assume here that rationality sometimes *requires* imprecise credences. I will, however, assume that credal imprecision is sometimes rationally *permissible*. In particular, I will assume that it is rationally permissible to have a vague credence in the proposition that God exists. As Alan Hájek (2000) points out, we can identify particular circumstances in which credal precision is rationally required. Such requirements might be generated by, for example, the Principal Principle (which says, very roughly, that, in a certain restricted range of cases, if I know that the objective chance of P is r, then my credence in P should be equal to r), suitably qualified versions of the Reflection Principle, various Deference principles (if I regard some other agent as an expert in the relevant sense, and know that their credence in P is r, then my credence should be equal to r), and perhaps others. But we can imagine possible agents – we ourselves likely among them – for whom no such principles apply to the proposition that God exists, and thus for whom a sharp credence in God's existence is not rationally mandated.

Assuming, then, that credal imprecision is rationally permissible, how can we model it formally? The most common approach is to represent an agent's doxastic state with a set of probability functions, rather than a single function (see, e.g., Joyce 2010, van Fraassen 1990, Jeffrey 1983b, and many others). This can be thought of as analogous to the supervaluationist account of vagueness (see, e.g., Hájek, 2000). Each function in an agent's set can be thought of as one admissible precisification of their doxastic state. Propositions about the agent's doxastic state about which all functions in their set agree are determinately true of the agent. Propositions true according to some, but not all, functions in their set are *indeterminately* true. For example, suppose that all of the functions in my set have Pr(aliens exist) > Pr(pigs fly). Then, it is determinately true that I am more confident that aliens exist than that pigs fly. Suppose that some of my functions have Pr(aliens exist) > Pr(humans go extinct in the next 6,000 years), but others have Pr(humans go extinct in the next 6,000 years) > Pr(aliens exist). Then, it is indeterminate whether I'm more confident that aliens exist than that humans will go extinct in the next 6,000 years, or vice versa. Or, suppose that, for each number r in the interval [c, d], there is some function in my set with Pr(aliens exist) = r, but for each number r outside that interval, there is no function in my set with Pr(aliens exist) = r. Then my credence that aliens exist is vague over the interval [c, d].[3]

This supervaluationist interpretation of the set of functions model generalizes to confirmation, decision-making, and more. For example, it is determinate that E confirms H just in case all functions in my set have Pr(H|E) > Pr(H). Or, if action A has highest expected value according to every function in my set, then it's determinate that I ought to choose A, and determinately impermissible for me to choose any alternative to A. If, however, some functions have EV(A) > EV(B), while others have EV(B) > EV(A), then it's indeterminate whether it's permissible to choose A, and indeterminate whether it's permissible to choose B.

This last result might strike some as odd. How can it be *indeterminate* whether it's permissible to do something? There are cases in which, for each possible option, there is at least one function in your set according to which that option has highest expected value, and at least one function in your set according to which some alternative option has higher expected value. If so, then, according to the supervaluationist interpretation of the set of functions model, for every possible option, it is indeterminate whether it's permissible for you to perform that option. This might seem like a problematic result. Isn't the role of decision theory to help guide you in making decisions? But if what it says is simply that each of your options is indeterminately permissible, that is no help at all.

Surely it is right that one important role for decision theory is to guide our decision-making. But we cannot demand that decision theory tell us precisely what to do in any possible situation. After all, everyone will agree that there are possible cases in which multiple options are all determinately permissible. In such a case, decision theory will not tell us exactly what to do – we will have to decide among the permissible options with no help from it. So, at most

---

[3] Some cases may seem to present a challenge to the idea that all determinate differences in confidence should be reflected in an agent's credences. For example, suppose I am throwing an infinitely fine dart (with no extension) at the [0, 1] interval in such a way that it is equally likely to hit any particular point in that interval. For any such point p, surely one should be determinately more confident that the dart will land on p than that a contradiction is true. However, according to some authors, one should have the same credence in both of these propositions, namely, 0. If so, then there are differences in comparative confidence that are not reflected in the agent's credences. My own preferred way of handling these cases is to insist that all rational confidence differences show up in the agent's credences. We can accommodate this by supposing that rational agents will have an infinitesimal credence in the proposition that the dart will land precisely on p, for any p in the unit interval. As noted in fn. 1, it is beyond the scope of this chapter to consider credences in God's existence that either precisely are, or imprecisely range over, some infinitesimal value(s).

what we can expect from decision theory is that it will *sometimes* tell us what to do, and *sometimes* provide helpful input in decision-making. And, this is true of the decision theory that falls out of the supervaluationist interpretation of the set of functions model. Sometimes there is a unique determinately permissible option, and in such cases this decision theory tells us precisely what to do. Other times it at least labels certain options as determinately impermissible – so we know to avoid them. In short, the fact that the supervaluationist decision rule sometimes labels options as indeterminately permissible does not prevent it from sometimes playing the helpful role in decision-making that we reasonably expect from a decision theory, even if it doesn't do everything that we might unreasonably hope it would. Elsewhere (Rinard, 2015) I have argued that this supervaluationist decision rule is superior to other decision rules one might pair with the set of functions model.

3 Against the Wide Interval View and the Reverse Principal Principle

Let us suppose, then, that it can be rationally permissible to have vague credences, and that these states are best formalized with the set of functions model, which is best interpreted along supervaluationist lines, with the attendant supervaluationist approach to neighbouring issues, such as decision theory. In this section I will rely on these assumptions to argue against some claims that are commonly made in the literature, and which, we will see, are ultimately relevant to Pascal's Wager. First, I will argue against *the wide interval view*, the claim that complete ignorance is properly represented as vague credence over the entire [0, 1] interval. I will then explore the implications this argument has for the idea that one should sometimes have a credence that is vague over an interval that includes just one extreme endpoint (i.e., a vague credence that includes 0 but not 1, or 1 but not 0). Finally, I will show that similar lines of argument undermine a claim I'll call the Reverse Principal Principle, which says that if all you know about some proposition P is that its objective chance is somewhere in the interval [c, d], then your credence in P ought to be vague over precisely that interval.

The wide interval view can be motivated by the natural thought that if one is *maximally* ignorant with respect to some proposition P – for example, if one has no evidence whatsoever relevant to P – then there should be no constraints on one's credence in P. If there are no constraints, then no function is ruled out of one's set by virtue of what it says about P, and so, for every number r in the unit interval, there will be some function in one's set with $Pr(P) = r$, and so one's credence in P will be vague over the entire unit interval. The wide interval view is widely accepted, endorsed by, among others, Joyce (2005), Kaplan (1996), and Weatherson (2007).

However, I will argue that this line of thought is deeply mistaken. The core of my argument will be that maximal ignorance – having *no clue* whether P – requires one to be determinately far from certain that P is true, and determinately far from certain that P is false. If so, then functions that assign extreme values to P – such as 0, or 1 – must be ruled out of one's set, in order to reflect this.

To make the point vivid, let's consider a particular case. Suppose you are presented with two urns, labelled GREEN and MYSTERY. You are certain (or at least, as certain as a rational agent possibly could be) that all the marbles in GREEN are green. But you have no evidence relevant to the colours of the marbles in MYSTERY. You have no clue what their colours might be. Let G-green be the proposition that the next marble drawn at random from GREEN will be green, and let M-green be the proposition that the next marble drawn at random from MYSTERY will be green. A defender of the wide interval view will claim that your credence in M-green should be vague over the entire unit interval, since you have no evidence relevant to it, and no clue whether or not it is true.

Whatever else we want to say about the case, this much seems clear: you should be certain, or very nearly so, that G-green is true; but you should be far from certain that M-green is true. So, you should be determinately much more confident of G-green than M-green. If so, then this must be reflected in the functions we allow into your set. Each such function must have $Pr(\text{G-green}) > Pr(\text{M-green})$. If so, then no function can have $Pr(\text{M-green}) = 1$, since such a function could not have $Pr(\text{G-green}) > Pr(\text{M-green})$. So, your credence in M-green cannot range over the entire closed interval [0, 1]. The same argument goes for any proposition, P, for which you lack all relevant evidence, about whose truth you simply have no clue. The wide interval view is false. (Further arguments against the wide interval view are given in Rinard 2013.)

Note that the argument just given against the wide interval view proceeded by first arguing that, when you have no clue about whether P is true, no function in your set should assign $Pr(P) = 1$. From this premise we can in fact derive a stronger conclusion: that your credence, if rational, is not properly represented by *any* interval (or set) that contains 1. That is, a rational credence in propositions about which you have no clue is not represented by any interval of the form [c, 1], or (c, 1], where c is any real number in the unit interval.

Parallel reasoning can also be used to establish that your credence should not be vague over an interval that includes the other extreme value, 0. Consider G-red, the proposition that the next marble drawn from GREEN will be red. You should be certain, or very nearly certain, that G-red is false. You shouldn't be anywhere near certain that M-green is false. So, you should be determinately more confident of M-green than G-red. So, every function in your set must have Pr(G-red) < Pr(M-green). So, no function in your set can have Pr(M-green) = 0, because such a function cannot have Pr(M-green) > Pr(G-red). So, no interval of the form [0, c] or [0, c), for any c in the unit interval, represents the credence of a rational agent in a proposition about which they have no clue. One upshot of this is that Bas van Fraassen's proposed probabilistic analysis of agnosticism – as vagueness over an interval that contains 0 – is mistaken. (Hájek, 1998 also argues against van Fraassen's analysis of agnosticism.)

But what if you *do* have a clue about whether P? You have some relevant evidence, say, or *a priori* considerations that bear on the matter. One possibility, of course, is that you have a precise credence in P, and perhaps that precise credence is 0 or 1. But what if the conditions under which credal precision is rationally required don't obtain, and you, compatible with perfect rationality, have an imprecise credence. Is it *ever* rationally permissible to have a vague credence that includes either of the extreme values 0 or 1, even if you do have relevant evidence or other relevant considerations?

I will suggest that this is never rationally permissible, for any contingent proposition P. If your credence in P is vague, then, I suggest, it should not include either of the extreme values. However, it is not my aim here to give a conclusive argument for this claim that ought to convince all possible opponents. Rather, what I will do is describe considerations that, to my mind, render this view plausible, and invite the reader to agree.

The heart of the matter, as we can see from the above discussion, is that, for any proposition P, as long as there is some proposition T such that you should be determinately more confident of T than P, and some other proposition F such that you should be determinately more confident of P than F, then your credence in P cannot be vague over either extreme value. This is because every function in your set must have Pr(T) > Pr(P) > Pr(F), and if some function has Pr(P) = 1, or Pr(P) = 0, then it cannot exhibit these inequalities.

This gives us a procedure for testing, for any proposition P for which your credence is rationally vague, whether your credence in P is rationally vague over either of the extreme values. For any such P, we can ask whether we can find some propositions T and F such that you're rationally required to be determinately more confident of T than P, and determinately more confident of P than F. If we can find two such propositions, then your credence in P is not rationally vague over either of the extreme values.

It is clear that, even if we have some idea about whether P is true, and have a considerable amount of evidence relevant to P, as long as we have a fair amount of uncertainty about P, it will be easy to find suitable values for T and F. For example, consider propositions like these: the average annual return of the stock market over the next decade will be over 5 percent; the incidence of anxiety and depression in American teenagers will continue to rise in the near future; most institutions of higher education in the United States will go bankrupt before 2035. I have a fair amount of evidence relevant to each of these three propositions, and I have some idea of whether they'll turn out true or not, but my credence in each is vague. I think my evidence on balance supports the first two claims, and I'm fairly confident they're true; I think my evidence on balance does not support the third claim, and I'm fairly confident it's false. In all three cases, though, my uncertainty is non-trivial, and I wouldn't be shocked to learn that I'm wrong about all three. So, it's easy to find values for T and F such that rationality requires me to be determinately more confident of T than any of those three propositions, and determinately more confident of any of them than F. For example, let T be some simple tautology, like P → P, and let F be some simple contradiction, like P & ~P. It would be irrational for an agent to fail to be determinately more confident of P → P than any of these three claims, or to fail to be determinately more confident of each of those claims than P & ~P. We could also use, as a value for T, some obvious mathematical truth, like 1 + 1 = 2, and, as a value for F, an obviously false claim like 1 + 1 = 27. There are many candidates for T and F that would do the job.

If we are looking for a proposition in which we could rationally have a vague credence that is vague over an extreme value, then we will have to look to propositions in which we rationally have less uncertainty: propositions such that we are extremely confident that they are true, or false, but in which we nonetheless have a vague credence. Consider the proposition that someone somewhere in the world will drink some water sometime in the next week. I am extremely confident that this proposition is true; I would be deeply shocked (and horrified) to discover that it's false. And yet, I am still, it seems, rationally required to be determinately even more confident of a simple tautology like P → P. Or, consider the proposition that I will find 28 four-leaf clovers in the next 10 minutes. I am extremely confident that this proposition is false; I haven't found a single four-leaf clover in years, and I fully intend to spend the next 10 minutes sitting at my desk writing this chapter, where there are no clovers of any kind to be found. And yet, surely rationality compels me to be determinately more confident that I will find those clovers than that P & ~P is true.

In general, for any contingent proposition P, it is quite plausible that I should be determinately more confident of a simple tautology than P, and determinately more confident of P than a simple contradiction. If so, then there is no contingent proposition such that rationality permits me to have a credence in that proposition that is vague over an interval that includes either extreme value.

Here's another route to the same conclusion. For any contingent proposition P, we can find some logically stronger proposition S, and some logically weaker proposition W. Plausibly, one should be determinately more confident of P than S, and determinately more confident of W than P. If so, then, for any such P, one's credence in P cannot rationally be vague over an extreme value.

Before concluding this section, I will draw one final lesson from the strategy of argument that I've been employing here. Consider the following claim, which I call the Reverse Principal Principle: if all you know about P is that the objective chance of P is somewhere in the [c, d] interval, then your credence in P should be vague over that same interval. Even where this principle is not explicitly stated, it is common in the imprecise credence literature to reason in a way that seems to presuppose it. And, as the name suggests, it can seem to nicely complement the Principal Principle, some version of which virtually everyone accepts, which is roughly as follows: if you know that the objective chance of P is r, then (in a suitably restricted class of cases), your credence in P should be equal to r.

However, the considerations raised in this section show that the Reverse Principal Principle is false. Consider again the example of our mystery urn. For every real number r between 0 and 1, for all we know, the proportion of green balls in MYSTERY is r, and so, for all we know, the objective chance of M-green is r. But it does not follow, as the Reverse Principal Principle would have it, that for each such r, some function in our set must have Pr(M-green) = r. That would entail the wide interval view, which, we have seen, is false.

Even though the Reverse Principal Principle is false, there is a truth in the vicinity. What is true is that, for each such r, we should not rule out the possibility that the objective chance of M-green is r. For example, all of the functions in our set should have Pr(the chance of M-green is 1) > 0. But this does not mean that some function in our set should have Pr(M-green) = 1. It is important, here, to keep firmly in mind the distinction between objective chance and subjective level of confidence. From the fact that, for all we know, the objective chance of P is r, it does not follow that our subjective level of confidence in P should be vague over an interval that includes r.

One way to bring this out vividly is to consider a case like the following. Suppose a fair coin has just been flipped. If it landed heads, then a marble will soon be drawn from the all-green urn. If it landed tails, then a marble will soon be drawn from the all-red urn. We have no other relevant information. Now, in this case, we know that the objective chance that the marble will be green is either 0 or 1. We have no idea which it is. But it certainly doesn't follow from this that our subjective level of confidence in the marble's being green should be vague over a set that contains 0, and 1. On the contrary, our credence that the marble is green should, in this case, be precisely 0.5. This example helps bring out the importance of clearly separating the following two questions: (1) What, for all we know, is the objective chance of P? (2) What is the rational subjective credence to have in P?

It has been my aim in this section to cast doubt on the idea that it is ever rational to have a credence in any proposition that is vague over a range that includes either of the extreme values, 0 or 1. I began by arguing against the wide interval view, the idea that in cases of maximal ignorance, in which you have no clue about whether a proposition is true, you should be vague over [0, 1]. Then I showed that, similarly, in such cases it is also not rational to be vague over a narrower interval with either extreme endpoint – i.e., you should not be vague over an interval with any of the following forms (where c is any real number between 0 and 1): [0, c], [0, c), (c, 1], [c, 1]. Next I suggested that even if you do have some clue about whether P is true, and a substantial amount of evidence bearing on the matter – even if you are fairly confident that P is true, or fairly confident that it's false – still, if P is contingent, it is not rational to have a credence in P that is vague over an interval that contains 0 or 1, since we can always find some proposition T such that you ought to be determinately more confident of T than P, and some proposition F such that you ought to be determinately more confident of P than F. Finally, I observed that one upshot of these conclusions is that the Reverse Principal Principle is false.

4 Imprecise Probability and the Wager

It's time to get back to Pascal's Wager. Now that we have in place the set of functions formalism for modelling imprecise credences and the supervaluationist interpretation of that formalism, and have investigated to some extent what these vague credences will rationally look like in a variety of different contexts, we can address the main question of the chapter: Does anything important change for Pascal's argument if we suppose that, instead of having a precise credence in the proposition that God exists, the agent has an imprecise credence? That is, assuming that the Wager goes through on the assumption that the agent has a non-zero precise credence, does it still go through if we suppose that the agent has an imprecise credence instead?

Let's first consider the case in which the agent's credence is vague over a range, or a set, that does not include 0. That is, different functions in their set will assign different values to the probability that God exists, but there is no function in their set with Pr(God exists) = 0. For such an agent, the Wager argument still goes through (assuming it was successful before). Recall that, as described in section 2, we can apply the supervaluationist interpretation of the set of functions model to matters of decision theory. It is determinate that an agent ought to choose A over B just in case all functions in their set agree that the expected value of A is higher than the expected value of B. As shown in section 1, any function that assigns a non-zero value to the probability that God exists will have the expected value of believing higher than the expected value of not believing. So, all functions in the agent's set will agree that the expected value of believing is higher than the expected value of not believing, so, they will all agree that the agent should believe in God; so, it will be determinately true that the agent ought to so believe.

The situation is importantly different, though, if the agent has a credence for God's existence that is vague over a set that includes 0; that is, if some function in their set has Pr(God exists) = 0. As noted in Hájek 2000, in such a case the Wager argument does not go through as before. What are the expected values of believing, and not believing, according to this function? We will have: EV(believe) = (∞)(0) + (b)(1), and EV(don't believe) = (h)(0) + (a)(1). Since zero times any value (whether infinite or finite) is still zero, we will have EV(believe) = b and EV(don't believe) = a. So, unlike in the previous cases, where believing had the highest expected value regardless of the comparative values of a and b, for the function with Pr(God exists) = 0, whether or not believing has higher expected value depends crucially on the comparative values of a and b. The original argument, then, does not go through as before.

Of course, there is potential room for a different argument – a dominance argument. For all that's been said here, it may be that we should assign higher value to believing than not believing, even on the assumption that God does not exist. Perhaps believing offers substantial benefits not available to the non-believer, enough to outweigh whatever pleasures they may forgo. What little empirical data we have that is relevant to this question does not rule out the possibility that religious people may in general be better off. If such a case could be made, then one could give a dominance argument in favor of belief that makes no assumptions at all about one's credence that God exists. According to this argument, one would be better off believing in God regardless of whether or not God exists, and so, one should believe, even if one is absolutely certain that God does not exist. Such an argument would be quite powerful, if its premises – including, crucially, the assumption that $b > a$ – could be adequately defended.

For the purposes of this chapter, though, my focus will be on the question of whether Pascal's reasoning goes through *without* making any assumptions about the comparative value of b and a. It is often touted as a great benefit of Pascal's argument that no such assumptions are made; let's see whether such an argument can be made to work, or at least, whether rational agents with imprecise credences are just as susceptible to such an argument as rational agents with precise credences.

The upshot so far is that the argument does not work for an agent whose credence in God is vague over a set that includes 0. The central question now, though, is whether it is *rational* to have a credence in the existence of God that is vague over 0. In the previous section I cast doubt on whether it is ever rational to have such a credence in any contingent proposition. Let us consider the specific case of the proposition that God exists.

First, is this proposition contingent? Some have thought that it is necessary that God exists. Others have thought that the concept of God is inconsistent and, so, that it is necessary that God does not exist. However, I will set aside such possibilities here. Presumably Pascal wants his argument to be persuasive to those who are not convinced that God exists necessarily, and so he wouldn't want to use that claim as a premise in his argument. And if it is *a priori* that the concept of God is inconsistent, and, thus, *a priori* that God does not exist, this would make trouble for Pascal before we even get to imprecise credences, since many hold that a rational agent should have precise probability 0 in *a priori* false propositions.

Let's suppose, then, that God's existence is contingent: there are possible worlds in which God exists, and possible worlds in which God does not exist. If so, then, applying the line of thought presented in the previous section, I would like to suggest that no rational agent should have a credence in God's existence that is vague over an interval that includes 0. Even if you are quite confident that God does exist, or quite confident that God does not exist, surely you ought to be determinately even more confident of a simple tautology, like P → P, and determinately less confident of a simple contradiction, like P & ~P. If so, then all functions in your set must have Pr(P → P) > Pr(God exists) > Pr(P & ~P), and so no function can have Pr(God exists) = 0. If so, then no *rational* agent will have a credence in God's existence that is vague over an interval that includes 0, and so, Pascal's argument, if it goes through for an agent with a precise non-zero credence in God's existence, will also go through for any *rational* agent with an imprecise credence in God's existence.

The crucial assumption on which this argument hangs is that a rational agent would be determinately more confident that God exists, if this is a contingent matter, than that a simple contradiction is true. But there are other routes to the conclusion that a rational agent's credence in God would not be vague over an interval that contains 0.

One such route, explored by Duncan (2003), begins with the common and plausible assumption that an agent's set of functions is updated by conditionalizing each function individually on whatever new evidence is learned. We then observe that, if one initially has a function in one's set with Pr(God exists) = 0, that function will continue to have Pr(God exists) = 0, no matter what evidence one updates on. Conditionalization can never raise a credence above zero. But then an agent who starts out vague over zero will *always* be vague over 0, and so will never, for example, come to believe that God exists, no matter what evidence they might possibly receive. But this seems irrationally dogmatic. Suppose, for example, that one hears a booming voice from the clouds announcing that it is the voice of God, which then makes all kinds of predictions about what will happen next, each of which comes true. Surely there are ways of filling out this case such that any rational agent, after all this takes place, would be determinately more confident that God exists than that he does not. But if their credence is still vague over an interval that contains 0 – which it will be if they started out vague over an interval that contains 0 – then no possible amount of evidence could lead them to be determinately more confident that God exists than not. Surely it is not rational to take such a dogmatic attitude, and so, not rational to have a credence in God's existence that is vague over 0.

Another route to this same conclusion focuses on rational action. Suppose you are offered a choice between two options: (1) receive $10 billion if God exists, nothing if God does not; or (2) receive $1 if God exists, nothing if God does not. Surely any rational agent would determinately prefer (1) over (2). It would be determinately irrational to choose (2), if (1) is an option. But if some function in your set has Pr(God exists) = 0, then you do not determinately prefer (1) over (2), since your functions do not agree that (1) has higher expected value that (2). According to the function with Pr(God exists) = 0, (1) and (2) have the same expected value, namely, 0. If it is rational to have such a function in your set, then it is indeterminate whether you ought to prefer (1) over (2), or whether you ought to be indifferent between them. But surely this is not indeterminate. Surely, it's determinate that you rationally ought to prefer (1) over (2). If so, then no function in your set should have Pr(God exists) = 0.

To summarize: I have given three different arguments for the claim that, if it is a contingent matter whether God exists, then no rational agent has a credence in God's existence that is vague over an interval that contains 0. I argued, first, that a rational agent ought to be determinately more confident that God exists than that P & ~P is true; second, that a rational agent ought to be such that some possible evidence would render them determinately more confident that God exists than that he does not; third, that a rational agent would determinately prefer a bet that pays a huge sum if God exists, and nothing if not, over a bet that pays a paltry sum if God exists, and nothing if not. In each case, I showed that, if this claim is right, then, given other plausible assumptions, it follows that no rational agent has a credence for God's existence that is vague over a range or set that contains 0.

If even one of these arguments is successful, then any rational agent with an imprecise credence in God's existence will have only functions that assign non-zero values to God's existence in her set of probability functions. If so, then if Pascal's argument is successful when aimed at an agent with a non-zero precise credence in God's existence, then it is equally successful when aimed at a rational agent with an imprecise credence in God's existence. Although other objections might undermine Pascal's Wager, going (rationally) imprecise does not.

15 Do Infinitesimal Probabilities Neutralize the Infinite Utility in Pascal's Wager?

Sylvia Wenmackers*

> Of these two infinites of science, that of greatness is much more obvious, ... the infinitely small is much harder to see.
>
> [L199/S230]

In the "Infinity – nothing" passage of *Pensées* (L418/S680), Pascal considers wagering for or against the existence of God, taking into account both probability ("chance") and utility ("happiness," "gain"). Jordan (2006) reconstructs the passage as presenting four related arguments: the first is based on weak dominance, the second is based on maximizing expected utility assuming the same probability for or against the existence of God, the third allows these probabilities to be different, and the fourth is based on strong dominance. Jordan calls the third part, which is the most discussed in contemporary philosophy, the "Canonical Wager."

It is only in the context of this Canonical Wager (henceforth simply referred to as "the Wager") that bringing up infinitesimal probabilities is relevant. Pascal explicitly excluded this possibility, assuming that there is "one chance of winning against a finite number of chances of losing," so that "there are not infinite chances of losing against that of winning" (L418/S680). In this chapter, we investigate whether Pascal was right in excluding infinitesimal probabilities. In other words: can infinitesimal probabilities be used to neutralize the infinite utility in the Wager, blocking the conclusion stating that it's prudent to wager for the existence of God? We will study this using a formal framework that enables us to represent both infinitesimal probabilities and infinite utilities and to combine them algebraically.

1 Infinities in the Wager

For our purposes, we can build on Hájek (2003)'s "perfectly standard interpretation" of the Wager passage in the *Pensées*, which is "cast in the

---

* I am grateful to Paul Bartha and Lawrence Pasternack for very helpful feedback on this chapter.

Table 15.1 Pascal's Wager

	Probabilities:	p Possible state 1:	$1 - p$ Possible state 2:
		God exists	God does not exist
Action 1:	Wager for God	S	f_2
Action 2:	Wager against God	f_1	f_3

anachronistic terminology of modern Bayesian decision theory (and that casting too is standard)" (p. 52).

1.1 Decision Matrix for the Wager

Hájek (2003, pp. 27–28) takes the argument to have three premises. Slightly rewriting:

(1) Rationality requires the agent to assign a positive, non-infinitesimal probability to God's existence, p.
(2) The decision matrix is as laid out in Table 15.1 with S the infinite utility associated with salvation and f_1, f_2, and f_3 finite utility values.
(3) Rationality requires the agent to opt for the action that has the highest expected utility (when there is one).

From the decision matrix, it follows (Hájek, 2003, p. 30) that the expected utility associated with wagering for God's existence is $S \times p + f_2 \times (1 - p)$, which is infinite given that S is. By comparison, the expected utility of wagering for God's non-existence is $f_1 \times p + f_3 \times (1 - p)$, a finite value. Maximizing the expected utility, this leads to the conclusion that it is rational to wager for God's existence.

The utilities in the decision matrix are based on Pascal's description in L418/S680 that "there is an infinity of an infinitely happy life to be won" (if you wager for the existence of God and this turns out to be correct), whereas "what you are staking is finite." To decide the best wager, we have to compare a finite probability value times an infinite utility value to a finite probability value times a finite utility value. Pascal writes (L418/S680): "The finite is annihilated in the presence of the infinite and becomes pure nothingness."

Hájek (2003)'s decision matrix leaves the finite utility values unspecified, which is consistent with L418/S680 as well as another fragment, in which Pascal writes (L199/S230): "In the perspective of these infinites, all finites are equal and I see no reason to settle our imagination on one rather than another." Although this is written in a different context (pertaining to the size of the human body as compared to the whole of the universe), this may be applied to the utility values in the Wager argument as well, making the argument very robust (f_1, f_2, and f_3 can be any real numbers). As long as neither probability value (for or against God's existence) is zero or some non-zero infinitesimal, the contrast between an infinite product and a finite product persists either way. So, also in this sense, the argument is very robust (p can be any real number in the open interval (0,1), i.e., $0 < p < 1$).

Since the Wager mentions infinite utility (a feature shared by all versions of the argument), that raises the question how to represent it formally. There are multiple options for doing so. In addition, we want to investigate the effect of allowing infinitesimal probabilities, so the formal model should allow us to represent infinitesimals as well. And since this argument involves expected utility, we should adopt a formal system that allows us to algebraically combine infinitesimal probabilities and infinite utilities. Otherwise the product and addition necessary for computing the expected utility associated with wagering for God (i.e., $S \times p + f_2 \times (1 - p)$, where we will allow p to be an infinitesimal) may turn out to be undefined.

Before looking into this, we first define infinite and infinitesimal numbers and consider the Wager in the context of real analysis.

1.2 Infinitely Large and Infinitely Small Numbers

In this chapter, we use the following definitions for infinite numbers and infinitesimals.

Infinite numbers are numbers of which the absolute value is larger than n for any natural number n. Within the fields of rational and real numbers, there are no infinite numbers.

Infinitesimals are numbers of which the absolute value is smaller than $1/n$ for any natural number n. Within the fields of rational or real numbers, the only infinitesimal is zero; such fields are called Archimedean. Fields of numbers that do contain non-zero infinitesimals are called non-Archimedean: we will see examples below. Sometimes the word infinitesimal is reserved for non-zero numbers, but here we will stick to the definition just given, which includes zero.

1.3 Infinity in Real Analysis

Hájek (2003)'s formalization of the Wager uses the lemniscate symbol for infinity, ∞, rather than S. Although it isn't meant to refer to a particular mathematical theory, the symbol is most commonly used in real analysis. Since both the symbol and the theory are well known, let's briefly elaborate. The values of real-valued sequences and functions that diverge become larger (or smaller) than any given finite value: we say that the limit "at infinity" is plus (or minus) infinity. This is represented in the affinely extended real number system by two symbols added to the set of reals, $\mathbb{R} \cup \{-\infty, +\infty\}$. This allows us to extend some arithmetic operations on the reals to sums and products involving infinite and finite limits (making $+\infty \times p + f_2 \times (1-p) = +\infty$ true for non-zero p and any f_2). Yet, some algebraic combinations cannot be defined in any consistent way, such as the indeterminate form $+\infty - +\infty$. So, unlike the real numbers, the extended reals do not form a field and these lemniscates should be recognized for what they are: shorthand notations that do not signify real numbers. By definition, divergent sequences and functions do not have a limit in the real numbers.

Pascal's L418/S680 description of the utility of eternal bliss as "an infinity of infinitely happy life" suggests that the infinite utility is a product of two infinities (infinite happiness per time and infinite duration) and also in L199/S230 Pascal discusses the "double infinity" of nature and science. To formalize this part of the argument, we have to take into account that standard measure theory doesn't deal with infinite quantities. It does allow us to state that a function diverges or, in other words, has a limit of infinity. It also allows us to express the rate of divergence, by considering the derivative of the function. A product of diverging functions diverges faster than its factors: the product function approaches infinity faster. Although the original argument is very robust, in the sense that any rate of divergence will do, it will be crucial to return to this issue once we consider infinitesimal probabilities.

If we look outside the field of real numbers of the standard calculus, there are multiple ways to represent infinitesimals and infinite numbers formally, although not all of them have been used explicitly to represent probabilities and utilities. We elaborate on this in the next section.

2 Harmonious Number Systems

In the context of a discussion on Pascal's Wager, Oppy (1991, p. 163) considered the epistemic possibility "that the probability that God exists is infinitesimal," in which case "the calculation of the expected return of a bet on [the existence of] God is no longer as straightforward as the initial argument suggested." Oppy (1991, p. 163) writes that "it is incorrect to suppose that 'non-zero' and 'finite' are coextensive; for it is epistemically possible that the probability that God exists is infinitesimal." Thereby, he raises the possibility that the agent's degree of belief might be infinitesimal rather than some non-zero real number. (Unlike Oppy, I will not use "finite" as a contrary to zero and/or infinitesimal numbers.)

Following up on this, Hájek (2003, p. 38) wrote: "Once infinitesimal probabilities are allowed, the reformulated argument no longer goes through automatically: the infinitesimal probability can 'cancel' the infinite utility so as to yield a finite expectation for wagering for God; and this may be exceeded by the expectation of wagering against God."

In this section, we investigate a necessary condition that the formal system needs to satisfy in order to allow for this "cancelling" of infinite quantities by infinitesimals. Two systems that satisfy it will be briefly introduced and some reasons for preferring one over the other will be evaluated.

2.1 "Cancelling" Requires a Harmonious Number System

Some well-known theories that include infinity do not include non-zero infinitesimals. Cantor's theory of cardinality, for instance, contains an infinite hierarchy of transfinite cardinalities, but not a single non-zero infinitesimal. Likewise for Cantor's theory of ordinal numbers. In fact, Cantor was strongly opposed to the idea of infinitesimals. He tried to prove their inconsistency (Cantor, 1966 [1932], pp. 407–9), but Zermelo pointed out that Cantor's purported proof relies on the multiplication of an infinitesimal by a transfinite ordinal number, which is undefined (Cantor, 1966 [1932], pp. 439). It seems more accurate to say that infinitesimals are simply incommensurable with transfinite cardinal and ordinal numbers (Benci, Horsten, and Wenmackers, 2018, § 4.1.1). Also Sobel (1996, endnote 8) observed that the multiplicatory inverse of Cantorian cardinalities are undefined and hence cannot be used to represent infinitesimals.

So, some ways of formalizing Pascal's Wager with infinitesimal probabilities simply will not work. If we use a theory that doesn't allow us to multiply transfinite numbers by non-zero infinitesimals, then we cannot investigate the question whether infinitesimal probabilities block the Wager's conclusion. Other approaches are more promising: there are number systems in which each infinite number is the multiplicatory inverse of a particular infinitesimal number, and vice versa. We will call number systems that have this property "harmonious." This property is suggested by the etymology of "infinitesimal": the word is formed in analogy to "decimal" (meaning tenth or one divided by ten), which suggests it is equal to one divided by infinity. For our purposes, we require this relation to be symmetric: each infinitesimal is equal to the inverse of an infinite number and vice versa. So, neither the infinite nor the infinitesimal numbers are conceptually prior to or privileged over the other in any way.

We now introduce two types of harmonious number systems. A first type of harmonious number system is given by non-standard models of rational or real closed fields, which are used in non-standard analysis. A second type of harmonious number system is given by Conway's surreal numbers. Both systems have also been discussed by Hájek (2003) in the context of the Wager, e.g., to tell apart various infinite utilities.

2.2 Robinson's Hyperrational and Hyperreal Numbers

A well-developed route for handling infinitesimal and infinite numbers is that of non-standard models of rational or real closed fields, due to Robinson (1966). The numbers in such non-Archimedean, ordered fields are known as hyperrational ($\in {}^*\mathbb{Q}$) and hyperreal numbers ($\in {}^*\mathbb{R}$). While I will not go into the technical details, the next subsection gives a sketch of the construction. If you decide to skip section 2.2.1, the main thing you need to know to follow the discussion in section 4 is this: all algebraic operations that are defined on the real numbers are also defined on the hyperreal numbers.

Non-standard analysis is a branch of mathematics geared toward proving results in the standard calculus via the non-standard domain: this is the main, but not the only application. Rounding off the infinitesimal part of a finite hyperreal number k yields a standard real and is called taking the standard part, written st(k). Authors like Edward Nelson and Vieri Benci have defended taking infinitesimals more seriously; in particular, by investigating hyperreal-valued probability functions in their own right (Nelson, 1987; Benci, Horsten, and Wenmackers, 2013).

2.2.1 Harmony of Non-Standard Models

I will briefly sketch how to construct a non-standard model of the real numbers, to illustrate how the aforementioned property of harmony comes into play. One way of constructing the hyperrational or hyperreal numbers is by taking the ultrapower of the rationals or reals with respect to a free ultrafilter or, equivalently, by taking the quotient ring of the rationals or reals under a maximal ideal. To explain this, let's start with Peano arithmetic, which gives the properties of natural numbers in first-order logic. Starting from a model of Peano arithmetic (which includes an infinite alphabet, with a name for each natural number), one builds a non-standard model by introducing a new symbol, for instance N, declaring it larger than any number already in the model. The proof relies on the Compactness theorem; for details see Boolos et al. (2007, pp. 302–18). N is the first example of an infinite number in the non-standard model (by the definition in section 1.2). Since the first-order properties are required to hold in this non-standard model, like any standard natural number, N will have a successor, which itself has a successor, and so on. So, by introducing N, we allow an infinite number of infinite numbers. All the numbers in a non-standard model of Peano arithmetic (finite and infinite ones) are called hypernatural numbers ($\in {}^*\mathbb{N}$).

Building non-standard models of rational or real closed fields can be done in similar fashion: starting from a model of the first-order properties of the rational or real field, one adds a new symbol. In this case we have two options for how to proceed, but the results will turn out to be equivalent. Either we add a symbol R for a new number larger than any rational or real number, or we add a symbol ε for a number larger than zero but smaller than $1/r$ for any positive rational or real number r. Since the non-standard model has to be closed under the same first-order operations as the standard model, it has to be closed under taking the reciprocal, so adding the infinite number R ensures that the non-zero infinitesimal $1/R$ is in the model as well, whereas adding the non-zero infinitesimal ε ensures that the infinite number $1/\varepsilon$ is in the non-standard model. This illustrates the harmony of the number system.

It is possible to embed the standard rationals and reals in a non-standard model in a canonical way. However, there is not just a single or even a canonical non-standard model of the real numbers. There are infinitely many fields of hyperreal numbers, with different properties. It is even possible to iterate the procedure of building a non-standard model starting from a non-standard model: simply add yet a new symbol and declare it bigger than any previous one. This yields hyperhyperreal numbers ($\in {}^{**}\mathbb{R}$).

2.2.2 Hyperreals and the Wager
Herzberg (2011) uses hyperreal numbers to represent utilities in the Wager, but he does not deal with infinitesimal probabilities in the main text. In a footnote,

he does remark (p. 71): "Zero probabilities would make the argument invalid straightaway, and infinitesimal probabilities would require a sufficiently high utility of salvation in order to preserve the validity of the argument," and he cites Oppy (1991) and Hájek (2003). Oppy's and Hájek's idea of cancelling is indeed what a field of hyperrationals or hyperreals allows us to formalize.

This was discussed by Sobel (1996): though the main text deals with representing the infinite utility in Pascal's Wager by Cantorian infinities, he included an appendix on Robinsonian hyperreals. In the endnotes, Sobel discusses the interplay between infinite utility and infinitesimal probabilities, when both are taken to be hyperreal numbers. In endnote 6, Sobel considers using an infinite hyperreal number to represent the utility of eternal bliss in a "hyperreal decision theory." He says this is relevant only if one is "supposed to prefer only every non-infinitesimal chance of eternal bliss even to a certainty of every worldly loss" (p. 55). According to Sobel, it makes no sense to apply hyperreals, however, if one is "supposed to prefer absolutely every chance of eternal bliss, even to a certainty of every worldly loss" (p. 55). This observation is related to the fact that for a given infinite hyperreal value, S (used here to represent the infinite utility of salvation), there is exactly one infinitesimal that is its inverse, $1/S$, but for any infinitesimal hyperreal there at least continuum many smaller ones (and similarly for larger ones). We will return to this is section 4.1, to see for which values of $p(S)$ expected utility does not favor belief over disbelief and the Wager fails.

2.3 Conway's Surreal Numbers

A decade after Robinson, Conway developed a system aimed to represent all numbers, now known as the surreal numbers (Knuth, 1974). The construction of the proper class of surreal numbers is done inductively, in stages. The first infinite surreal number is constructed at the same stage as the first non-zero infinitesimal and they are each other's multiplicatory inverse. Hence, harmony is built into the number system.

In a reconstruction of Pascal's Wager argument, Hájek (2003, p. 35) favors the surreal numbers over hyperreals, citing their ingenuity and user-friendliness. Also Easwaran (2014) favors Conway's system.

To the best of my knowledge, however, there is not yet a well-developed probability theory or decision theory based on the surreal numbers.[1] Hence

---

[1] Shortly before this book was finalized, I discovered that Chen and Rubio (forthcoming) have developed such a theory and that they applied it to Pascal's Wager. The interested reader is encouraged to read their excellent article.

my pragmatic choice for the hyperreals. Since there exists an isomorphism between the field of Conway numbers and a field of hyperreals that is maximal in a particular sense, this choice need not be an exclusive one.

2.4 Toward a Harmonious Decision Theory

Combining standard probability theory (that uses the real numbers, which do not include any non-zero infinitesimals) with infinite utilities is notorious for producing puzzling and even paradoxical results, of which the St. Petersburg paradox is the most famous example.

Using a harmonious number system seems to be a necessary but not a sufficient condition to formalize Pascal's Wager with infinitesimal probabilities: we need to apply it in the context of a theory that is capable of representing probabilities as well as utilities. I take this as a reason to prefer the non-standard approach – at least at this point in time. The situation may be assessed differently once a decision theoretic framework is built around the Conway numbers, or maybe a different system altogether.

There are several ways to base a probability theory on non-standard fields of numbers. Some approaches are geared toward obtaining standard probability functions (without infinitesimals), such as Loeb's (1975), but others embrace infinitesimal probabilities and other values in the hyperreal unit interval. An important example is due to Nelson (1987); more recently, Benci, Horsten, and Wenmackers (2013). There have been counterarguments against the notion of infinitesimal probabilities. Criticisms include worries about the non-uniqueness of the underlying hyperreal fields, the observation that infinitesimal chances are incompatible with symmetry considerations (such as temporal translation and rotation), and an argument that suggests physical agents cannot hold hyperreal credences (Easwaran, 2014). We will not engage with these criticisms here; for an overview and replies, see Benci, Horsten, and Wenmackers (2018).

Non-Archimedean probabilities do not mix well with real-valued utilities. Hence, to deal adequately with infinitesimal probabilities in the context of decision theory, a non-Archimedean utility theory is needed, such as the one developed by Pivato (2014): a recent theory that both represents probabilities and utilities based on such non-standard fields of numbers. This can be used to model choice under uncertainty, as is the case in Pascal's Wager.

We will not present the theory in detail here, but the crucial part is that, in order to ensure harmony, the construction of a non-Archimedean field of numbers has to be one and the same to represent the infinitesimal probabilities as well as the infinite numbers. In what follows, we assume this theory, without spelling it out here.

On the other hand, we could also wonder if it is necessary to model infinite utilities and infinitesimal probabilities at all. Bartha (2007) focuses on relative utilities as a way to represent preferences without having to represent infinite utilities explicitly. Advantages are that he doesn't have to "pre-suppose a definite number of dimensions or to attach a definite value to the utility of salvation" (p. 29). He merely raises the option of infinitesimal probabilities at the end, suggesting that they may be helpful to deal with more complex problems. Indeed, there are non-Archimedean orders that do not presuppose a field of numbers, which suffice for a theory of qualitative probability (Narens, 1980), but for computing expected utilities some field of numbers is required.

3 But Is It Sufficiently Pascalian?

Before we continue to bring these recent formal systems to bear on the case of infinitesimal probabilities in a wager argument, we briefly pause to reflect on the question of whether these approaches are sufficiently Pascalian.

The formal systems described in sections 2.2 and 2.3 were not developed in Pascal's time. What is more, one should keep in mind that the same holds for standard decision theory. One of its components, contemporary probability theory, has its roots in the correspondence between Pascal and Fermat, but also encompasses many mathematical notions not available in Pascal's own time, such as functions and the field of real numbers, of which the unit interval is now used to represent probability values. Due to the axiomatization by Kolmogorov (1956 [1933]), probability theory has since been incorporated into standard measure theory: the theory that uses real numbers to assign values to lengths, areas, and volumes is also used to express probabilities. Hence, our standard decision theory built on this framework can only be applied to Pascal's Wager anachronistically.

3.1 Pascal on Infinities and Infinitesimals

Some seminal ideas concerning infinity and infinitesimals were definitely known to Pascal.

3.1.1 Pascalian Harmony

In light of what I called the harmony of some number systems, it is interesting that Pascal (L199/S230) clearly saw a symmetry between the infinitely large and the infinitely small, writing: "These extremes touch and join by going in opposite directions, and they meet in God and God alone." He regarded human bodies as placed between these two extremes, "between these two abysses of infinity and nothingness," neither of which we can perceive or truly grasp. For instance, our finite size is nothing compared to the whole of the infinite universe, but on the other hand he suggested that small things or animal bodies aren't simply infinitely divisible, but infinitely structured into the small (suggesting a fractal structure *avant-la-lettre*).

3.1.2 Pascalian Infinity

A generation before Pascal, Galileo (1954 [1638]) had written on a paradox that arises if we try to compare infinite sizes (in particular, the collection of whole numbers and that of perfect squares; see also Mancosu, 2009). Galileo had concluded from this that it is meaningless to determine (in-)equality of sizes of infinite collections. Yet, in his *Pensées*, Pascal (L199/S230) did write of infinites (in plural) and said that "mathematics, for instance, has an infinity of infinities of propositions to expound," which may suggest that he did allow distinctions among infinite sizes.

Numerosity theory (Benci and Di Nasso, 2003; see also Mancosu, 2009) is a particular motivation and axiomatic approach for the hypernatural numbers, which is closely related to the non-Archimedean probability (NAP) theory mentioned in section 2.4. According to numerosity theory, on which NAP is based, there is a numerosity of natural numbers, and this infinite number – usually called α – is either even or odd. Adding a unit to α, or any other numerosity, yields another value. Although NAP theory can be derived from axioms and need not be derived from numerosity theory or any other form of non-standard analysis, it is precisely this Euclidean part–whole property that powers the regularity of the NAP-function.

Textual evidence of the passage just before the Wager argument suggests Pascal might have objected to such a theory (L418/S680): "[W]e know that it is untrue that numbers are finite. Thus it is true that there is an infinite number, but we do not know what it is. It is untrue that it is even, untrue that it is odd, for by adding a unit it does not change its nature." This suggests that Pascal had a Galilean concept of infinity, or possibly proto-Cantorian (given the cited part of L199/S230), both of which are incompatible with infinite hypernatural (and by extension, hyperreal) numbers. Pascal (L418/S680) also wrote: "Unity added to infinity does not increase it at all, any more than a foot added to an infinite measurement: the finite is annihilated in the presence of the infinite and becomes pure nothingness." Although this goes in the same direction, it can be understood as a merely comparative claim, unlike the previously cited passage.

Hájek (2003) worries that assigning an infinite hyperreal to the value of salvation is insufficiently Pascalian, because salvation is supposed to be the best thing possible, whereas for each infinite hypernatural number, there are infinitely many larger ones. But it is not clear that the existence of larger numbers on the utility scale entails the existence of states or rewards corresponding to them. Rota (2017, endnote 4) also comments that this need not be a decisive reason to reject the use of hyperreals in a reconstruction. In addition, he suggests that Hájek's underlying worry may be that there is arbitrariness "in selecting a particular [infinite] number to represent the value." Although Hájek wrote with the surreal numbers in mind, the same worry may be flagged for hyperreal numbers. This interpretation seems plausible to me, since similar worries have been raised in arguments geared against infinitesimal probabilities: there the worry is related to assigning a particular infinite number to represent the size of an infinite sample space (for instance by numerosities; Benci and Di Nasso, 2003), rather than to an infinite utility as is the case for the Wager. Rota (2017) does not find the objection a strong one, however, since any choice for the infinite value yields a valid argument.

3.1.3 Pascalian Infinitesimal Probabilities

All formalizations that include infinitesimal probabilities are more recent than the foundations of standard probability theory. Nevertheless, the notion of infinitesimals in general, and of infinitesimal probabilities in particular, is certainly not alien to Pascal's time. The notion of infinitesimals was prominent in the European mathematics of the time. Bonaventura Cavalieri had rediscovered forgotten ideas of Archimedes: areas and volumes can be computed as sums of infinitesimally thin cross-sections. This idea was crucial for the development of the calculus by Newton and, even more explicitly so, by Leibniz. This use of infinitesimals was famously criticized by Berkeley, who called them "Ghosts of departed Quantities" (Berkeley, 1734; section XXXV, p. 59). But it is undeniable that they were at the roots of our calculus, which has been called "infinitesimal calculus" for a long time. It was only after the nineteenth-century formalization of the calculus in terms of real numbers and epsilon-delta definitions that the notion of infinitesimals was banished from mathematics. To this day, mathematicians continue to use the word "infinitesimal," but not in the sense defined in section 1.2. As a result, the notion of infinitesimal probabilities in the original sense became suspect as well – especially after the twentieth-century incorporation of probability theory in measure theory (itself firmly rooted in the calculus).

As the quote in the second paragraph of this chapter shows, Pascal explicitly mentions infinitesimal probabilities, albeit only to exclude this case from his argument. Interestingly, Hájek (2003) writes that Pascal "deserves considerable credit for apparently having a notion of infinitesimal probability years ahead of his time" (p. 39). In light of the history of the concept, which I just briefly sketched, however, I believe this should not be so surprising. Nevertheless, when we apply the very recent formalizations of the concept to Pascal's argument, we should be aware that we approach it very differently from anyone in Pascal's time.

See section 5.1 for another aspect that may be non-Pascalian about any reconstruction with infinitesimal probabilities.

3.2 Pascalian Probability and Cromwell's Rule

3.2.1 Epistemic versus Personal Probabilities

It may also be worthwhile to reflect on Pascal's interpretation of probabilities in general. It is (pre-)classical: probabilities are related to (lack of) knowledge, but they are epistemic, not personal probabilities. They are much closer to Laplacian chances (on which all agents can agree), than to the contemporary (neo-)Bayesian interpretation of probabilities as credences: rational degrees of belief that depend on a person's knowledge, not all of which needs to be common knowledge. Regarding the existence or non-existence of God, Pascal remarks that neither possibility can be ruled out *a priori*. In the classical setting, both agents would have to agree on the (prior) probabilities.

In the contemporary Bayesian approach to probability, probability assignments are personal. In the objective Bayesian approach, priors have to be shared, whereas they may differ for subjective Bayesians. For all Bayesians, probabilities should be updated upon receiving new information (which may be different across agents). Based on different life experiences, someone may have become an atheist and someone else a theist. In a Bayesian formalization, this is reflected in these persons having different posterior probabilities. If we grant Pascal that there is no decisive knowledge to be gained either way, then we may assume that the posterior probability that the atheist assigns to "God exists" is not *zero*, neither is that of the theist *one*.

3.2.2 Cromwell's Rule
It is in this context that the requirement of "regularity" (Carnap, 1950b) or "strict coherence" (Shimony, 1955) is usually discussed, which requires us to assign non-zero (prior) probability to all elementary possibilities. (This requirement is violated in standard probability for uniform probability distributions over infinite sample spaces.) Avoiding the extreme probability values (zero and one) allows the agents to remain within reach of each other's arguments and of evidence contrary to their current highest degree of belief: they may be "converted" by Bayesian conditionalization. Pascal's Wager argument doesn't take this dynamical aspect in account, but for his argument to be effective, he too needs to rule out probability zero (the only standard infinitesimal) as well as other infinitesimals.

Hájek (2003, § 3.2) discussed regularity in the context of Pascal's Wager. In the literature on the foundations of statistics, Lindley (1991, p. 104) coined the term "Cromwell's rule" for essentially the same idea: probability unity should only be assigned to tautologies, and probability zero should be reserved for contradictions. The name refers to Oliver Cromwell, who in 1650 wrote a letter to the General Assembly of the Church of Scotland in which he urged: "I beseech you, in the bowels of Christ, think it possible that you may be mistaken." However, Cromwell doesn't seem to have applied this plea to himself: he did not consider the possibility that his own belief could be mistaken, and the suggestion was rejected. In the Wager argument, Pascal tries to start from a more open-minded position: although he is a theist, he considers the option favored by an atheist *a priori* possible too. In doing so, he puts himself in a vulnerable position, within reach of arguments from atheists, but of course, his goal is opposite: to make them consider and embrace theism.

We already quoted Hájek commenting that if we allow infinite utilities, we should also allow infinitesimal probabilities in the formalism. Oppy made a similar remark, that also touches upon regularity:

> Note, by the way, that it is no objection to observe that infinitesimals are somewhat dubious entities. The same point could be made in the language of measure theory – i.e., in the mathematical theory which is appropriate

for dealing with probabilities in the case in which there are infinitely many options. The set of worlds in which the Christian God exists may have measure zero, and yet be non-empty. [1991, p. 168]

In other words, standard probabilities violate the requirement of regularity; infinitesimals provide a way to restore it (Benci, Horsten, and Wenmackers, 2013; albeit at the cost of other problems see Benci, Horsten, and Wenmackers, 2018).

4 Applying Hyperreal Decision Theory to the Wager

Recall from section 2.2.1 that all algebraic operations that are defined on the real numbers are also defined on the hyperreal numbers. One effect of this is that formulas involving infinite utility values and infinitesimal probabilities will look exactly the same as formulas that only involve finite values. Using the decision matrix of section 1.1, but allowing both probabilities and utilities to take values in some fixed non-Archimedean field of hyperreal numbers $(p, f_1, f_2, f_3, S \in {}^*\mathbb{R})$, the Wager is successful (as an argument for the existence of God) if

$$S \times p + f_2 \times (1 - p) > f_1 \times p + f_3 \times (1 - p)$$

and unsuccessful if

$$S \times p + f_2 \times (1 - p) \leq f_1 \times p + f_3 \times (1 - p).$$

Assuming the utilities are fixed, this can be rewritten as a condition on the value of p. The Wager is successful iff (if and only if):

$$p > (f_3 - f_2)/(S - f_1 - f_2 + f_3). \tag{1}$$

Now we can reformulate the main question of this chapter as follows: does it suffice to demand regularity in the probability assignments of the agent in order for the argument to keep being successful? As is known from the literature, the answer is: it depends.

For instance, against Rescher's (1985, pp. 16–17) claim that the Wager is favorable "as long as the chance of winning is nonzero," Sobel (1996, fn. 10) offered a counterexample, which we can now check against our success condition. Sobel's counterexample corresponds to setting $f_3 - f_2 = 1$ and $p = 1/(2(S - f_1))$. Substituting this in condition (1), the Wager is successful iff $f_1 > S - 1$. This cannot hold, since we assume f_1 to be finite and S infinite. Hence, Sobel (1996) indeed provided a counterexample to Rescher's claim that regularity suffices.

The fraction on the right-hand side of inequality (1) is small in the order of $1/S$, so there is some infinitesimal threshold on p above which the Wager is successful. We will investigate this dependence.

4.1 Dependence of the Wager's Success on the Incredulity

In general, the larger the infinite utility S is compared to $1/p$, the easier it is for the Wager to be successful. We illustrate this by considering two special cases.

4.1.1 Probability Inversely Proportional to Utility of Salvation

Let us first consider a skeptical agent whose degree of belief is inversely proportional to the magnitude of the promised salvation. In such a case, we have that the infinitesimal probability p is equal to f/S, with f some positive hyperreal number that is neither infinitesimal nor infinite, such that $p \times S$ is finite and non-infinitesimal. This case is of special interest, not because it is especially well motivated, but because (i) it encompasses the switch between successful and unsuccessful wagering and (ii) it seems not well motivated to rule it out.

Because of (i), the case where $p = f/S$ is very similar to a decision problem in which all utilities and probabilities are finite. In particular, the values of the three finite utilities in the decision matrix do become crucial for the conclusion based on utility maximalization. If someone could argue, based on the interpretation of p and S, that this case is irrelevant, we would only have to deal with cases for which it is clear whether the Wager is successful or not, irrespective of the values of f_1, f_2, and f_3. Moreover, if we could rule out this case, we would probably not need hyperreal analysis at all (see also section 4.2). However, I see no obvious way of ruling it out, so let's proceed with the detailed analysis.

Substituting $p = f/S$ in inequality (1) and rewriting, we see that the Wager is successful iff

$$f_3 < f_2 - 1/(S/f - 1) \times f_1 + f/(1 - f/S).$$

Taking the standard part (cf. section 2.2), we obtain $st(f_2) \geq st(f_3) - st(f)$ as a necessary (though not a sufficient) condition for a successful wager. In any case, the seemingly irrelevant matter of how to assign the finite utilities in the Wager (cf. section 1.1), turns out to be crucial – even sensitive up to infinitesimal differences – for determining the successfulness of the argument once p is allowed to be an infinitesimal. This result is consistent with Oppy (1991), who already wrote that the expected utility will depend upon the exact values of the infinite utility and the infinitesimal probability.

Of course, whether any valuation of f_1, f_2, and f_3 that makes a crucial difference between successful and unsuccessful wagering makes sense given the meaning of the utilities can still be debated, but I will not pursue this matter too far. After all, setting p proportional to $1/S$ in the first place was just meant as an illustration. But it is interesting to observe that in the fourth part of the Wager (in Jordan 2006's reconstruction), Pascal goes on to argue that one gains even in this life by living piously (argument from strong dominance). In other words, he argues that $f_2 > f_3$, which is indeed a sufficient condition for a successful wager in the presence of infinitesimal probabilities.

4.1.2 Probability Inversely Proportional to Either Duration or Intensity of Salvation

In section 1.3 we already indicated that Pascal characterized S as "an infinity of an infinitely happy life" suggesting it to be a product of two infinities. Now, it is true for *any* infinite hyperreal, that it can be regarded as a product of two infinite hyperreals (e.g., the initial number's square root). But if $1/p$ were to scale with either infinite dimension of salvation (either its duration, or its intensity, but not both), the Wager would still be successful. Could this "doubly infinite" utility still be an ace up Pascal's sleeve, even in the presence of some "moderate" infinitesimal probabilities?

Consider an agent to which the existence of a being with an infinite capacity is infinitely implausible, but additional infinite capacities do not increase the implausibility so strongly. This suggests substituting $p \sim 1/\sqrt{S}$. Let's be slightly more general, however, by substituting in inequality (1): $p = f/S^a$, with f and $a < 1$ positive hyperreal numbers that are neither infinitesimal nor infinite. (Hence, $p \times S$ is infinite, as is the case in a wager without infinitesimal probabilities.) After rewriting, we see that the Wager is successful iff:

$$f_3 < f_2 - 1/(S^a/f - 1) \times f_1 + f/(1 - f/S^a) \times S^{1-a}.$$

Observe that the righthand-side has a positive infinite value (the first and second term are finite and the third term is positive infinite). Given that f_3 is finite, this inequality always holds, so the Wager is always successful when, e.g., $p \sim 1/\sqrt{S}$.

4.2 Keeping It Real

In section 1.3 we already remarked that sequences and functions may diverge at different rates. Here we can develop this observation and connect it to the hyperreal framework. It will also help us to answer the following question: did we really need decision theory with hyperreal numbers to get to this result? Or can we regard non-standard methods, once again, merely as a tool to find a result within the standard domain?

To address this question, it is helpful to consider the following interpretation of infinitesimals: the limit of a function that converges to zero also represents information about the rate of convergence. An example may be helpful: consider the functions $1/x$ and $1/x^2$. Both converge to zero if we take the limit of x to infinity, but they do so at different rates. The non-standard limit of these functions are infinitesimals: let's call them ε_1 and ε_2. Both infinitesimals are infinitely close to the same real number: zero. In addition, the fact that $1/x^2$ converges to zero much faster than $1/x$ is represented in the relationship between them: it holds that $\varepsilon_2 = \varepsilon_1^2 < \varepsilon_1$. Infinite hyperreal numbers can be understood in a similar vein: as limits of diverging functions, encoding additional information about the rate of divergence. For instance, the functions x and x^2 both diverge, but at different rates. The non-standard limits associated with them are, respectively: $1/\varepsilon_1$ and $1/\varepsilon_2 = 1/\varepsilon_1^2 > 1/\varepsilon_1$.

Applying this interpretation to Pascal's Wager is not straightforward, however, since it involves no limit processes associated with S and p. For the special cases in the previous section, where we assumed a functional dependence of p on S, we can apply this interpretation.

Consider inequality (1), but now interpret all quantities appearing in it as real-valued functions of S. We are interested in the limit as S goes to infinity. The boundary case for an unsuccessful wager implies that the left- and right-hand side of (1) are "equivalent infinitesimals" (with infinitesimal *not* in the sense of section 1.2): this means that the limit of their ratio equals one. (Computing this limit will typically involve applying L'Hôpital's rule.) This is equivalent to taking the standard part of both sides of (1) in its original interpretation (i.e., with symbols representing hyperreal values).

So, the move to real numbers is slightly less precise (yielding a necessary but not a sufficient success condition) and presupposes a functional dependence of p on S. Considering whether such an assumption is always sensible leads to a further complication, to which we turn next.

4.3 Independent Disbelief and Practical Limitations on Incredulity?

Oppy (2006a, p. 257) wrote: "While one now has a neat mathematical apparatus for handling infinities and infinitesimals, it is not clear that we have any way of understanding how this apparatus can be applied to the case at hand." Although he wrote this with Conway's numbers in mind, it seems applicable here, too.

We stressed the importance of employing a decision theory that is built upon a harmonious number system. This harmony is necessary to ensure that the relevant expected utilities can be computed. Now, the special cases discussed in section 4.1 may raise a worry: unless the size of the utility of eternal bless S directly influences the agent's probability assignment p, it may not be clear what the relation between the two values is. Yet, as long as it is transparent to the agent what the values of p and S are, they are representable by some tractable relation $p(S)$ and it can be determined by inequality (1) whether the Wager fails or not.

On the other hand, we might deny the existence of any such equation, expressible within the number system the agent uses, to represent the relation between p and S. In other words, we may deny the harmony of the number system to the relevant quantities. For instance, if S belongs to a non-standard model of the non-standard model that the agent uses ($S \in {}^{**}\mathbb{R}$), it is larger than any number representable by the agent and the Wager goes through (because $p \times S$ is infinite). Since presumably S is fixed at the dawn of time, this would require foreknowledge of what are the largest and smallest positive numbers ever constructed in human mathematics. Without such foreknowledge, S has to be some kind of absolute infinity, to which none of the formal systems herein discussed applies.

4.4 Ruling Out Mixed Strategies

Hájek (2003) holds that the Wager is an invalid argument and considers four valid reformulations, finding fault with all of them. To argue that Pascal's Wager is invalid, Hájek points out that mixed strategies have infinite utility too and hence are not to be preferred over the pure strategy of wagering for God's existence directly. For example, letting one's wager depend on the outcome of a coin toss, a die toss, or some other random event, no matter how unlikely, will still yield infinite expected utility.

Using hyperreals to represent the expected utilities, however (and assuming larger than infinitesimal probabilities), it can be shown that that of the pure strategy is the largest. For example, if an agent considers a chance process to decide between wagering for or against God, which leads to wagering for God with some probability q, the expected utility of this mixed strategy is $q \times (S \times p + f_2 \times (1 - p)) + (1 - q) \times (f_1 \times p + f_3 \times (1 - p))$, which is strictly smaller than the expected utility of the pure strategy of wagering for God. If either p or q or both are allowed to take infinitesimal values, as before, the argument's success may co-depend on the finite utilities.

5 Final Thoughts

The Wager fragment doesn't motivate why Pascal excludes infinitesimal probabilities: Is it because he (correctly) intuited that the argument may fail or would he (also) deny that a rational agent can have such credences in the first place? We already discussed in section 3.2.1 that Pascal's proto-classical interpretation would probably require the probabilities to be equal for all agents, but we set aside that aspect here.

5.1 Assigning Probability to the Inconceivable

Since Pascal wrote that humans can neither grasp the infinitely large nor the infinitely small, he could indeed deny that a finite human can hold an infinitesimal credence in any proposition. Whereas God is capable to provide salvation of any utility, there is a limit to the fineness of difference a human can perceive and reason about: we are "infinitely remote from an understanding of the extremes" (L199/S230). In this sense, the harmony of the numbers may speak *against* allowing infinitesimal probabilities in a wager argument, for compared to God's, human capacities are imperceptibly small. Also, from a naturalistic perspective, Easwaran (2014, section 5.4) has argued that actual agents cannot hold hyperreal credences (though I worry that this argument is too strong, for a similar argument would also rule out irrational probabilities).

Of course, some people might set their degree of belief inversely proportional to the estimated utility value of salvation, but then it seems Pascalian to point out that any such estimation of salvation will fall short as a finite value ($p = 1/S'$ with S' some finite estimation of salvation, such that section 4.1 does not apply). The Wager may then fail, not because the probability assigned to the hypothesis of God's existence is infinitesimal, but simply because the utility value is very large but finite (S underestimates the true value of salvation by an infinite factor).

Both observations can motivate why Pascal excluded infinitesimal probabilities in the first place.

5.2 Infinitely Many Alternative Hypotheses

As a final point, we remark that infinitesimal probabilities may be introduced into the Wager in a different way. In order to apply our probability theory, we first need to check whether the possibilities are mutually exclusive and jointly exhaustive. It isn't clear that this is the case: the negation of the proposition "God exists" may include possibilities other than "God doesn't exist," such as the concept of God being contradictory or meaningless, but Pascal doesn't mention these. Moreover, it is clear that Pascal only discussed the existence or non-existence of the God of Christianity, but there are infinitely many alternative deities conceivable. If any non-zero probability is assigned to the non-elementary premise that a supernatural being exists, a uniform and regular probability assignment will assign an infinitesimal probability to each elementary premise, including to "Pascal's Christian God exists." Hence, the many-gods-objection to the Wager may pose a context in which infinitesimal probabilities arise (more) naturally.

Oppy (1991, p. 166) also considered an agent who deliberates on "an infinite range of possible deities about whose existence he acknowledges that his reason is impotent to decide" and writes that such an agent "ought to assign no more than an infinitesimal value to the subjective probability that any one of these deities exists." Likewise, Oppy (2006a, pp. 249–50) asks us to "[c]onsider the hypothesis that there is a source of infinite utility that one will obtain just in case one forms a correct belief about the identity of a natural number that is intimately associated with admission to this source of infinite utility." For instance, the agent has to guess a deity's favorite natural number to receive salvation. Distributing probability over the natural numbers in a regular and uniform way requires infinitesimal probabilities (Wenmackers and Horsten, 2013).

Alternatively, and similarly to section 4.1, we could consider agents whose degree of belief is a function of the utility provided by each deity. Then, distinguishing between a deity that can grant an eternal life or one infinitely happy day, on the one hand, and Pascal's God who can grant eternal infinite happiness, on the other hand, becomes relevant. The double infinity of salvation may be a double-edged sword: if infinitesimal probabilities are allowed, it raises the utility S, but it might lower the associated degree of belief *p* proportionately. Yet, excluding infinitesimal probabilities in such a context may guarantee success once more.

However, as Oppy (2006a, pp. 250–51) remarked, for each hypothesis, there exist alternatives involving a malevolent deity that assigns salvation the opposite way: rewarding disbelievers (switching the rows in the decision matrix of a benevolent deity) or punishing believers (allotting utility -*S* instead of *S*). Successful wagering in the presence of symmetric hypotheses would require an additional argument for assigning higher probability to a benevolent deity than a malevolent one, but then the infinite utility of salvation is no longer at the heart of the argument. So, the symmetry of the hypotheses neutralizes Pascal's argument completely, irrespective of whether we allow infinitesimal probabilities.

Bibliography

Alston, W. P. (1991). *Perceiving God*. Cornell University Press.
Anderson, E. (1993). *Value in Ethics and Economics*. Harvard University Press.
Anderson, R. (1995). Recent criticisms and defenses of Pascal's Wager. *International Journal for Philosophy of Religion*, 37: 45–56.
Anon. (2015). Which party is better for the economy? April 14. politicsthatwork.com/blog/which-party-is-better-for-the-economy.php
Aquinas, T. (1992). *Super Boethium De trinitate*. In Leonina (ed.), *Opera omnia*. Vol. 50.
Ariew, R. (2005). Introduction. In B. Pascal, *Pensées* (pp. xi–xvi). Hackett.
Arnobius (1907–1911). *Adversus Gentes*, in *Ante-Nicene Fathers: Translations of the Writings of the Fathers down to AD 325*. Vol. 5. Trans. H. Bryce and H. Campbell. Scribner's.
Bartha, P. (2007). Taking stock of infinite value: Pascal's Wager and relative utilities. *Synthese*, 154: 5–52.
Bartha, P. (2012). Many Gods, Many Wagers: Pascal's Wager meets the replicator dynamics. In J. Chandler and V. Harrison (eds.), *Probability in the Philosophy of Religion* (pp. 187–206). Oxford University Press.
Bartha, P. (2016). Probability and the Philosophy of Religion. In A. Hájek and C. Hitchcock (eds.), *The Oxford Handbook of Probability and Philosophy* (pp. 738–71). Oxford University Press.
Bartha, P. and DesRoches, C. T. (2016). The relatively infinite value of the environment. *Australasian Journal of Philosophy*, 95: 328–53.
Benci V. and Di Nasso, M. (2003). Numerosities of labelled sets: a new way of counting. *Advances in Mathematics*, 173: 50–67.
Benci, V., Horsten, L., and Wenmackers, S. (2013). Non-Archimedean probability. *Milan Journal of Mathematics*, 81: 121–51.
Benci, V., Horsten, L., and Wenmackers, S. (2018). Infinitesimal probabilities. *The British Journal for the Philosophy of Science*, 69: 509–52.
Berkeley, G. (1734). *The Analyst*. J. Tonson in the Strand.
Betty, L. S. (2001). Going beyond James: a pragmatic argument for God's existence. *International Journal for Philosophy of Religion*, 49: 69–84.
Birault, H. (1988). Nietzsche and Pascal's Wager. *Man and World*, 21: 261–85.
Blackburn, S. (1999). *Think: A Compelling Introduction to Philosophy*. Oxford University Press.

Blanchet, L. (1919). L'Attitude religieuse des jésuites et les sources du pari de Pascal. *Revue de métaphysique et de morale*, 26: 477–516, 617–47.

Boolos, G. S., J. P. Burgess, and R. C. Jeffrey. (2007). *Computability and Logic*, 5th ed. Cambridge University Press.

Bostrom, N. (2009). Pascal's mugging. *Analysis*, 69: 443–45.

Buben, A. (2011). Christian hate: death, dying, and reason in Pascal and Kierkegaard. In P. Stokes and A. Buben (eds.), *Kierkegaard and Death* (pp. 65–80). Indiana University Press.

Buben, A. (2013). Neither irrationalist nor apologist: revisiting faith and reason in Kierkegaard. *Philosophy Compass*, 8: 318–26.

Buben, A. (2016). *Meaning and Mortality in Kierkegaard and Heidegger: Origins of the Existential Philosophy of Death*. Northwestern University Press.

Buchak, L. (2012). Can it be rational to have faith? In J. Chandler and V. Harrison (eds.), *Probability in the Philosophy of Religion* (pp. 225–47). Oxford University Press.

Burns, J. P. (1999). Grace. In A. Fitzgerald and J. C. Cavadini (eds.), *Augustine through the Ages: An Encyclopedia* (pp. 391–98). W. B. Eerdmans.

Cantillon, A. (2014). *Le Pari-de-Pascal: étude littéraire d'une série d'enonciations*. Vrin.

Cantor, G. (1966 [1932]). *Gesammelte Abhandlungen mathematischen und philosophischen Inhalts. Herausgegeben von Ernest Zermelo, nebst einem Lebenslauf Cantors von A. Fraenkel*. Springer. Reprinted by Hildesheim, Olms.

Cargile, J. (1966). Pascal's Wager. *Philosophy*, 41: 1–18.

Carnap, R. (1950a). Empiricism, semantics, and ontology. *Revue Internationale de Philosophie*, 4: 20–40. Reprinted in *Meaning and Necessity*. University of Chicago Press, 1956.

Carnap, R. (1950b). *Logical Foundations of Probability*. University of Chicago Press.

Catholic Church (1992). *Catechism of the Catholic Church*. Libreria Editrice.

Casey, Patricia. (2009). *The Psycho-social Benefits of Religious Practice*. The Iona Institute.

Chen, E. and Rubio, D. (forthcoming). Surreal decisions. *Philosophy and Phenomenological Research*.

Chillingworth, W. (1840 [1638]). *The Religion of Protestants a Safe Way to Salvation*, in *The Works of William Chillingworth*. Hooker.

Clifford, W. K. (1879). The Ethics of Belief. In *Lectures and Essays*, Vol. 2 (pp. 177–211). Macmillan.

Colyvan, M. and Hájek, A. (2016). Making ado without expectations. *Mind*, 125: 829–57.

Connor, J. A. (2006). *Pascal's Wager: The Man Who Played Dice with God*. HarperOne Publishers.

Conway, J. H. (1976). *On Numbers and Games*. Academic Press.

Craig, W. (1994). *Reasonable Faith: Christian Truth and Apologetics*. Crossway Books.

Davis, S. (2004). It is rational to believe in the resurrection. In M. L. Peterson and R. J. VanArragon (eds.), *Contemporary Debates in Philosophy of Religion* (pp. 164–74). Blackwell.

Decety, J., Cowell, J. M., Lee, K., Mahasneh, R., Malcolm-Smith, S., Selcuk, B., and Zhouet, X. (2015). The negative association between religiousness and children's altruism across the world. *Current Biology*, 22: 2951–55.

DeGeneres, E. (1995). *My Point ... and I Do Have One*. Doubleday.

Deleuze, G. (1983). *Nietzsche and Philosophy*. Trans. H. Tomlinson. Continuum.

Denning, J. (1947). *Miller v Minister of Pensions*, 2 All England Reports 372.

Diderot, D. (1875–77 [1746]). J. Assézar and M. Tourneux (eds.) *Pensées philosophiques*. Vol. 1. Garnier (fr.wikisource.org/wiki/Pensées_philosophiques/Addition_aux_Pensées_philosophiques).

Diderot, D. (1875). Addition aux Pensées philosophiques. In J. Assézat (ed.), *Œuvres complètes de Diderot*. Vol. 1 (pp. 158–70). Garnier Frères.

Diderot, D. (2009). *Diderot's Early Philosophical Works* (reprint). Trans. M. Jourdain. BiblioBazaar.

Diener, E. and Seligman, M. (2002). Very happy people. *Psychological Science*, 13: 81–84.

Diener, E., Tay, L., and Myers, D. (2011). The religion paradox: if religion makes people happy, why are so many dropping out? *Journal of Personality and Social Psychology*, 101: 1278–90.

Dreier, J. (1996). Rational preference: decision theory as a theory of practical rationality. *Theory and Decision*, 40: 249–76.

Duff, A. (1986). Pascal's Wager and infinite utilities. *Analysis*, 46: 107–9.

Duncan, C. (2003). Do vague probabilities really scotch Pascal's Wager? *Philosophical Studies*, 112: 279–90.

Duncan, C. (2007). The persecutor's wager. *The Philosophical Review*, 116: 1–50.

Duncan, C. (2013). Religion and secular utility: happiness, truth, and pragmatic arguments for theistic belief. *The Philosophy Compass*, 8: 381–99.

Easwaran, K. (2014). Regularity and hyperreal credences. *Philosophical Review*, 123: 1–41.

Ecklund, E., Johnson, D., Scheitle, C., Matthews, K., and Lewis, S. (2016). Religion among scientists in international context. *Socius*, 2: 1–9.

Edgell, P., Gerteis, J., and Hartmann, D. (2006). Atheists as "other". *American Sociological Review*, 71: 211–34.

Edwards, W. (1982). Conservatism in human information processing (excerpted). In D. Kahneman, P. Slovic, and A. Tversky (eds.), *Judgment under Uncertainty* (pp. 359–69). Cambridge University Press.

Ehrman, B. (2003). *Lost Christianities: The Battles for Scripture and the Faiths We Never Knew*. Oxford University Press.

Ehrman, B. (2009). *Jesus, Interrupted: Revealing the Hidden Contradictions in the Bible (And Why We Don't Know About Them)*. Harper Collins.

Ehrman, B. (2014). *How Jesus Became God: The Exaltation of a Jewish Preacher from Galilee*. Harper Collins.

Ehrman, B. (2016). *Jesus Before the Gospels: How the Earliest Christians Remembered, Changed, and Invented Their Stories of the Savior*. Harper Collins.

Eid, M. and Larsen, R. (eds.) (2008). *The Science of Subjective Well-being*. Guilford.

Elster, J. (2003). Pascal and decision theory. In N. Hammond (ed.), *Cambridge Companion to Pascal* (pp. 53–75). Cambridge University Press.

Fagan, P. (2006). Why religion matters even more. Backgrounder #1992, Heritage Foundation (www.heritage.org/civil-society/report/why-religion-matters-even-more-the-impact-religious-practice-social-stability).

Feinberg, J. (1974). Noncomparative Justice. *Philosophical Review*, 83:297–338.

Flew, A. (1976). *The Presumption of Atheism*. Elek.

Foley, R. (1994). Pragmatic reasons for belief. In J. Jordan (ed.), *Gambling on God: Essays on Pascal's Wager*. Rowman and Littlefield.

Fouke, D. C. (1989). Argument in Pascal's *Pensées*. *History of Philosophy Quarterly*, 6: 57–68.

Franklin, J. (1998). Two caricatures, I: Pascal's Wager. *International Journal for Philosophy of Religion*, 44: 109–14.

Franklin, J. (2015). *The Science of Conjecture: Evidence and Probability before Pascal*. Johns Hopkins University Press.

Frey, P. (ed.) (1983). *Chess Skill in Man and Machine*. Springer.

Frierson, P. R. (2012). Two concepts of universality in Kant's moral theory. In S. M. Shell and R. Velkley (eds.), *Kant's Observation and Remarks: A Critical Guide*. Cambridge University Press.

Fugate, C. (2014). The highest good and Kant's proof(s) of God's existence. *History of Philosophy Quarterly*, 31: 137–58.

Gale, R. (1991). *On the Nature and Existence of God*. Cambridge University Press.

Galen, L. and Beahan, J. (2013). A skeptical review of religious prosociality research. *Free Inquiry*, 33: 14–24.

Galileo (1954 [1638]). *Discorsi e Dimostrazioni Intorno a Due Nuove Scienze*. Translated as: *Dialogues Concerning Two New Sciences*. Northwestern University.

Garber, D. (2007). Religio Philosophi. In L. M. Antony (ed.), *Philosophers without Gods: Meditations on Atheism and the Secular Life*. Oxford University Press.

Garber, D. (2009). *What Happens after Pascal's Wager: Living Faith and Rational Belief*. Marquette University Press.

Gauthier, D. (1986). *Morals by Agreement*. Oxford University Press.

Gebauer, J., Sedikides, C., and Neberich, W. (2012). Religiosity, social self-esteem, and psychological adjustment: on the cross-cultural specificity of the psychological benefits of religiosity. *Psychological Science*, 23: 158–60.

Golding, J. (1990). Toward a pragmatic conception of religious faith. *Faith and Philosophy*, 7: 486–503.

Golding, J. (1994). Pascal's Wager. *The Modern Schoolman*, 71: 115–43.

Golding, J. (2003). *Rationality and Religious Theism*. Ashgate Academic Press.

Golding, J. (2018). *The Jewish Spiritual Path: The Way of the Name*. Urim.

Greco, J. and Turri, J. (2016). Virtue Epistemology. *The Stanford Encyclopedia of Philosophy*. Winter 2016. Ed. Edward N. Zalta. (plato.stanford.edu/archives/win2016/entries/epistemology-virtue/).

Habermas, J. (1990). *The Philosophical Discourses of Modernity*. Trans. F. Lawrence. MIT Press.

Hacking, I. (1975). *The Emergence of Probability*. Cambridge University Press.
Hacking, I. (1994 [1972]). The logic of Pascal's Wager. *American Philosophical Quarterly*, 9 (1972): 186–92. Reprinted in J. Jordan (ed.), *Gambling on God: Essays on Pascal's Wager*. Rowman & Littlefield.
Hájek, A. (1998). Agnosticism meets Bayesianism. *Analysis*, 58: 199–206.
Hájek, A. (2000). Objecting vaguely to Pascal's Wager. *Philosophical Studies*, 98: 1–16.
Hájek, A. (2003). Waging war on Pascal's Wager. *Philosophical Review*, 12: 27–56.
Hájek, A. (2012a). Blaise and Bayes. In J. Chandler and V. S. Harrison (eds.), *Probability in the Philosophy of Religion* (pp. 167–86). Oxford University Press.
Hájek, A. (2012b). Pascal's Wager. *The Stanford Encyclopedia of Philosophy*. Winter 2012. Ed. Edward N. Zalta (plato.stanford.edu/archives/win2012/entries/pascal-wager/).
Hájek, A. (manuscript). Staying regular? (hplms.berkeley.edu/HajekStayingRegular.pdf).
Hall, D., Matz, D., and Wood, W. (2010). Why don't we practice what we preach? A meta-analytic review of religious racism. *Personality and Social Psychology Review*, 14: 126–39.
Halpern, J.Y. (2010). Lexicographic probability, conditional probability, and non-standard probability. *Games and Economic Behavior*, 68: 155–79.
Hamilton, M. A. (2005). *God vs. the Gavel: Religion and the Rule of Law*. Cambridge University Press.
Harman, G. (1986). *Change in View*. MIT Press.
Hausner, M. (1954). Multidimensional Utilities. In R. M. Thrall, C. H. Coombs, and R. L. Davis (eds.), *Decision Processes* (pp. 167–80). John Wiley.
Haybron, D. (2008). Philosophy and the science of subjective well-being. In M. Eid and R. Larsen (eds.), *The Science of Subjective Well-being*. Guilford.
Helman, C. (2011). The world's happiest countries. Forbes, Jan. 19 (www.forbes.com/2011/01/19/norway-denmark-finland-business-washington-world-happiest-countries.html).
Henrich, J. (2005). "Economic man" in cross-cultural perspective. *Behavioral and Brain Sciences*, 28: 795–855.
Herbermann, C. (ed.) (1913). *The Catholic Encyclopedia*. Vol. 10. Robert Appleton Company.
Herzberg, F. (2011). Hyperreal expected utilities and Pascal's Wager. *Logique et Analyse*, 213: 69–108.
Hewstone, M., Rubin, M., and Willis, H. (2002). Intergroup bias. *Annual Review of Psychology*, 53: 575–604.
Hick, J. (1990). *Philosophy of Religion*. Prentice Hall.
Himma, K. (2002). Finding a high road: the moral case for Salvific Pluralism. *International Journal for the Philosophy of Religion*, 52: 1–33.
Hintikka, J. (1962). Cogito Ergo Sum: inference or performance? *Philosophical Review*, 71: 3–32.
Hofstadter, D. (1983). Metamagical Themas. *Scientific American*, 248.6: 14–29.
Holyer, R. (1989). Scepticism, evidentialism and the Parity Argument: a Pascalian perspective. *Religious Studies*, 25: 191–208.

Hume, D. (2000 [1748]). *An Enquiry Concerning Human Understanding*. Ed. T. L. Beauchamp. Oxford University Press.

Hume, D. (1998 [1779]). *Principal Writings on Religion, including Dialogues Concerning Natural Religion and the Natural History of Religion*. Ed. J. Gaskin. Oxford University Press.

Hunter, G. (2013). *Pascal the Philosopher*. University of Toronto Press.

IHEU (2016). Freedom of Thought Report. International Humanist and Ethical Union (freethoughtreport.com/).

James, W. (1936 [1902]). *Varieties of Religious Experience*. Longmans, Green, and Co.

James, W. (1956 [1896]). The Will to Believe. In *The Will to Believe and Other Essays in Popular Philosophy*. Dover Publications.

Janzen, G. (2011). Pascal's Wager and the Nature of God. *Sophia*, 50: 331–44.

Jeffrey, R. (1983a). *The Logic of Decision*. University of Chicago Press.

Jeffrey, R. (1983b). Bayesianism with a Human Face. In J. Earman (ed.), *Minnesota Studies in the Philosophy of Science: Testing Scientific Theories*. Vol. 10. University of Minnesota Press.

Jones, W. E. (1998). Religious conversion, self-deception, and Pascal's Wager. *Journal of the History of Philosophy*, 36: 167–88.

Jordan, J. (ed.) (1994a). *Gambling on God: Essays on Pascal's Wager*. Rowman and Littlefield Publishers.

Jordan, J. (1994b). The many-gods objection. In J. Jordan (ed.), *Gambling on God: Essays on Pascal's Wager* (pp. 101–13). Rowman and Littlefield Publishers.

Jordan, J. (2006). *Pascal's Wager: Pragmatic Arguments and Belief in God*. Oxford University Press.

Joshanloo, M. and Weijers, D. (2014). Aversion to happiness across cultures. *Journal of Happiness Studies*, 15: 717–35.

Joyce, J. (2005). How probabilities reflect evidence. *Philosophical Perspectives*, 19: 153–78.

Joyce, J. (2010). A defence of imprecise credences in inference and decision making. *Philosophical Perspectives*, 24: 281–323.

Kant, I. (1991). *Practical Philosophy*. Ed. and trans. M. Gregor. Cambridge University Press.

Kant, I. (1996). *Religion and Rational Theology*. Ed. and trans. A. W. Wood and G. Di Giovanni. Cambridge University Press.

Kant, I. (1997 [1785]). *Groundwork of the Metaphysics of Morals*. Ed. and trans. P. Guyer and A. W. Wood. Cambridge University Press.

Kant, I. (1998 [1781/1787]). *Critique of Pure Reason*. Ed. and trans. P. Guyer and A. W. Wood. Cambridge University Press.

Kant, I. (2007). *Anthropology, History, and Education*. Ed. and trans. R. B. Louden and G. Zöller. Cambridge University Press.

Kaplan, M. (1996). *Decision Theory as Philosophy*. Cambridge University Press.

Kasser, J., and Shah, N. (2006). The metaethics of belief: an expressivist reading of *The Will to Believe*. *Social Epistemology*, 20: 1–17.

Kauffman, W. (1978). *Critique of Religion and Philosophy*. Princeton University Press.

Keynes J. M. (1921). *A Treatise on Probability*. Macmillan.
Kierkegaard, S. (1967–1978). *Søren Kierkegaard's Journals and Papers*. 7 vols. Trans. H. V. Hong and E. H. Hong, assisted by G. Malantschuk. Indiana University Press.
Kierkegaard, S. (1988). *Stages on Life's Way*. Trans. H. V. Hong and E. H. Hong. Princeton University Press.
Kierkegaard, S. (1989). *The Concept of Irony*. Trans. H. V. Hong and E. H. Hong. Princeton University Press.
Kierkegaard, S. (1990). *For Self-Examination* and *Judge for Yourself!* Trans. H. V. Hong and E. H. Hong. Princeton University Press.
Kierkegaard, S. (1997–2013). *Søren Kierkegaards Skrifter*, 28 vols. (plus corresponding commentary vols.) Ed. N. J. Cappelørn et al. Gads.
Klick, Jonathan (2004). Econometric Analyses of U.S. Abortion Policy: A Critical Review. *Fordham Urban Law Journal*, 31: 751–82.
Knuth, D.E. (1974). *Surreal Numbers: How Two Ex-Students Turned on to Pure Mathematics and Found Total Happiness*. Addison-Wesley.
Koenig, H. G., King, D. E., and Carson, W. B. (2010). *Handbook of Religion and Health*. Oxford University Press.
Koenig, H. G., McCullough, M. E., and Larson, D. B. (2001). *Handbook of Religion and Health*. Oxford University Press.
Kolmogorov, A. N. (1956 [1933]). *Grundbegriffe der Wahrscheinlichkeitrechnung. Ergebnisse der Mathematik*. Springer. Trans. N. Morrison, *Foundations of Probability*. Chelsea Publishing Company.
Kolokowski, L. (1995). *God Owes Us Nothing*. University of Chicago.
Lahn, A. (2015). Gender equality gives men better lives (sciencenordic.com/gender-equality-gives-men-better-lives).
Larson, E. and Witham, L. (1998). Leading scientists still reject God. *Nature*, 394.6691: 313.
Lawes, M. J. and Perrin, M. R. (1995). Risk sensitive foraging behaviour of the round-eared elephant shrew. *Behavioral Ecology and Sociobiology*, 37: 31–37.
Lehrer, K. and Wagner, C. (1981). *Rational Consensus in Science and Society*. D. Reidel.
Leith, J. H. (1982). *Creeds of the Churches*. John Knox Press.
Levi, A. (1995). Introduction. In A. Levi (ed.), *Pensées and Other Writings* (pp. vii–xxxvii). Oxford University Press.
Levitt, S. and Dubner, S. (2005). *Freakonomics*. Harper Collins.
Lewis, C. S. (2001). *The Problem of Pain*. Harper Collins.
Lewis, D. (1969). *Convention*. Harvard University Press.
Lindley, D. (1991). *Making Decisions*. Wiley.
Loeb, P. A. (1975). Conversion from nonstandard to standard measure spaces and applications in probability theory. *Transactions of the American Mathematical Society*, 211: 113–22.
Luce, R. D. and Raiffa, H. (1957). *Games and Decisions*. John Wiley and Sons.

Lun, V. M. C. and Bond, M. H. (2013). Examining the relation of religion and spirituality to subjective well-being across national cultures. *Psychology of Religion and Spirituality*, 5: 304–15.

Luther, M. (1959). The last sermon in Wittenberg. In J. W. Doberstein (ed. and trans.), *Luther's Works: Sermons I* (pp. 371–80). Muhlenberg.

Luzzatto, M. C. (1978). *The Way of God*. Feldheim.

Lycan, W. and G. Schlesinger (1996). You bet your life: Pascal's Wager defended. In J. Feinberg (ed.), *Reason and Responsibility* (pp. 119–27). Wadsworth Publishing.

Lynn, Rev. B. (2015). *God and Government*. Prometheus Books.

Lyubomirsky, S. (2007). *The How of Happiness*. Penguin.

McBrayer, J. P. (2014). The Wager renewed: believing in God is good for you. *Science, Religion and Culture*, 1: 130–40.

McClennen, E. (1994). Finite Decision Theory. In J. Jordan (ed.), *Gambling on God: Essays on Pascal's Wager* (pp. 115–38). Rowman and Littlefield.

Mackie, J. (1982). *The Miracle of Theism*. Clarendon Press.

Maia Neto, J. R. (1991). The christianization of Pyrrhonism: scepticism and faith in Blaise Pascal, Søren Kierkegaard, and Lev Shestov. PhD. diss., Washington University.

Maimonides, M. (2010). *Mishneh Torah (Code of Jewish Law): Sefer Hamadah-Book of Knowledge*. Moznaim.

Mancosu, P. (2009). Measuring the size of infinite collections of natural numbers: was Cantor's theory of infinite number inevitable? *The Review of Symbolic Logic*, 2: 612–46.

Mang, E. (2009). How religion influences Canadian politics (rabble.ca/blogs/bloggers/ericmang/2009/12/how-religion-influences-canadian-politics).

Marks, G. and Miller, N. (1987). Ten years of research on the false-consensus effect: an empirical and theoretical review. *Psychological Bulletin*, 102: 72–90.

Martin, M. (1983). Pascal's Wager as an argument for not believing in God. *Religious Studies*, 19: 57–64.

Martin, M. (1990). *Atheism: A Philosophical Justification*. Temple University Press.

Martin, M. (2004). It is not rational to believe in the resurrection. In M. L. Peterson and R. J. VanArragon (eds.), *Contemporary Debates in Philosophy of Religion* (pp. 174–84). Blackwell.

Martin, M. and Monnier, R. (eds.) (2003). *The Impossibility of God*. Prometheus Books.

Maslen, C. (2016). Pragmatic decisions about God from different points of view: the costs of apostasy. *International Journal of Philosophy of Religion*, 80: 103–13.

Mellor, D. (2005). *Probability: A Philosophical Introduction*. Routledge.

Miles, T. (2011). Friedrich Nietzsche: rival visions of the best way of life. In J. Stewart (ed.), *Kierkegaard Research: Sources, Reception and Resources: Kierkegaard and Existentialism* (pp. 263–98). Ashgate.

Mochon, D., Norton, M. I., and Ariely, D. (2011). Who benefits from religion? *Social Indicators Research*, 101: 1–15.

Mooney, C. (2015). New study reaffirms the link between conservative religious faith and climate change doubt. (www.washingtonpost.com/news/energy-environment/wp/2015/05/29/this-fascinating-chart-on-faith-and-climate-change-denial-has-been-reinforced-by-new-research).
Morewedge, C. and Giblin, C. (2015). Explanations of the endowment effect: an integrative review. *Trends in Cognitive Sciences*, 19: 339–48.
Moriarty, M. (2003). Grace and religious belief. In N. Hammond (ed.), *Cambridge Companion to Pascal* (pp. 152–57). Cambridge University Press.
Moser, P. K. (2008). *The Elusive God*. Cambridge University Press.
Moser, P. K. (2013). *The Severity of God*. Cambridge University Press.
Moser, P. K. (2017). *The God Relationship: The Ethics for Inquiry about the Divine*. Cambridge University Press.
Mougin, G. and Sober, E. (1994). Betting against Pascal's Wager. *Noûs*, 28: 382–95.
Murray, D. B. and Teare, S. W. (1993). Probability of a tossed coin landing on its edge. *Physical Review*, 48: 2547–52.
Myers, D. (2000). The funds, friends, and faith of happy people. *American Psychologist*, 55: 56–67.
Myers, D. (2012). Reflections on religious belief and prosociality: comment on Galen (2012). *Psychological Bulletin*, 138: 913–17.
Myers, D. (2013). Religious engagement and well-being. In S. David, I. Boniwell, and A. Ayers (eds.), *Oxford Handbook of Happiness*. Oxford University Press.
Narens, L. (1980). On qualitative axiomatizations of probability theory. *Journal of Philosophical Logic*, 9: 143–51.
Nathanson, S. (1994). *The Ideal of Rationality*. Open Court.
Neander, A. (1847). *Über die geschichtliche Bedeutung der Pensées Pascal's für die Religionsphilosophie insbesondere* (Ein zur Feier des Geburtstages Seiner Majestät des Königs in der öffentlichen Sitzung der Akademie am 16. Oktober 1846 gehaltener Vortrag, 2. opl.).
Nehr, A. (2012). Pascal's Wager, *Free Inquiry*, 32: 38.
Nelson, E. (1987). *Radically Elementary Probability Theory*. Princeton University Press.
Nemoianu, V. M. (2010). The insufficiency of the many gods objection to Pascal's Wager. *American Catholic Philosophical Quarterly*, 84: 513–30.
Nemoianu, V. M. (2015). Pascal on divine hiddenness. *International Philosophical Quarterly*, 55: 325–43.
Newport, F., Witters, D., and Agrawal, S. (2012). Religious Americans enjoy higher well-being. February 16 (gallup.com/poll/152723/Religious-Americans-Enjoy-HigherWellbeing.aspx).
Nietzsche, F. (1968). *The Will to Power*. Trans. W. Kaufmann and R. J. Hollingdale. Vintage Books.
Nietzsche, F. (1990). *Twilight of the Idols* and *The Anti-Christ*. Trans. R. J. Hollingdale. Penguin Books.
Nietzsche, F. (1992). *Ecce Homo*. Trans. R. J. Hollingdale. Penguin Books.
Nietzsche, F. (1997). *Daybreak*. Trans. R. J. Hollingdale. Cambridge University Press.

Nietzsche, F. (2002). *Beyond Good and Evil*. Trans. J. Norman. Cambridge University Press.
Nietzsche, F. (2003). *Writings from the Late Notebooks*. Trans. K. Sturge. Cambridge University Press.
Nisbett, R. (2003). *The Geography of Thought*. Free Press.
Nishimura, K. (1992). Foraging in an uncertain environment: patch exploitation. *Journal of Theoretical Biology*, 156: 91–111.
O'Connell, M. (1997). *Blaise Pascal: Reasons of the Heart*. Eerdmans.
O'Connell, R. J. (1984). *William James on the Courage to Believe*. Fordham University Press.
Oishi, S. and Koo, M. (2008). Two new questions about happiness: "is happiness good" and "is happiness better?" In M. Eid and R. Larsen (eds.), *The Science of Subjective Well-being*. Guilford.
Oppy, G. (1991). On Rescher on Pascal's Wager. *International Journal for Philosophy of Religion*, 30: 159–68.
Oppy, G. (2006a). *Arguing About Gods*. Cambridge University Press.
Oppy, G. (2006b). *Philosophical Perspectives on Infinity*. Cambridge University Press.
Pascal, B. (1670). *Pensées de M. Pascal*. Guillaume Desprez.
Pascal, B. (1777). *Gedanken Paskals*. Trans. J. F. Kleuker. Johann Heinrich Cramer.
Pascal, B. (1840). *Gedanken über die religion und einige andern Gegenstände*, 2 vols. Trans. K. Adolf Blech. Wilhelm Besser.
Pascal, B. (1919 [1659]). Polemical fragments. In *Pensées*. Trans. W. F. Trotter. Collier.
Pascal, B. (1963 [1653]). On the conversion of the sinner. Trans T. P. Johnston. In L. Lafuma (ed.), *Pascal: Œuvres Complètes*. Macmillan.
Pascal, B. (1966 [1670]). *Pensées*. Trans. A. J. Krailsheimer. Penguin Books.
Pascal, B. (1982). *The Provincial Letters*. Trans. A. J. Krailsheimer. Penguin.
Pascal, B. (1995 [1655]). Letter on the possibility of the commandments. In A. Levi (ed. and trans.), *Pensées and Other Writings* (pp. 205–12). Oxford University Press.
Pascal, B. (1995 [1656]). Treatise concerning Predestination. In A. Levi (ed. and trans.), *Pensées and Other Writings* (pp. 213–26). Oxford University Press.
Pascal, B. (1995a). *Pensées and Other Writings*. Ed. and trans. H. Levi. Oxford University Press.
Pascal, B. (1995b). Writings on Grace. In H. Levi (ed. and trans), *Pensées and Other Writings*. Oxford University Press.
Pascal, B. (2005). *Pensées*. Trans. R. Ariew. Hackett.
Pasternack, L. (2011). The development and scope of Kantian belief: the highest good, the practical postulates and the fact of reason. *Kant-Studien*, 102: 290–315.
Pasternack, L. (2012). The many gods objection to Pascal's Wager: a decision theoretic response. *Philo*, 15: 158–78.
Pasternack, L. (2014). Kant's touchstone of communication and the public use of reason. *Society and Politics*, 8: 78–90.
Pasternack, L. (2015). Kant's "appraisal" of Christianity: biblical interpretation and the pure rational system of religion. *Journal of the History of Philosophy*, 53: 485–506.

Pasternack, L. (2017). Restoring Kant's conception of the highest good. *Journal of the History of Philosophy*, 55: 435–68.
Penelhum, T. (1971). *Religion and Rationality: An Introduction to the Philosophy of Religion*. Random House.
Peterson, M. L. and VanArragon R. J. (eds.) (2004). *Contemporary Debates in Philosophy of Religion*. Blackwell.
Pew Research Center (2009). Scientists, politics, and religion. July 9 (www.people-press.org/2009/07/09/section-4-scientists-politics-and-religion/).
Pew Research Center (2010). Jesus Christ's return to earth. July 14 (www.pewresearch.org/fact-tank/2010/07/14/jesus-christs-return-to-earth/).
Pew Research Center (2012). The world's Muslims, ch. 3. August 9 (www.pewforum.org/2012/08/09/the-worlds-muslims-unity-and-diversity-3-articles-of-faith/).
Pew Research Center (2014). Religious landscape study (www.pewforum.org/religious-landscape-study/religious-family/atheist/).
Pew Research Center (2015). Religion and views on climate and energy issues. Oct. 22 (www.pewinternet.org/2015/10/22/religion-and-views-on-climate-and-energy-issues/).
Pew Research Center (2017). Sharp partisan divisions in views of national institutions. July 10 (www.people-press.org/2017/07/10/sharp-partisan-divisions-in-views-of-national-institutions/).
Pirolli, P. (2007). *Information Foraging Theory: Adaptive Interaction with Information*. Oxford University Press.
Pivato, M. (2014). Additive representation of separable preferences over infinite products. *Theory and Decision*, 77: 31–83.
Putnam, H. (1981). *Reason, Truth and History*. Cambridge University Press.
Putnam, R. and Lim, C. (2010). Religion, social networks, and life satisfaction. *American Sociological Review*, 75: 914.
Qin, J., Kimel, S., Kitayama, S., Wang, X., Yang, X., and Han, S. (2011). How choice modifies preference. *NeuroImage*, 55: 240–46.
Quine, W. (1953). Two Dogmas of Empiricism. In *From a Logical Point of View*. Harvard University Press.
Quine, W. (1976 [1936]). Truth by Convention. Reprinted in *The Ways of Paradox*. Harvard University Press.
Quine, W. (1976 [1963]). Carnap and Logical Truth. Reprinted in *The Ways of Paradox*. Harvard University Press.
Quine, W.V. and Ullian, J.S. (1978). *The Web of Belief*. Random House.
Quinn, P. (1994). Moral objections to Pascalian Wagering. In J. Jordan (ed.), *Gambling on God: Essays on Pascal's Wager* (pp. 61–81). Rowman and Littlefield.
Radcliff, B. (2013). *The Political Economy of Human Happiness*. Cambridge University Press.
Rescher, N. (1985). *Pascal's Wager: A Study of Practical Reasoning in Philosophical Theology*. University of Notre Dame.
Rescher, N. (1988). *Rationality*. Oxford University Press.
Resnik, M. D. (1987). *Choices*. University of Minnesota Press.

Reuchlin, H. (1839). *Geschichte von Port Royal*, 2 vols. Gotha.
Reuchlin, H. (1840). *Pascals Leben und der Geist seiner Schriften zum Theil nach neu aufgefundenen Handschriften mit Untersuchungen über die Moral der Jesuiten*. Cotta'sche.
Rinard, S. (2013). Against radical credal imprecision. *Thought*, 2: 157–65.
Rinard, S. (2015). A decision theory for imprecise probabilities. *Philosophers' Imprint*, 15: 1–16.
Rinard, S. (2017). No exception for belief. *Philosophy and Phenomenological Research*, 94: 121–43.
Robinson, A. (1966). *Non-Standard Analysis*. North-Holland Publishing Co.
Rohde, H. P. (ed.) (1967). *Auktionsprotokol over Søren Kierkegaards Bogsamling*. Det Kongelige Bibliotek.
Rota, M. (2016a). *Taking Pascal's Wager: Faith, Evidence and the Abundant Life*. InterVarsity Press.
Rota, M. (2016b). A better version of Pascal's Wager. *American Catholic Philosophical Quarterly*, 90: 415–39.
Rota, M. (2017). Pascal's Wager. *Philosophy Compass*, 12:e12404.
Rothstein, B. (2010). Happiness and the welfare state. *Social Research*, 77: 441–68.
Royden, H. L. (1968). *Real Analysis*. 2nd ed. Macmillan.
Ryan, J. (1945). The argument of the wager in Pascal and others. *New Scholasticism*, 19: 233–50.
Saka, P. (2001). Pascal's Wager and the Many Gods objection. *Religious Studies*, 37: 321–41.
Saka, P. (2002). Pascal's Wager. Internet Encyclopedia of Philosophy (www.utm.edu/research/iep).
Saka, P. (2007). *How to Think about Meaning*. Springer.
Saroglou, V., Delpierre, V., and Dernelle, R. (2004). Values and religiosity: a meta-analysis of studies using Schwartz's model. *Personality and Individual Differences*, 37: 721–34.
Savage, L. J. (1954). *Foundations of Statistics*. John Wiley.
Scherman, N. (2001). *The Complete Artscroll Siddur*. Mesorah Publications.
Schlesinger, G. (1994). A Central Theistic Argument. In J. Jordan (ed.), *Gambling on God: Essays on Pascal's Wager* (pp. 83–100). Rowman and Littlefield.
Schopenhauer, A. (1958). *The World as Will and Representation*. Vol. 2. Trans. E. F. J. Payne. The Falcon's Wing.
Searle, J. (1983). *Intentionality*. Cambridge University Press.
Sellier, P. (1999). *Port-Royal et la littérature, I: Pascal*. Honoré Chamion.
Shimony, A. (1955). Coherence and the axioms of confirmation. *The Journal of Symbolic Logic*, 20: 1–28.
Shontell, A. (2016). 69-year-old monk who scientists call the worlds happiest man. Business Insider, Jan. 27 (www.businessinsider.com/how-to-be-happier-according-to-matthieu-ricard-the-worlds-happiest-man-2016-1).
Sirmond, A. (1635). *De immortalitate animae Demonstratio physica et Aristotelica Adversus Pomponatium et asseclas*.

Sirmond, A. (1637). *Démonstration de l'Immortalité de l'Ame, Tirée des Principes de la Nature, Fortifiée de ceux d'Aristote.*
Skyrms, B. (1990). *The Dynamics of Rational Deliberation.* Harvard University Press.
Skyrms, B. (1996). *Evolution of the Social Contract.* Cambridge University Press.
Skyrms, B. (2000). *Choice and Chance: An Introduction to Inductive Logic.* 4th ed. Wadsworth.
Slater, M. (2009). *William James on Ethics and Faith.* Cambridge University Press.
Sloan, R. (2006). *Blind Faith.* St. Martin's Press.
Smart, J. J. C. (1986). Utilitarianism and its applications. In J. P. DeMarco and R. M. Fox (eds.), *New Directions in Ethics* (pp. 24–41). Routledge and Kegan Paul.
Snoep, L. (2008). Religiousness and happiness in three nations. *Journal of Happiness Studies*, 9: 207–11.
Sobel, H. (1996). Pascalian Wagers. *Synthese*, 108: 11–61.
Sober, E. (1981). Revisability, *a priori* truth, and evolution. *Australasian Journal of Philosophy*, 59: 68–85.
Sober, E. (1990). Contrastive empiricism. In W. Savage (ed.), *Minnesota Studies in the Philosophy of Science, XIV: Scientific Theories* (pp. 392–412). University of Minnesota Press.
Sober, E. (2008). Empiricism. In S. Psillos and M. Curd (eds.), *The Routledge Companion to Philosophy of Science* (pp. 129–38). Routledge.
Sorensen, R. (1994). Infinite decision theory. In J. Jordan (ed.), *Gambling on God: Essays on Pascal's Wager.* Rowman and Littlefield
Stavrova, O., Fetchenhauer, D., and Schlosser, T. (2013). Why are religious people happy? The effect of the social norm of religiosity across cultures. *Social Science Research*, 42: 90–105.
Stephen, L. (1898). Pascal. In *Studies of a Biographer* (pp. 277–93). Duckworth and Co.
Suh, E. and Koo, J. (2008). Comparing subjective well-being across cultures. In M. Eid and R. Larsen (eds.), *The Science of Subjective Well-being.* Guilford.
Swinburne, R. (1979). *The Existence of God.* Clarendon Press.
Talbott, T. (2004). No Hell. In M. L. Peterson and R. J. VanArragon (eds.), *Contemporary Debates in Philosophy of Religion* (pp. 278–88). Blackwell.
Thirouin, L. (1991). *Le Modèle du jeu dans la Pensée de Pascal.* Vrin.
Titelbaum, M. (2013). *Quitting Certainties – A Bayesian Framework Modeling Degrees of Belief.* Oxford University Press.
UK Office of National Statistics. (2016). Measuring national well-being. Feb. 2 (www.ons.gov.uk/peoplepopulationandcommunity/wellbeing/datasets/measuringnationalwellbeinghappiness).
Van Fraassen, B. (1990). Figures in a probability landscape. In M. Dunn and A. Gupta (eds.), *Truth or Consequences.* Kluwer.
Van Fraassen, B. (1998). The agnostic subtly probabilified. *Analysis*, 58: 212–20.
Voltaire, F. M. A. (1763). *The Works of M. de Voltaire.* Vol. 26. Trans. T. Smollett et al. Newbery, Baldwin et al.
Voltaire, F. M. A. (1856). *A Philosophical Dictionary.* Vol. 1. Trans. unknown. J. P. Mendum.

Voltaire, F. M. A. (1961 [1778]). *Philosophical Letters (Letters Concerning the English Nation)*. Trans. E. Dilworth. Bobbs-Merrill.

Voltaire, F. M. A. (1971 [1764]). *Philosophical Dictionary*. Ed. and trans. T. Besterman. Penguin Books.

Voltaire, F. M. A. (1994 [1734]). Pascal's thoughts concerning religion. (Letter XXV). In N. Cronk (ed.), *Letter Concerning the English Nation* (p. 127). Oxford University Press.

Von Neumann, J. von and O. Morgenstern. (1944). *Theory of Games and Economic Behavior*. Princeton University Press.

Wainwright, W. (1995). *Reason and the Heart*. Cornell University Press.

Walker, D. (1964). *The Decline of Hell: Seventeenth-Century Discussions of Eternal Torment*. University of Chicago Press.

Walls, J. (2004). Eternal Hell and the Christian concept of God. In M. L. Peterson and R. J. VanArragon (eds.), *Contemporary Debates in Philosophy of Religion* (pp. 268–78). Blackwell.

Warda, A. (1922). *Immanuel Kants Bücher*. Martin Breslauer.

Weatherson, B. (2007). The Bayesian and the Dogmatist. *Proceedings of the Aristotelian Society*, 107: 169–85.

Weeden, J. and Kurzban, R. (2014). *The Hidden Agenda of the Political Mind*. Princeton University Press.

Wenmackers, S. and Horsten, L. (2013). Fair infinite lotteries. *Synthese*, 190: 37–61.

Wernham, J. C. S. (1987). *James's Will-to-Believe doctrine: a heretical view*. McGill-Queen's College Press.

Wetsel, D. (1994). *Pascal and Disbelief: Catechesis and Conversion in the Pensées*. Catholic University of America Press.

Wetsel, D. (2003). Pascal and Holy Writ. In N. Hammond (ed.), *The Cambridge Companion to Pascal* (pp. 162–81). Cambridge University Press.

Williams, B. (1973a). Deciding to believe. In *Problems of the Self*. Cambridge University Press.

Williams, B. (1973b). The Makropulos Case: reflections on the tedium of immortality. In *Problems of the Self*. Cambridge University Press.

Wing, Nick. (2017). There are still no open atheists in Congress. Huffington Post, Jan. 4 (www.huffingtonpost.com/entry/no-atheists-in-congress_us_586c074ae4b0 de3a08f9d487).

Wood, W. (2013). *Blaise Pascal on Duplicity Sin and the Fall: The Secret Instinct*. Oxford University Press.

Yeniaras, V. and Akarsu, T. N. (2016). Religiosity and life satisfaction: a multi-dimensional approach. *Journal of Happiness Studies*, 18: 1815–40.

Zuckerman, M., Silberman, J., and Hall, J. A. (2013). The relation between intelligence and religiosity: a meta-analysis and some proposed explanations. *Personality and Social Psychology Review*, 17: 325–54.

Index

acceptance of a hypothesis, 107, 182
Agnostic Rule, 19, 103–4, 107, 116, 117, 119, *See also* evidentialism
agnosticism, 23, 85, 115, 116, 186, 195, 225, 251–53, 285
altruism, 60, 206
Anderson, Robert, 168
Anti-Wager, 202, 205–8
apology. *See* Pascal's Wager, as a theological or apologetic argument
Aquinas, Saint Thomas, 110, 174
Archimedes, 304
Arnobius, 27–28, 37
atheism, 20, 85, 116, 126, 141, 170, 171, 172, 173, 176, 185, 186, 192, 193, 195, 199, 202, 203, 205, 208, 225, 233, 234, 251–53
Augustine, Saint, 47, 54, 64, 66, 82, 89, 164, 170, 204
authentic religious commitment, 14, 21, 168, 209–21
  and artificially induced belief, 214, 221

Bartha, Paul, 33, 39, 43, 143–45, 255, 265–77, 302
Bayesian belief updating, 43, 229, 278
Bayesian representation of belief, 43, 229, 278, 280, 294, 305
Benci, Vieri, 298
Berkeley, George, 182, 304
Better-Chances condition, 145, 241
Better-Prizes condition, 145
Birault, Henri, 98
Buchak, Lara, 237
Buridan's ass, 130, 136, 232

Calvinism, 67, 69
Cantillon, Alain, 45
Cantor, Georg, 297
cardinal numbers, 297, 300
Carnap, Rudolf, 235

cartel of deities, 253–54
  grouchy/perverse, 253, 271
  jealous, 271
  nice, 251
Cavalieri, Bonaventura, 304
Chillingworth, William, 28–29, 39
Christianity, 45, 89, 91, 97, 110, 150, 170, 173, 204, 210
  and slave-morality, 97–98
  evidence for and against, 49, 72, 156–57
Clement of Alexandria, 176
Clifford, William, 19, 103, 211
common knowledge assumption, 232
Continuity Axiom, 241, *See* preference, Continuity Axiom
conventionalism, 235
Conway, John, 137, 265, 300
Council of Trent, 51, 53, 54, 56, 57, 78
Craig, William Lane, 165
credences
  definition, 5
  imprecise/vague, 22, 280–83, *See also* Wide internal view, Supervaluationism, Reverse Principal Principle
  infinitesimal. *See* infinitesimal credence/probability
  precise, 278, 280
Cromwell's Rule, 306

damnation, 47, 53, 66, 67, 86, 126, 141, *See also* exclusivism
  annihilation, 149, 158, 170
  eternal punishment, 29, 52, 79, 157, 170
decision matrix. *See* decision table
decision table
  definition, 3
  with relative utilities. *See* relative decision table
decision table for Pascal's Wager
  as argument from dominance, 10, 124

decision table for Pascal's Wager (cont.)
   as argument from dominating expectation. *See* decision table for Pascal's Wager, as Canonical Wager
   as argument from expectation, 126
   as argument from long-run average utility, 140, 144
   as argument from strict dominance, 11
   as argument from superdominance, 125
   as argument from superduperdominance, 146
   as Canonical Wager, 7, 13, 36, 148, 169, 236, 261, 294
   finite utility version, 40, 141
   for contemporary individual, 42
   for post-Christian intellectual, 41
   hybrid version, 189
   many-gods version, 150, 151, 155, 171, 175, 177, 178, 180, 240, 244
   metaphysical version, 188
   mundane version, 189
   relative utility version, 144, 244
   surreal utility version, 137
   vector-valued utility version, 139
   with many theologies, 227, 228
decision theory, 2, 128, 148, 196, 282
   and realistic decision-making, 27, 31, 34
   and religious belief, 168
   finite. *See* standard decision theory
   harmonious, 301–2
   hyperreal, 23, 307–12
   infinite. *See* infinite decision theory
   non-standard. *See* infinite decision theory
decision under ignorance, 20, 169, 172
Deference principles, 281
Deleuze, Gilles, 98
Denning, Lord Alfred, 32
Diderot, 32, 84, 109, 149, 168
discounted present value, 39, 193
divine commandments, 209, 217
divine foreknowledge, 46, 52
divine hiddenness, 18, 71–75, 76, 82
dominance, 3, 124, 170, *See also* strict dominance, superdominance and weak dominance
Dominance Principle, 3
dominance reasoning, 125, 193, 290
double predestination, 53
doxastic voluntarism, 107
Dreier, Jamie, 241

Duff, Antony, 239, 262, 266
Duncan, Craig, 291
dynamics of rational deliberation, 21, 230–32
   attractors, 195
   equilibrium. *See* equilibrium for rational deliberation

Elster, Jan, 58
epistemic peers, 172
epistemic virtue, 20, 162, 164, 167
equilibrium for rational deliberation, 195, 230, 238
Escobar y Mendoza, Antonio, 34
ethics for inquiry, 64, 81
evidentialism, 15, 21, 103–4, 187, 199, 251, 257
evolutionary game theory, 22, 239, 247, 255
exclusivism, 70, 114, 149, 156, 170
   moral objections to, 70, 71, 80, 157–63
expectation, 5, 128, *See also* expected utility
   indeterminate, 133
   infinite, 38, 133, 135
   undefined, 133, 175
expected monetary value (EMV), 4, 128
   maximization of (EMV maximization), 4
expected utility (also expected value), 4, 5, 132, 279
   Expected Utility Theorem, 6
   infinite, 152
   maximization of (EU Maximization), 5, 15, 130, 148, 289
expected utility theory, 126, 127, 132
extended number system, 263–65
extended real numbers, 8, 38, 152, 154, 260, 296
external costs. *See* religion, external costs of

Fagan, Patrick, 201
faith. *See also* religion
   role of heart and reason in, 2, 68, 91
finite utility, 29, 265
   arbitrarily large, 19, 153–54
Flew, Anthony, 32
Foley, Robert, 196
forced choice, 31, 105, 111, 116
freedom, 46, *See also* grace, and human freedom

Gale, Richard, 168
Galileo, 303
gamble, 102, 127, 241
Garber, Daniel, 117–19

global warming, 28
God, 225
   concepts of, 65
   evidence for existence or non-existence, 66
   goodness of, 65
   justice of, 66, 67
   love of, 67, 80, 82
   mercy/forgiveness of, 66, 67, 69, 70, 71, 75, 163
   perfection of, 69, 70, 80, 81, 154, 158
Goldbach's Conjecture, 160
grace, 1, 17, 31, 46, 66, 67, 77, 169
   and human freedom, 46, 51, 54–58, 67, 78, 164
   and justification, 54–58
   sweetness/delight/inspiration of, 54–55, 58, 60, 67, 79
   universality of, 53, 70
Gregory of Nyssa, 176

Habermas, Jürgen, 178
Hacking, Ian, 9, 19, 123, 124, 126, 129, 130, 169, 236
Hájek, Alan, 13, 14, 27, 40, 154, 239, 262, 265, 277, 281, 285, 289, 293, 297, 298, 300, 304, 305, 306
harmonious number system, 296–98, 299
Harmony Condition, 23
Hausner, Melvin, 242
Heaven, 155
Hell, 170
Herzberg, Frederik, 137, 299
Hintikka, Jaakko, 178
Holyer, Robert, 168
Horsten, Leon, 275
human condition
   and love of diversions, 50
   and self-love, 50
   greatness and wretchedness, 48
human nature, 46, 47
Hume, David, 151, 157
Hunter, Graeme, 70, 75, 80
hybrid wager, 102, 111–17, 119, 189
hyperreal numbers, 137, 298–99, 307

ignorance
   maximal, 283
   representation of. *See* credences, imprecise/vague

imprecise credences. *See* credences, imprecise/vague
infinite decision theory, 9, 38, 241
   naïve, 9, 15, 22, 38
infinite number, 295, 303
infinite series, 7
infinite utility, 7–9, 14, 22, 36, 76, 102, 109, 126, 136, 139, 241, 261, 263, 293, 296, 300
   limit argument against, 193
infinitesimal credence/probability, 12, 22, 23, 113, 137, 154, 251, 263, 267, 274–75, 276, 278, 293–314
   Pascalian, 304–5
infinitesimals, 38, 295
infinity, 260, 296
   reflexivity under addition, 136, 304
   reflexivity under multiplication, 136
information-foraging decisions, 42
invalidity of Pascal's Wager, 15, 19, 123–47, 210
Islam, 41, 49, 84, 170, 204

James, William, 14, 19, 101–8, 111–17, 119, 189, 198, 221
   "The Will to Believe", 19, 101–8
   Will to Believe (WTB) argument. *See* Jamesian Wager
Jamesian Wager, 19, 101–8
Jansenism, 1, 17, 18, 31, 46, 54, 62, 64, 66, 67, 69, 70, 71–75, 89, 170, 183
Jeffrey, Richard, 134, 140, 239
Jesuits, 34, 88
Jesus Christ, 46–47, 72, 89
Jordan, Jeff, 34, 151, 153, 168, 172–74, 191, 198, 201, 293
Joyce, James, 284
Judaism, 21, 49, 170, 209
   Kedusha, 219
   Shema, 219
   spiritual virtues, 219
   Torah, 210, 217
   traditional, 216–20

Kant, Immanuel, 18, 85–88, 100, 182, 196
   Highest Good, 86–88, 182, 184, 185
   Practical Postulates (God and Immortality), 86–88
Kaplan, Mark, 284
Kierkegaard, Søren, 18, 88–95, 100

Kolmogorov axiomatization, 302
Kurzban, Robert, 195

Lehrer, Keith, 196
Leibniz, Gottfried Wilhelm, 304
lexicographic ordering, 138
lexicographic utility, 138, 242
live hypothesis, 19, 105, 112, 116, 168, 172–74
Loeb, Peter, 301
Luther, Martin, 89, 99
Lyubomirsky, Sonja, 192, 195

Maimonides, Moses, 217
many-gods objection, 14, 17, 19, 20, 32, 33, 84, 108–17, 148, 169–71, 172, 189, 198–99, 225, 238, 240, 262, 266, 276, 313, *See also* theological hypotheses, religion, philosophers' fictions, many-wagers model, decision table for Pascal's Wager, exclusivism
  ambitious version, 149–52
  modest version, 163–67
  response based on appeal to authentic religion, 168, 172, 198
  response based on appeal to finite utility, 152–54
  response based on appeal to live/realistic hypotheses, 32–36, 111–17, 168, 172–74, 198
  response based on appeal to tradition, 168, 172
  response based on constraints of rationality. *See* theological hypotheses, substantive constraints on
  response based on dynamics, 238, 245–50
many-wagers model, 245–50, 268
  challenges to, 250–59, 271–77
  equilibrium distribution for, 248, 268
  stable equilibrium, 248, 268–69, 276
  strongly stable equilibrium for, 256, 270, 276
  updating rule for, 247
Martin, Michael, 110, 114
McClennen, Edward, 8, 125, 140, 241
Miles, Thomas, 95, 99
mixed strategy, 6, 38, 131, 133, 138, 139, 140
Molina, Luis de, 46, 69
Molinism, 46, 52, 69
Mougin, Gregory, 21, 225
Myers, David, 200, 201
Mynster, Jacob Peter, 95

Narens, Louis, 302
Nathanson, Stephen, 197
negative infinite utility, 133, 141
Nelson, Edward, 298
Newton, Sir Isaac, 304
Nietzsche, Friedrich, 18, 95–100
non-standard analysis, 137, 265, 298, 303
numerosity theory, 137, 303

O'Connell, Marvin, 69
objections to Pascal's Wager, 14
  from arbitrary prudentialism. *See* prudentialism, arbitrary
  from divine hiddenness, 18, 78
  from evidentialism, 15, 21, 211, 261
  from imprecise credences, 289–92
  from inauthenticity, 14, 209, 211–14
  from invalidity. *See* objections to Pascal's Wager, mixed-strategies objection, invalidity of Pascal's Wager
  many-gods objection. *See* many-gods objection
  mixed-strategies objection, 15, 38, 130–32, 169, 239, 240, 244, 262, 266, 267, 311
  problems with infinite utility, 14, 22, 140, 194, 261
  unjustified positive probability of God's existence, 15, 36, 261
Oppy, Graham, 33, 36, 154, 250, 296, 300, 306, 309, 311, 313
ordinal numbers, 137, 297
Origen, 176
original sin. *See* sin, original
Ovid, 191

Paley, William, 110
Panshin, Alexei, 195
Pantheon of deities, 246, 275
Pascal
  Pensées, 1, 89
  Provincial Letters, 29, 34, 46
Pascalian belief revision, 237–39
Pascal's Wager
  and contemporary philosophy, 45
  and Judaism, 21, 216–20
  as a theological or apologetic argument, 45, 46, 51, 63, 94
  as argument from dominance, 10, 20, 124

as argument from dominating expectation,
    12–14, 130, 169
as argument from expectation, 11–12, 126–29
as argument from finite utility, 19, 29, 30, 39,
    140–41, 153–54
as argument from generalized expectation,
    129–30
as argument from long-run average utility,
    139–40
as argument from relative utility, 143–45
as argument from strict dominance, 10, 20,
    108
as argument from superdominance, 38,
    123–26
as argument from superduperdominance,
    145–46
as argument from weak dominance of
    expectation, 135
Canonical Wager, 12–14, 169, 188, 236, 260,
    278, 293
certainty of gain from, 77
decision tables for. *See* decision table for
    Pascal's Wager
dynamical models, 21, 22, 232–34, 237, 238,
    246
hybrid wager. *See* hybrid wager
hyperreal version, 23, 299–300, 304, 307–12
metaphysical version of, 20, 188, 190, 193, 198
mundane version of, 20, 170, 189, 191, 194,
    199, 207
objections to. *See* objections to Pascal's Wager
Pensées text and analysis, 9–14
pre-Pascalian versions
    Arnobius, 27–28
    Chllingworth, William, 28–29
    Sirmond, Antoine, 29–30
surreal version, 136–38, 300
vector-valued version, 138–39
passions, 2, 51, 76, 238
Pasternack, Lawrence, 196
payoff matrix. *See* decision table
Peano arithmetic, 299
Pelagianism, 51, 52, 57, 67, 70
Penelhum, Terence, 158
performative contradiction, 177–79, 180, 181–86
philosophers' fictions, 20, 27, 33, 35, 168–69,
    179–86
piety, 170, 211–12
Pivato, Marcus, 301

pragmatic assumption, 21, 211, 214–15, 221
pragmatic justification, 1, 19, 88, 102, 170
Precautionary Principle, 28
precisification of an indeterminate value, 134,
    136, 281
predestination, 18, 46, 51, 52, 53, 57, 69, 85, 164
preference
    axioms, 6, 145
    Continuity Axiom, 145
    non-Archimedean, 143, 301
    ordering, 6
    transitivity of, 6
Principal Principle, 281
Prisoner's Dilemma, 201, 207
probabilistic independence, 3, 125
probability. *See also* credences
    Bayesian updating. *See* Bayesian
        representation of belief
    epistemic, 160, 305
    infinitesimal. *See* infinitesimal credence/
        probability
    non-Archimedean, 301, 303
    objective, 35, 160
    subjective, 5, 35
proof beyond reasonable doubt, 32
prudential argument. *See* pragmatic
    justification
prudentialism, 21
    arbitrary, 21, 227–29, 252
    constrained, 234–35
    unbridled, 234–35
pure strategy, 6, 144
Putnam, Hilary, 178, 182

Quine, W.V.O., 234
Quinn, Philip, 158–61

rational choice theory. *See* decision theory
rationality
    classical, 202, 205–7
    cooperative, 202, 205–7
    epistemic, 15, 161, 207
    pluralism of, 20, 195, 208
    prudential, 15, 192
    relativity of. *See* relativity, of rationality
reason, 65
    and consequences of the *Fall*, 48–49
reconciliation and redemption, 66, 67, 71
Reflection Principle, 281

Reformation, 54, 176
regularity, 306, See also Cromwell's Rule
relative decision table, 144, 244, 266
relative utilities, 22, 39, 143, 242–45, 265–68
   infinite, 144, 242
   relative utility matrix. See relative decision table
relativity
   of belief, 190–92
   of decision-rule, 200–2, 205–7
   of perspective, 196–99, 205–7
      consensus, 196
      external, 196
      internal, 196
      intersubjective, 196
      objective, 196
   of rationality, 188
   of value, 192–96
religion
   and happiness, 115, 192, 195, 199, 200–5
   empirical studies of, 20, 117, 189, 195, 200–2
   external costs of, 201, 202–5
   political dimensions of, 194, 203
   popular, 27
   rationality of, 117, 192
religious experience and knowledge of God, 211, 221
Replicator Dynamics, 22, 247
representation theorem
   for hyperreal expected utilities, 137
   for relative utilities, 242
   for standard (finite) utilities, 6, 143, 145
Rescher, Nicholas, 34, 168, 197, 307
Resnik, Michael, 145
Reverse Principal Principle, 287–88
Robinson, Abraham, 137, 265, 298
Rota, Michael, 38, 201, 304

Saka, Paul, 189
salvation, 124
   and exclusivist deity. See exclusivism
   and human desert. See grace
   and predestination. See predestination
   and universalist deity. See universalism
salvific exclusivism. See exclusivism
satisficing, 154
Schlesinger, George, 16, 38, 143, 153, 198, 240

Schlesinger's Principle, 16, 38, 143, 153, 156, 240, 244
Schopenhauer, Arthur, 96–97
self-knowledge and self-deception, 50, 118
Sellier, Phillipe, 45
semi-Pelagianism, 51, 57, 70
sin, 46, 165
   original, and the *Fall*, 46–48
Sirmond, Antoine, 29–30, 31, 34, 37
skeptic-favoring god. See theological hypotheses, skeptic-favoring god
skepticism
   brain-in-vat, 33, 182
   demon, 33
   in Pascalian society, 1, 45
Skyrms, Brian, 21, 22, 229–33, 239
Slater, Michael, 101
Sobel, Howard, 297, 300, 307
Sober, Elliott, 21, 235
Socrates, 90
St. Petersburg game, 138
St. Petersburg paradox, 301
standard decision theory, 3, 6, 140, 241
Stephen, Leslie, 151
strict dominance, 3, 125
strong dominance. See strict dominance
superdominance, 38, 125
supervaluation, 134, 281–83, 289
supervaluationism, 281–83
surreal numbers, 137, 265, 298, 300
surreal utility, 137

theism, 102, 106–7, 109–10, 116, 151, 158, 164, 168, 169, 176, 185, 189, 192, 193, 195, 199, 202, 205, 208, 234
   ecumenical, 170, 173
   rationality of. See religion, rationality of
   traditional, 171, 173
theological hypotheses, 32–34, 168–69, 170
   detached god, 249
   Evil Deceiver, 178
   grouchy/perverse god, 249, 262
   jealous god, 240, 249
   nice god, 249, 262
   perverse master, 110, 114
   pointless god, 249
   skeptic-favoring god, 151, 238
   substantive constraints on, 20, 169, 174–86
      anti-skepticism constraint, 177–79

outcome plurality constraint, 176–77
practical reason constraint, 181–86
stability constraint, 175–76
trickster deity, 175
universalist god, 249, *See* universalism
very grouchy/perverse god, 249, 262
very nice god, 249, 262
virtuous god, 257, 276
theology, 21, 45, 225
in Pascal's work, 45, 46, 75, 77
natural, 73, 164
P theology, 225
X theology, 226, 252
Thirouin, Laurent, 45
Titelbaum, Michael, 229
Tragedy of the Commons, 201, 205

universalism, 18, 53, 58, 82, 114, 176–77, 249, 251
diabolical version, 177
utility, 5
as representation of preferences, 5
finite. *See* finite utility
incomparably good. *See* finite utility
infinite. *See* infinite utility
lexicographic, 265
long-run average, 140
negative infinite. *See* negative infinite utility
relative. *See* relative utilities

surreal infinite. *See* surreal utility
utility functions, 5, 39, 145
vector-valued. *See* lexicographic utility

vague credences. *See* credences, imprecise/vague
van Fraassen, Bas, 285
Voltaire, 31, 84–85, 100, 111, 168

Wagerer's Dilemma, 201, 207
wagering for God, 31, 169, 261
Wagner, Carl, 196
Wainwright, William, 108
weak dominance, 3, 125
weak precisificational dominance rule, 135, 136
Weatherson, Brian, 284
Weeden, Jason, 195
Wenmackers, Sylvia, 275
Wetsel, David, 45, 49, 66, 74
Wide Interval View, 283–87
Williams, Bernard, 107
Winding Road game, 230
wishful thinking, 107
worship and religious practice, 212, 218
WTB argument. *See* James, William, Will to Believe argument

Zermelo, Ernst, 297